Resolution of the Twentieth Century Conundrum in Elastic Stability

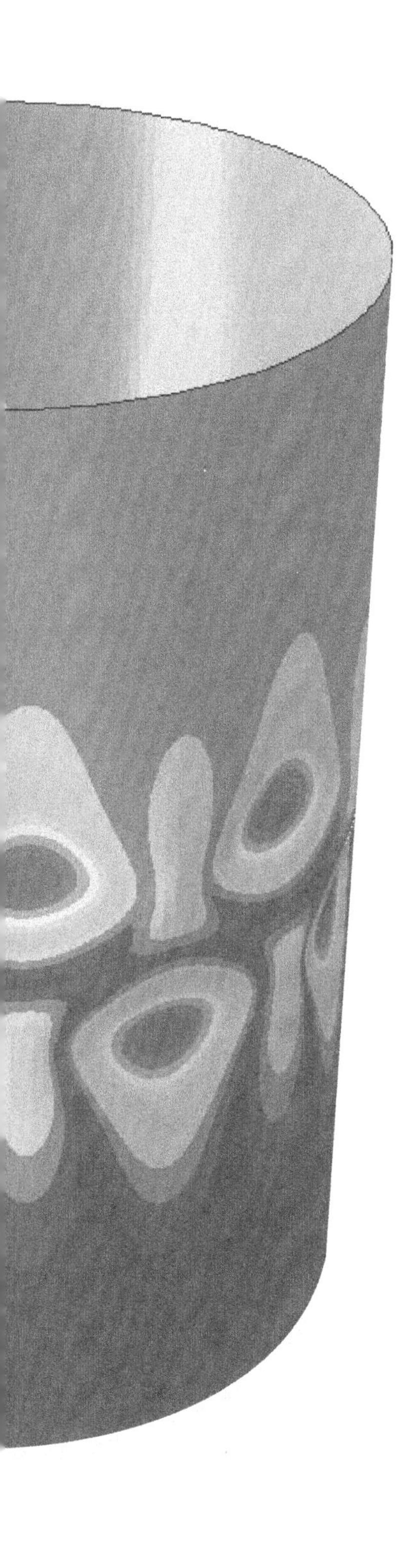

Resolution of the Twentieth Century Conundrum in Elastic Stability

Isaac Elishakoff
Florida Atlantic University, USA

World Scientific

NEW JERSEY • LONDON • SINGAPORE • BEIJING • SHANGHAI • HONG KONG • TAIPEI • CHENNAI

Published by

World Scientific Publishing Co. Pte. Ltd.
5 Toh Tuck Link, Singapore 596224
USA office: 27 Warren Street, Suite 401-402, Hackensack, NJ 07601
UK office: 57 Shelton Street, Covent Garden, London WC2H 9HE

Library of Congress Cataloging-in-Publication Data
Elishakoff, Isaac.
Resolution of the 20th century conundrum in elastic stability / by Isaac Elishakoff
(Florida Atlantic University, USA).
pages cm
Includes bibliographical references.
ISBN 978-9814583534 (hardcover : alk. paper)
1. Elastic analysis (Engineering) 2. Elastic plates and shells. I. Title.
TA653.E45 2014
620.1'1232--dc23

2014007315

British Library Cataloguing-in-Publication Data
A catalogue record for this book is available from the British Library.

In-house Editor: Amanda Yun

Typeset by Stallion Press
Email: enquiries@stallionpress.com

Printed in Singapore

Dedication

This monograph is dedicated to Professor John Hutchinson (Harvard University), correspondence with whom convinced me of the advisability of initiating this monograph: "I agree that it is one of the most fascinating phenomena in mechanics in the last half century or so. I'll be interested to see your take on this," he wrote to me (Hutchinson, 2011). In other words, he is the one, in addition to this author, to whom the reader ought to feel indebted for this excursion into the exciting 20th-century history of elastic buckling of "prima-donna" shells and bridging of the yawning chasm between theory and practice.

Preface

Shells were first used by the Creator of the Earth and its inhabitants. The list of natural shell-like structures is long, and the strength properties of some of them are remarkable. Egg shells range in size from those of the smallest insects to the large ostrich eggs, and cellular structures are the building blocks for both plants and animals. Bamboo is basically a thin-walled cylindrical structure, as is the root section of a bird's feather. The latter structural element develops remarkable load-carrying abilities.

E. E. Sechler (1974)

. . . Imperfections distributions are likely to be random in nature. . .

J. Hutchinson, R.C. Tennyson and D.B. Muggeridge (1971)

To a mature, well-educated scientist a discrepancy means an opportunity. It shows there is something to be discovered.

J.M.T. Thompson (2013)

This disagreement was much debated and was not explained until von Kármán and Tsien (1941) in their seminal paper found the explanation. . .

Z.P. Bažant (2000)

In apparently perfect resonance with the above introductory notes of Sechler (1974), Ramm and Wall (2004), stress:

> Shell structures are the most often used structural elements in nature and technology. The outstanding success is due to the curvature allowing a thin shell to carry transversal loading primarily by inplane action. As a consequence shell structures can be designed with an extreme slenderness leading to a high efficiency often simultaneously satisfying also a shel(l)ter function. The underlying concept has for example been exploited by the small scale building blocks of nature, namely the cells, prestressed by turgor pressure. But it is also carried over to macroscopic natural structures like blood vessels, bones, petals or egg and nut shells. The ancient master builders were aware of the magnificent load carrying features of shells when they designed and built large scale vault structures like the Roman Pantheon still existing today. Modern technology makes use of the shell principle in innumerable applications like pipes, containments, cooling towers, membrane roofs, aircraft fuselages or car bodies, to mention only a few, and our daily life is full of smaller and larger shell structures.

This monograph overviews the efforts that led to resolution of the 20th-century conundrum in elastic stability of shells. In particular, the dramatic disagreement between theoretical and experimental results and the subsequent introduction of the empirical knockdown factor are discussed in detail. The mismatch between theory and experiment was qualitatively explained by Warner Tjardus Koiter, in his now-famous thesis, as well as in the papers by von Kárman and Tsien, and by Lloyd H. Donnell and C.C. Wan. However, these studies did not offer means for rigorous theoretical derivation of the knockdown factor for the shells with generic imperfection patterns encountered in practice. Numerous attempts to resolve the conundrum via deterministic theoretical, experimental and probabilistic analyses remained unsuccessful. The conundrum consists in two facts. On the one hand, in the humble opinion of this writer, there is impossibility of using hundreds and perhaps thousands of deterministic papers devoted to predicting the rigorous knockdown factor. On the other the fact that Wynstone Barrie Fraser and Bernard Budiansky (1969) and numerous other investigators, while recognizing the usefulness of the probabilistic approach to resolve the above conundrum arrived yet to another conundrum: they asserted that the buckling load of stochastic structures is a deterministic quantity. Some investigators recommended the latter as the design load.

In 1979, this author lucked out on reliability-based theoretical means for derivation of the knockdown factor and its judicious allocation, although Richard Hamming (1986) states that "The particular thing you do is luck, but that you do something is not," citing Louis Pasteur's "Luck favors the prepared mind." This book is dedicated to the story of how these ideas developed. Specifically, the early works are considered in some detail, for rightfully, according to Bernard of Chartres (attributed to him by John of Salisbury in 1159), "we are like dwarfs on the shoulders of giants, so that we can see more than they, and things at a greater distance, not by virtue of any sharpness of sight on our part, or any physical distinction, but because we are carried high and raised up by their giant size." A similar statement can be found in the works of the Jewish tosafist Isaiah di Trani (c. 1180–1250):

> The wisest of the philosophers asked: "We admit that our predecessors were wiser than we. At the same time we criticize their comments, often rejecting them and claiming that the truth rests with us. How is this possible?" The wise philosopher responded: "Who sees further, a dwarf or a giant? Surely a giant, for his eyes are situated at a higher level than those of the dwarf. But if the dwarf is placed on the shoulders of the giant, who sees further?... So we too are dwarfs astride the shoulders of giants. We master their wisdom and move beyond it. Due to their wisdom we grow wise and are able to say all that we say, but not because we are greater than they.

Having said this, one has to bear in mind what Rabbi Solomon Almoli [*d. circa* 1542] wrote (Leiman, 1993):

> It is possible for the later authorities to know and comprehend more than the earlier authorities on two accounts. First, on account of specialization... Second, on account of the cumulative effect of knowledge.

While writing this book, two goals were kept in mind. Due to the explosion in the output of scientific papers, people can no longer see the forest for the trees. It was imperative to stop and reflect on this grand topic, to provide the state of the art to the interested reader. The second goal is associated with one of the recommendations of Gian-Carlo Rota (1997): "You are more likely to be remembered by your expository work."

The story I have to tell is fascinating. Many great minds of the 20th century took part in it. Central to it were qualitative explanations of the chasm between theory and practice, offered by von Kármán and Tsien in 1941, and the general theory of elastic stability developed by Koiter in his thesis and published in 1945. These explanations were followed by a flood of others, but no rigorous answer was provided.

In my own case, the central inspiring role was played by the Harvard group. In numerous papers researchers belonging to the stochastic part of the Harvard group and their followers treated an infinitely long structures with random, homogeneous, and ergodic initial imperfections. They eventually arrived at the paradoxical conclusion that all structures fail at the same buckling load, depending on the value of the spatial spectral density at the wavelength corresponding to the classical buckling load. I have to admit that without the strumming effect of this conclusion, this book would have never been written.

This is a personal and inevitably biased account. But whose story ever was, is, or will be different? The facts are presented through my own prism. They are given, to quote the fitting words of Tony Judt (2012) (with whom I vehemently disagree on numerous other things), with

> a degree of engagement lacking in the detached scholar: I think this is what people mean when they describe my writing as "opinionated." And why not? A historian (or indeed anyone else) without opinions is not very interesting, and it would be strange indeed if the author of a book about his own time lacked intrusive views on the people and ideas who dominated it. The difference between opinionated book and one which is distorted by the author's prejudices seems to be this: the former acknowledges the source and nature of his views and makes no pretense at unmitigated objectivity...

Some anecdotes may seem as totally unnecessary or irrelevant to the overly critical reader. They were, however, deemed instructive in providing the general "landscape" in which the discussed ideas developed and matured. Many more anecdotes could have been included, but I wanted merely to bring out the "flavor" of things.

This book, therefore, is a *blend* between the research monograph and a personal memoir. It vastly differs from Timoshenko's book *As I Remember* since the latter is counted as an autobiography. It is closer to the essay *A Personal History of Random Data Analysis* by Bendat (2005). The book before you deals only with engineering scientists who dealt with the elastic stability conundrum.

It is hoped that you will enjoy this account of how the "monster" (in the terminology of Chuck Babcock of Caltech) of imperfection sensitivity was tamed.

Galley–proofs of this book were corrected while the author served as a Visiting Eminent Scholar at the Hunan University, and Visiting Professor at the National University of Defence Technology, both in the city of Changsha, People's Republic of China. Sincere thanks are expressed to Professor Chao Jiang, Dr. Yuan Li, Professor Jingwei Gao and Dean, Professor Xu Han for their warm and kind hospitality.

The author will welcome responses from readers at elishako@fau.edu, or by snail mail, on things that were unwittingly omitted or neglected, or on alternative approaches to the imperfection sensitivity.

Isaac Elishakoff
Boca Raton, U.S.A. and Changsha, P.R.C.
December 2013

Acknowledgments

First of all, I express my gratitude to the numerous colleagues who provided invaluable opinions on the deterministic, probabilistic, or bounded initial imperfections over decades. I still experience the sense of wonder I had felt on reading the paper by Fraser and Budiansky (1969). While some reviewers described it as "a rather complete analysis," I had the audacity to differ both with this view and with the extensions of the above work at various universities. I am grateful to Professors Warner T. Koiter (with whom I had the honor of collaborating in two papers), Bernie Budiansky, and Ari van der Neut for endorsing the research directions that I undertook over the years, to Professor Johann Arbocz for collaboration between 1979 and 1992; his experimental data and numerical techniques for buckling load evaluation proved indispensable for implementing my approaches; to Professors Yakov Ben-Haim, Xiaojun Wang and Zhi-Ping Qiu for collaborations in the nonprobabilistic, convex analysis of imperfection sensitive structures; to Dr. W. Jefferson Stroud of the NASA Langley Research Center, for collaboration in stochastic imperfection–sensitivity, during his yearlong sabbatical stay at Florida Atlantic University; to Professor Josef Singer for the sage advice; to Dr. James H. Starnes, Jr. for the support provided by the NASA Langley Research Center, to the Technion — Israel Institute of Technology for providing an incomparable scientific atmosphere for critical inquiry; to Delft University of Technology for the unforgettable scientific and social environment offered during my sabbatical stay there in 1979/80 and my numerous subsequent visits, culminating in the summer of 2000 with serving as inaugural holder of the W.T. Koiter Visiting Chair Professorship at the Koiter Institute Delft; to Dr. David Bushnell, formerly of Lockheed Martin, and Professor Zdenek P. Bažant of Northwestern University for their encouraging comments on the preliminary version of this essay. Thanks to M. Yohann Miglis of Florida Atlantic University; Mr. Wim Verhaeghe of the Katholieke Universiteit Leuven, Belgium; M. Axel Delmas and M. Y. Bekel of the Ecole Centrale Paris, France; M. Thomas Gomez of the Polytech' Engineering School in Clermont-Ferrand, France; M. Clément Soret and especially Mr. Baptiste Ducreux, Etienne Archaud (who also kindly created figures using freely available Clipart gallery from Microsoft Word) and Nicolas Sarlin of the

Institut Français de Mécanique Avancée in Clermont-Ferrand, France, for typing the numerous versions of the manuscript. Without the reliable and meticulous help most kindly provided by Mr. Ducreux, it would have taken many more months to complete this book project. Sincere thanks to Professor Ken P. Chong, Amercian editor of the journal *Thin-Walled Structures* for enthusiastic encouragement in preparing this monograph. Author expresses gratitude to Dr. Benedikt Kriegesmann of Airbus-Hamburg, for kind permission to include our joint paper (Elishakoff, Kriegesmann, Rolfes, Hühne and Kling, 2012) as the basis for Chapter 6. Thanks also to Professor Igor Andrianov of the RWTH, Aachen University, Federal Republic of Germany for providing some Russian references. Some of the research reported herein was supported by the NASA Langley Research Center with Dr. James H. Starnes, Jr. as program manager. Special gratitude in expressed to Ms. Barbara Steinberg of Florida Atlantic University and Mr. Siman Yang of Hunan University, P.R.C. for scanning numerous pages. Last but not least I would like to express my sincere gratitude to Ms. Rochelle Kronzek and Ms. Amanda Yun of World Scientific for their most helpful and extremely dedicated cooperation.

The first part of this book dealing with solely the probabilistic treatment appeared earlier as a journal essay (2012). It was gratifying to receive positive responses from eminent scientists: (1) "Just read your rather remarkable survey. I loved it! Thanks"; (2) "I have skimmed through it and it looks quite a brilliant treatise." (3) "Fantastic contribution. It will be my top reference to all students starting their degree program. In fact, I will enforce it as a requirement to start a degree program" ; (4) "Instructive and thought provoking article . . . it was a real pleasure to read the paper and note that we share common points of view on this topic . . .". I am extremely thankful to them for their solidarity, not listing their names with a view that this book ought to stand on its own legs.

Contents

Chapter 1

Introduction

To the layman buckling is a mysterious, perhaps even awe-inspiring, phenomenon that transforms objects originally imbued with symmetrical beauty into junk.

D. Bushnell (1981)

... The performance of shell very much depends on how it is designed and how it is treated, or to phrase in a more colloquial word: A shell can be in a "good mood" or in a "bad mood."

E. Ramm and W.A. Wall (2004)

In this chapter we describe how the author got involved with the imperfection sensitivity concept, specifically how he got to be excited by the paradoxical result reported by Fraser and Budiansky (1969). Then rather brief historical excursion is made into the history of elastic stability that originated by the experiments of Pieter van Musschenbroek (1692–1761) and the theory of Leonhard Euler (1707–1783).

1. How I Got Involved with the Imperfect World of Imperfection Sensitivity

No matter how smart you are, there will always be probabilistic problems that are too hard for you to solve analytically.

P. Nahin (2008)

"Nillius in verba" (Take nobody's word for it)

The motto of the Royal Society

In the spring semester of 1977 (if my memory serves me well), Professor Bernard Budiansky of Harvard University was expected at the Department of Aeronautical Engineering at the Technion — Israel Institute of Technology, as part of his sabbatical. Professor Josef Singer (Fig. 1.1), head of the Structures Group, asked me to read some of Professor Budiansky's works, get interested in some of them, and discuss my ideas with him. I spent considerable time in the library on this assignment and eventually found the paper (Fraser and Budiansky, 1969) on imperfection sensitivity of an infinite column resting on a nonlinear elastic foundation. What attracted me most in it was the conclusion from which the following excerpt is reproduced word

Fig. 1.1. Professor Josef Singer (1923–2009) was an active contributor to the experimental analysis of shells.

for word, to avoid any impression of selective quoting:

> We consider the buckling of an ensemble of infinitely long columns, with initial deflection, resting on nonlinear elastic foundations. The initial deflections are assumed to be Gaussian, stationary random functions of known autocorrelations and the problem is solved by the method of equivalent linearization. *We find that each column in the ensemble has the same buckling load that depends only on autocorrelation of the initial deflection function* Results are presented for columns whose initial deflection functions have an exponential — cosine autocorrelation.

My first reaction was "Wow!" since I could not possibly visualize all columns having the same buckling load even though each of them necessarily possesses a different imperfection profile! The authors stated further: "... we show that each column in the ensemble has the same buckling load which depends only on the autocorrelation of the initial deflection and not on a particular realization of one of these functions."

This brought some relief: at least, the buckling load depends on the autocorrelation function of the initial imperfections, if not on a particular realization of them.

The authors themselves had some reservations. Choosing an exponential-cosine autocorrelation function, they remarked: "Whether or not structural imperfection can be validly represented by this autocorrelation function and spectral density is a question that can be only answered when appropriate experimental evidence is available."

They did not mention that prior to examining the validity (or lack thereof) of the chosen autocorrelation, the "experimental evidence" had to verify whether the initial imperfections constituted a stationary and Gaussian random field.

On another topic, the authors noted: "We hoped in solving this problem that the method used could be extended to solve the problem of the buckling of thin cylindrical shells with random imperfections. Unfortunately, the method of equivalent linearization which provided the key to this problem does not seem to work on the shell problem. However, recent work by Amazigo (1969) has shown that this problem and the shell problem can be solved by means of a modified truncated hierarchy technique."

The authors did not explain why stochastic linearization worked for an infinite beam, but failed for shells. The final sentence of their paper reads: "These results are very limited and do not offer a very substantial justification for our assumption that the method of equivalent linearization gives an accurate result for the column of infinite length. However, they do increase our confidence slightly." Again, they did not explain in what their confidence was increased. However, Fraser and Budiansky's work led me to that of Amazigo (1969). In fact, Amazigo performed series of works on this fascinating topic. About them the reader can consult Sec. 2.3.

Lewis Carroll in *Alice in Wonderland* advises: "Begin at the beginning . . . and go on till you come to the end; then stop." I shall follow this advice — more or less.

2. Digest of History of Elastic Stability from Musschenbroek and Euler to Koiter

It is tempting to begin this survey of the subject of buckling with a long backward look at its history and to muse on the special fascination it has held for so many engineers and scientists.

B. Budiansky and J.W. Hutchinson (1979)

A surprising phenomenon is that almost none of the values for the buckling of the bifurcation points predicted by the theory agree with those measured experimentally, the experimental values being systematically lower than the theoretical ones, with serious prejudice to the safety of engineering shell-type structures.

P. Villaggio (1997)

As Bushnell (1981) writes, ". . . to the layman buckling is a mysterious, perhaps even awe-inspiring, phenomenon that transforms objects originally imbued with symmetrical beauty into junk."

The history of elastic stability, as opposed to the much older general concept of stability, starts with the experiments of Musschenbroek (1729), the first author who "enunciated that the resistance of a beam under axial loads depends on the inverse of the square of its length" (Godoy, 2006) and the theoretical studies of Euler (1744) (see also Oldfather *et al.*, 1933; van den Broek, 1947) who gave us the now classical formula

$$P_{\text{cr}} = \frac{\pi^2 EI}{L^2} \tag{1.1}$$

for buckling of a uniform column of length L, moment of inertia I, and material with modulus of elasticity E. This formula was criticized in the beginning, so Euler had to defend it, stating that it was "not only entirely new but also most remarkable" (Truesdell, 1960). Bažant (2000a) notes:

> Experiments, however, could not verify the calculated critical loads. This fact was explained by Young (1807), who realized that imperfections such as initial curvature, initial bending moments or load eccentricity play an important role and derived a formula for what is known today as the magnification factor for deflections and bending moments in columns due to axial load.

Only about a century later, with the Industrial Revolution did the formula find wide application in the context of slender elements. Its initial criticism was associated with the fact that it was unsatisfactory for short columns, since the behavior of such columns was beyond the elastic range. According to van den Broek (1947),

> There is no question as to the correctness of the Euler formula. There is a serious question, however, as to its applicability to engineering design problems. The ambiguity arises from the fact that the Euler theory presupposes ideal pin-ended conditions, which are non-realizable in practice. This does not imply a conflict between theory and practice. It exemplifies the fact that a theory can be quite perfect and beautiful, and yet more or less futile. I have called the Euler formula a beacon in a sea of darkness. One beacon, however, is not enough. The engineering profession is still crying out for more light on the subject of column design.

In Koiter's (1976) words, "A cold war before the end of last (19th) century arose out of the circumstance that columns were hardly ever slender enough to allow the application of Euler's formula."

According to Lord Chilver (1976),

> . . . problems of "buckling", as such, became important design problems when, during the 19th century, the introduction of metallic structural materials made it possible to build structures of more extreme geometric *proportions*. During the early 19th century, slender cast-iron columns introduced a new element of clear, wide areas between the columns of a building. . . With the emphasis on extreme geometric proportions, it is not surprising that the field of structural stability theory, from the 19th century onwards, was concerned primarily with the study of "buckling."

Summarizing the whole century very roughly, one can say that it was devoted mainly to development of the theory of plates and plate behavior, vibration, and buckling. The 20th century witnessed the derivation of the classical buckling formula for the uniform axial compression stress σ_{cr}:

$$\sigma_{cr} = \frac{E}{\sqrt{3(1-\nu^2)}}\frac{h}{R} \approx 0.6E\frac{h}{R} \tag{1.2}$$

for a cylindrical shell with uniform thickness h, radius of the circular cross section R, and Young's modulus of the shell material E.

Equation (1.2) is due to the work of Lorenz (1908; 1911), Timoshenko (1910), and Southwell (1914). Hoff (1966) remarks on the priority:

> It appears... that Lorenz was the first to calculate the critical stresses of axially compressed circular cylindrical shells for both the axisymmetric and the general cases of buckling. Of course, his results were not quite accurate, and were later improved upon by Timoshenko in the case of axisymmetric buckling and by Southwell and Timoshenko in the general case. The classical formula of Eq. (1.2) was apparently first derived by Timoshenko.

In another paper Hoff (1967) calls Timoshenko "the Russian [Ukrainian] giant, under whom [he] had the good fortune of working for his doctor's degree."

The following quote is ascribed to Yogi Berra: "In theory there is no difference between theory and practice. In practice there is." Surely this statement is applicable to thin elastic shells. Indeed, shells fared no better than columns. This led Koiter (1976) to remark:

> To put it mildly, buckling theory and experiments have not always co-existed in harmony... Agreement between theory and experiment was hardly better for the buckling of plates and shells in the twenties and thirties of the present century. As usual, both sides were again to blame for their disagreement. Ill-conceived applications of linear theory to essentially nonlinear problems can hardly be expected to agree with the results of badly controlled experiments, no matter how much money is spent on both.

The difference between theory and experiment was also addressed by Nicholas Hoff in his above paper:

> In his classical *Habilitationsschrift*, that is a formal lecture presented when he was appointed Privatdozent in Göttingen, Flügge (1932) compared the results of his analysis with data on experiments described in the technical literature. As he was unable to find test results on axially compressed shells, he manufactured and tested a number of rubber and celluloid cylinders in the Institute for Applied Mechanics in the University of Göttingen. The length-to-radius ratio of the specimens varied between 1.76 and 5, and their radius-to-wall thickness ratio from 90 to 138. All the specimens buckled elastically, and the ratio of the experimental buckling stress to the buckling stress according to Eq. (1.2) ranged from 0.52 to 0.65 for the celluloid cylinders...
>
> But an experiment carried out in connection with the rapid development of thin-walled aluminum alloy airplane structures in the early nineteen thirties, shells having radius-to-thickness ratios up to 1500 were tested (Lundquist 1933; Donnell 1934). As these specimens often failed at stresses as low as 15% of the classical theoretical value, Flügge's explanation of causes of the disagreement was no longer sufficient.

In a later paper Hoff (1967) writes:

> Unfortunately serious doubts regarding the validity of the theory [Eq. (1.2)] arose when three experimental papers appeared in 1933 and 1934. In the first, a bright 20-year-old student, N. M. Newmark, working for his master's degree in civil engineering under the direction of W. M.Wilson, published the results of his tests carried out at the University of Illinois in Urbana. The radius-to-thickness ratios of Wilson and Newmark's specimens ranged from 28 to 1000. Thus earlier results obtained by Lilly in 1905, by Mason in 1909, and by Popplewel and Carrington in 1916, which characterized tubes of radius-to-thickness ratios less than 25, were extended significantly. Of course, during WWI Andrew Robertson of Bristol University had carried out tests with tubes and with cylindrical shells made of steel strip, with R/h values ranging from 5 to 500; but due to war conditions, his results did not become available to the engineering public until 1927 (references to these authors can be found in the paper by Wilson and Newmark, 1933).

In 1985, when I presented a seminar at the Department of Aerospace Engineering of the University of Notre Dame, and mentioned the seminal contributions of Koiter in imperfection sensitivity of shells, Professor Lawrence H.N. Lee and Professor Nai Chien Huang remarked that it were Donnell and Wan (1950) who first dealt with imperfection sensitivity of shells. However, as Sir Francis Darwin, the second son of Charles Darwin and a distinguished botanist in his own right (Rawson, 1997), stipulated, "in science, the credit goes to the man who convinces the world, not to the man to whom the idea occurs." More importantly, Koiter (1945) completed his seminal works much before the Donnell and Wan (1950) paper, although Donnell and Wan did not know of Koiter's work. Koiter mentions in his public lecture (1979, p. 243),

> In the spring of 1940 ... I read an existing article by Leslie Cox in the Journal of the Royal Aeronautical Society of March 1940, and his simple account of the behaviour of a column with a nonlinear elastic support suggested to me the basic idea that instability *at* the critical load might be the explanation of the disastrous effect of imperfections.
>
> This conjecture led to the development of the general nonlinear theory of elastic stability in my thesis. My idea came into full focus in the summer of 1940 ... The basic work, including the application to the baffling problem of buckling of cylindrical shells under axial compression, was completed in February 1942... I have a vivid recollection of the New Year's Eve in December 1942 when my wife and I sat in front of a very modest fire and discussed, in addition to the most acute war-time problems, what would happen to my work which I considered to be significant but questioned whether it would be recognized as such.

Referring to this period, Hutchinson and Koiter (1970) mention:

> The entirely different postbuckling behavior of thin shell structures came to light in the early 1940's when Kármán and Tsien (1939; 1940; 1941) showed that large discrepancies between test and theory for the buckling of certain types of thin shell structures was due to the highly unstable postbuckling behavior of these structures. At roughly the

> same time in war-time Holland, Koiter developed general theory of stability for elastic systems subject to conservative loading which was published as his doctoral thesis in 1945.

Professor Marcello Pignataro (1998) writes, in his obituary piece on Koiter, about that period:

> Professor Flügge had been sent from Germany to Delft to cover the Rector chair. According to the occupant law, Ph.D. students who were willing to discuss their thesis were obliged to take an oath of allegiance to the Nazi government. Koiter's thesis on the stability of elastic equilibrium was ready at that time, but the author, refusing this imposition, waited for the liberation of his country. The thesis thus appeared only in 1945.

After his now famous thesis was published, he did not return to this topic until about 15 years later. Many posed the following question: "Why didn't he publish results of his dissertation in English?" Koiter replied (1979, p. 244):

> . . . I considered I had already published my basic thinking on elastic stability, and that it would be improper to reiterate on the same topic, a curious mixture of modesty and immodesty, the latter because I took it for granted that my published work was accessible in principle to anyone actively interested in the field.

On the question, "How did he return to this topic?" Koiter (1979, p. 244) wrote:

> My unhurried return to the field of nonlinear elastic stability during the nineteen fifties was accelerated by a course I gave at Brown University in the academic year 1961–62. . . Interest in this field also mushroomed at that time, in particular at Harvard University where I fondly believe my work had some influence, and at University College London where a similar approach for discrete elastic systems was developed more or less independently, as described so eloquently in the preface by Thompson and Hunt in their monograph "A general theory of elastic stability."

Here is the quote from the definitive monograph by Thompson and Hunt (1973):

> For several years the first author was blissfully unaware of the classic dissertation of Professor W. T. Koiter which had surprisingly lain largely unknown since 1945, and has in fact only recently been translated into English by the National Aeronautics and Space Administration of America. This was indeed most fortunate, since the weight of Professor Koiter's contribution could well have discouraged him from proceeding with his own development of the subject. As it transpired, the full significance of Professor Koiter's work has filtered slowly into our consciousness in a gentle stream, moderated by the Dutch language and by our temperaments which have invariably preferred to explore the field for ourselves. This having been said, we must nevertheless hasten to admit our deep indebtedness to Professor Koiter's work . . . Although framed in generalized coordinates rather than in continuum terms, our approach owes much to the aforementioned work of Professor W. T. Koiter and we are pleased once again to acknowledge our indebtedness to this source. . .

Fig. 1.2. Professor Michael T. Thompson proposed probabilistic analysis of initial imperfections independently of V.V. Bolotin.

According to Bushnell (1985, p. 273),

> Koiter (1945; 1963) was the first to develop a theory which provides the most rational explanation of the large discrepancy between test and theory for the buckling of axially compressed cylindrical shells and externally pressurized spherical shells: the early collapse is due to small, unavoidable geometric imperfections.

3. Knockdown Factors

> *...the design of cylindrical shells is based on the theoretical critical load modified by empirical reduction, or knock down, factors for each kind of loading.*
>
> D.O. Brush and B.O. Almroth (1957)

> *If the answer is highly sensitive to perturbations, you have probably asked the wrong question.*
>
> L.N. Trefethen (1998)

Literature provides various analytical expressions for the knockdown factors. Upper bound is approximated by the following formula (Grigoliuk and Kabanov, 1978):

$$\gamma = 0.5 - 3.5 \cdot 10^{-3}(R/h)^{1/2} + 4.6(h/R)^{1/2}. \tag{1.3}$$

Lower bounds have been approximated by various expressions. Kanemitsu and Nojima (1939) proposed the following expression,

$$\gamma = 14.9(h/R)^{0.6}, \tag{1.4}$$

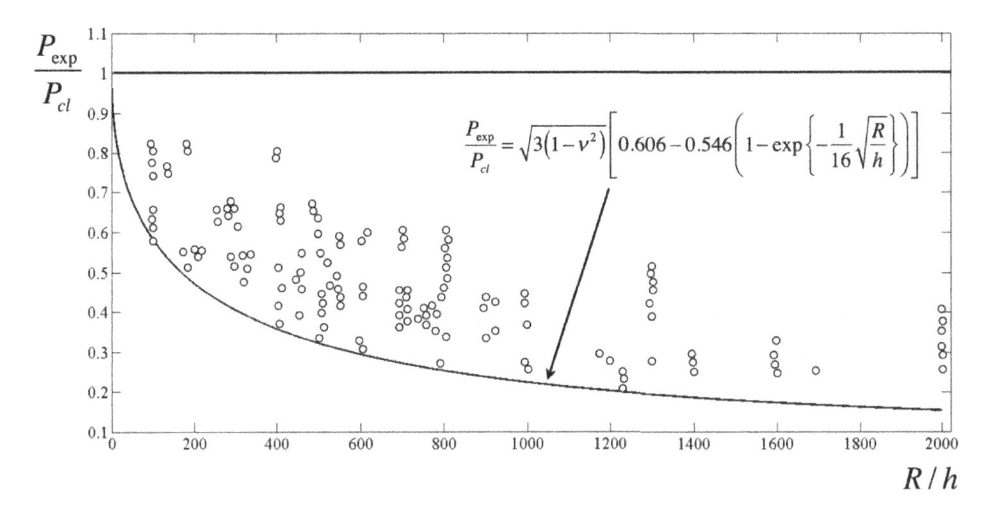

Fig. 1.3. Variation of the ratio P_{exp} over P_{cl} with the radius-to-thickenss ratio.

obtained with carefully prepared specimen in the range $R/h > 500$ and $0.5 < L/R < 5$. Grigoliuk and Kabanov (1978) state that a better limit for the range of values $0.03 < L/R < 5$ is given by Weingarten *et al.* (1965):

$$\gamma = 1 - 0.9(1 - \mathrm{e}^{-\sqrt{R/h}/16}) \tag{1.5}$$

Gerard and Becker (1957) proposed

$$\gamma = 6.35(h/R)^{0.54}. \tag{1.6}$$

Alekseev (see Grigoliuk and Kabanov, 1978; Eq. (6.2)) suggested

$$\gamma = 3.87\sqrt{h/R}. \tag{1.7}$$

Kirste (see Grigoliuk and Kabanov, 1969) suggested, via geometric considerations, $k = 0.308$; Pogorelov (1966) proposed the value $k = 0.296$ whereas Sachenkov (1962) argued for value $k = 0.352$. Kabanov (see Grigoliuk and Kabanov, 1978, Eq. (9.5)) introduced a corrective term in the above Alekseev approximation:

$$\gamma = 3.87(h/R)^{1/2} + 10^{-3}(R/h)^{1/2}. \tag{1.8}$$

Antona and Gabrielli (1966) proposed following dependence:

$$\begin{aligned}\gamma = 1.73[&0.235 \cdot 10^{-7}(R/h)^2 + 0.1755 \\ &-9.2 \cdot 10^{-5}(R/h) - 1.72 \cdot 10^{-2}(R/h)^3].\end{aligned} \tag{1.9}$$

The above formulas pertain to shells of average length. For short shells, the following formula is proposed by Kanemitsu and Nojima (1939):

$$\gamma = 1 - 0.9(1 - e^{-\sqrt{R/h}/16}) + 1.46(R/L)^2(h/R), \quad \text{for } R/h = 100, \quad (1.10)$$

whereas Weingarten *et al.* (1965) suggested,

$$\gamma = 14.9(h/R)^{0.6}[1 + 0.0177(R/L)^{1.3}(R/h)^{0.3}]. \quad (1.11)$$

As de Vries (2009) mentions, "in Europe the commonly used handbook of the European Conventional Steelwork, ECCS (1988) uses a similar equation." This knockdown factor is a combination of the two expressions

$$\begin{aligned} \gamma &= \frac{0.83}{\sqrt{1 + 0.01(R/h)}} \quad \text{for } \frac{R}{h} < 212 \\ \gamma &= \frac{0.70}{\sqrt{0.1 + 0.01(R/h)}} \quad \text{for } \frac{R}{h} > 212. \end{aligned} \quad (1.12)$$

These formulas are valid for cylinders with dimensions satisfying the following condition,

$$\frac{L}{h} \leq 0.95\sqrt{\frac{R}{h}}. \quad (1.13)$$

Comparison of the design buckling loads is conducted in Fig. 2.3 in the dissertation by de Vries (2009).

Discussing these and other formulas for knockdown factors, Grigoliuk and Kabanov (1978, p. 136) note, in a free translation from the Russian:

> There is a need in further improvement of the above formulas. In theory there is a need in more accurate solutions which will fully take into account the boundary conditions, initial imperfections of arbitrary form, plasticity of material and inhomogeneity of the state. In experiments there is a need in investigations at relatively constant conditions with use of the *statistical theory* [author's emphasis] in order to take into account influence of most of the parameters.

Whereas the need for probabilistic theory was acknowledged, no work appeared this probabilistic tool to be effectively utilized. Harris *et al.* (1957) advocated the "90% probability curve" reproduced as Fig. 58 in the book by Brush and Almroth (1975). They note: "The reduction factor for $R/h = 500$, for example, is seen to be 0.24. Multiplication of the coefficient 0.605 by this factor gives, for the design stress, the value $\sigma_{cr} = 0.15Eh/R$." Brush and Almroth (1975, p. 185) stress:

> Most industrial organizations establish their own design criteria for shell buckling and issue them in 'structural methods' handbooks. They frequently are looseleaf and are continually updated. Such books are in marked contrast to, say, the firmly established,

industry-wide Manual of Steel Construction (1970) for the design of columns and beams.

In spite of the wealth of test data for homogeneous isotropic cylinders, this semiempirical basis for design is only partially satisfactory." Bushnell writes (1985, p. 284): "As a consequence of the lack of adequate theoretical analysis, the designers of axially compressed cylindrical shells have been forced to use empirical methods. In 1957, Harris, Suer, Skene and Benjamin made the first attempt to devise a design limit by use of a statistical analysis of available test results. For different probability levels, a reduction factor is given as a function of the radius-to-thickness ratio... The disadvantage of this statistical analysis procedure is that some test results affect the design although they are really irrelevant because of the manner of fabrication or their size. There is no bonus for the manufacturer who can produce an almost perfect shell. For analysis such as that (by Harris *et al.*, 1957), it appears that the number of available tests constitutes a sufficient statistical background as long as the buckling coefficient for a fixed probability is a function of only one variable, R/h. This is the case for longer shells; however, for shorter shells the shell length becomes a substantial parameter and the number of available tests is really not satisfactory.

4. Studies by the Caltech Group

The thin shell is an intriguing device with many contradictions. Compared to flat plate and beam structures, great material savings are possible. On the other hand, the potential for great catastrophes is always present... the change in the solution because of a small change in the parameters can be quite large.

C.R. Steele (1989)

The shell turns out to be rather "phony" not at all reaching its seemingly high buckling load...

E. Ramm and W.A. Wall (2004)

To follow the contribution of the California Institute of Technology (Caltech) to shell buckling, it is instructive to read what Hoff (1966) had to write:

Between 1939 and 1941, Dr. Theodore von Kármán and the very capable and enthusiastic group of students and research men gathered around him in the Guggenheim Aeronautical Laboratory of the California Institute of Technology laid the foundations of the large-displacement theory of the buckling of thin shells. Kármán was puzzled on the one hand by the large difference between the complete predictability of the buckling stress of rods and thin plates, and on the other hand by the very substantial difference between theoretical and experimental values of the stress at which thin shells collapsed. He also observed that the development of buckles and bulges was gradual with elastic rods and plates while it was sudden, and even explosive, with shells. After buckling, columns and plates continued to carry the buckling load, and were even capable of supporting further increased loads, but with shells the load supported after buckling always dropped to a fraction of the buckling load. Yet the equations defining the equilibrium and the stability of rods, plates and shells were all based on the same well-proven hypotheses of the theory of elasticity.

...The only matter in which this shortcoming of shell theory could be remedied was to add terms to the equations that were nonlinear in the displacements. But relatively simple equations containing such terms had already been developed by a former collaborator of von Kármán, Lloyd H. Donnell, who had worked in the Guggenheim Aeronautic Laboratory of the California Institute of Technology between 1930 to 1933. To the linear terms of his small displacement equations (Donnell, 1933) which were to become famous later, Donnell added the nonlinear terms contained in the von Kármán large-deflection plate equation (von Kármán, 1910) (apparently Donnell was not aware of the existence of these equations when he derived his large-displacement theory) to obtain what are generally known today as the von Kármán–Donnell large-displacement shell equations (Donnell, 1934). Donnell also assumed that the shell was inaccurately manufactured...

...The first innovation in the analysis of von Kármán and Tsien (1941) was the introduction of a buckle pattern based on visual observation of the shells after failure.

... von Kármán and Tsien accomplished a great deal. They discovered the existence of three states of equilibrium corresponding either to a prescribed displacement of the loading head of the testing machine (loading in a rigid testing machine), or to a prescribed value of the load (so-called dead-weight loading).

...In the presence of initial deviations from uniformity the maximum value of the compressive stress would be smaller than that indicated by the letter C in Fig. 9, and the difference between the numerical value 0.605 for $(\sigma/E)(R/h)$ and the experimental value would increase with increasing values of the initial deviation. *However, the random nature of the initial deviations makes it very difficult to evaluate the practical maximal value of the buckling stress* [author's emphasis]. Hence, von Kármán and Tsien suggested that for design purposes the engineer use the minimal value of the equilibrium stress, that is $0.194E(h/R)$.

Professor Robert Jones (2006, pp. 535–536) writes about the importance of the paper by von Kármán and Tsien (1941):

The importance of initial geometric imperfections for shells was discussed by Donnell in 1934. However, significant numerical results were not obtained until 1950 for axially loaded shells by Donnell and Wan and until 1956 for shells under external pressure by Donnell. Those numerical results were obtained largely because of the work of von Kármán and Tsien (1941) *who... were the first* [author's emphasis] to demonstrate that equilibrium states other than the prebuckled state exist at stresses below the Euler stress.

The essence of Donnell's approach to the study of initial geometric imperfections is that he assumed the imperfections to be of the same shape as the mode into which the perfect shell buckled. Such geometric imperfection presumably would seem to be the most damaging to the shell. Accordingly, Donnell's approach must lead to an overestimate of the effect of initial geometric imperfections on buckling. That is, using the most damaging kind of geometric imperfections must lead to an upper bound on their effect.

It appears imperative to also quote from the book by Bažant and Cedolin (1991, p. 468):

The reason that many shells fail at a much smaller load than the classical critical load consists in the combined effect of nonlinearity and imperfection. This was discovered

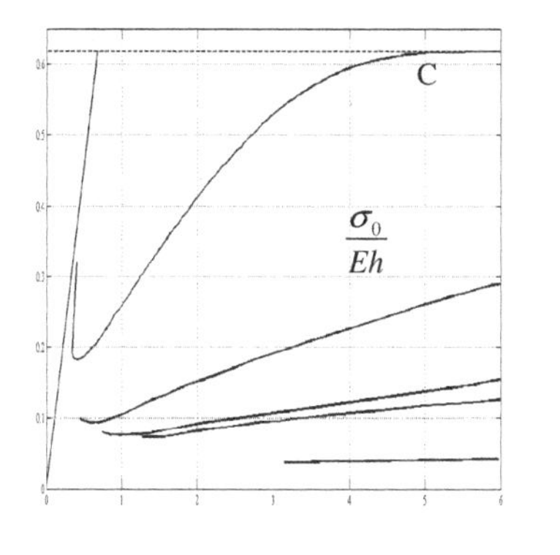

Fig. 1.4. Nondimensional stress vs. deviation.

by Kármán with Dunn and Tsien (1940) (see also von Kármán and Tsien, 1941). In their *revolutionizing* [author's emphasis] paper, they demonstrated by approximate nonlinear analysis that, after reaching the critical states, the load can rapidly decrease at increasing deflection, that is, the structure undergoes softening. Further they demonstrated that even a small disturbance can cause shell to jump to a postbuckling state at which the carrying capacity is greatly reduced.

Dryden (1963), then the Deputy Administrator of National and Space Administration (NASA), comments about these works as follows:

One of von Kármán's major contributions to solid mechanics is his analysis of the physical nature of the buckling of spherical shells and the development of a nonlinear theory of the phenomena which accounts for the great discrepancies between the experimental results and those obtained from linear elastics theory. The theory was first reported in a paper by von Kármán and Tsien in 1939. The new phenomenon is that commonly known as "oil canning" from the observed snapping of the bottom of an oil can between two positions, each of which is stable, but for a large range of deflection the sheet is unstable and moves to one or the other of the two stable positions. This phenomenon occurs when the bending stiffness is small as in thin sheets. The buckling loads attained are much smaller than those obtained by the classical linear theory.

A comprehensive discussion of the effects of the influence of curvature on the buckling characterization of structures was given by von Kármán, Dunn, and Tsien in 1940. Comparison was made of the discrepancies between the classical theory of buckling of cylindrical shells and the experimental evidence.

The buckling of thin cylindrical shells under axial compression was investigated with H.S. Tsien. The concepts of the nonlinear large deflection theory are applied to this case. The effect of the elastic characteristics of the testing machine on the buckling phenomena was discussed.

Koiter (1976, p. 12) referred to the above work as follows:

The pioneering investigations of nonlinear imperfection-sensitive column models by Cox (1940) and von Kármán *et al.* (1940) spawned many further researches, including the general theory (Koiter, 1945).

The works by von Kármán and his group were continued at Caltech by Sechler and Babcock and their collaborators and students. Here is an entry on Ernest Sechler (1905–1979) in the Graduate Laboratories of the California Institute of Technology (GALCIT) website:

> Guided by the observation of a failed rocket launch where asymmetrical forces prevailed without buckling of the external structure, he instituted intense efforts to understand the large difference between theoretical and measured buckling loads of cylindrical structures. Thus culminated in the findings of one of his students, C.D. Babcock. In the words of Y.-C. Fung, Sechler was a person who always found a simpler way to deal with a complex problem, guided and checked by experiments rather than theory alone. This philosophy has become a mainstray of GALCIT teaching and functions.

We also reproduce an excerpt from the entry on Charles D. Babcock (1935–1988):

> ... [He] reconciled theory and experiments on buckling of cylindrical shells through the use of exceedingly closely toleranced shells (electroplating process) and control of the edge boundary conditions. Used imperfection theory to place the problem of shell stability and reliability on a different level.

The following legitimate question arises: "Why did the Caltech scientists stress the experiments on shells?" The possible answers can be gained from the quotes by Jones (2006):

> It seems obvious that initial geometric imperfections are the most disruptive if they have the same shape as the shape into which the shell is attempting to buckle. However, at the same time, we must realize that such a situation is not very representative of actual geometric imperfections. Moreover, actual geometric imperfections would appear to be extremely difficult to characterize, much less account for in a theoretical approach.

It appears that the Caltech work on imperfection sensitivity was culminated by the NASA Report by Arbocz and Babcock (1968), which presented "... the first known results of complete imperfection surveys carried out on cylindrical shells before and during the loading process up to the buckling load."

Referring to the experimented work, Koiter (1976, p. 12) wrote:

> Considerable experimental difficulties had still to be overcome in the buckling of shells, and the first reliable verification of theory was achieved by Roorda (1965a; 1965b) (see also Koiter, 1966a), somewhat ironically in a simple model which might also have been investigated a long time ago. (Later, Hunt (2006) referred to these experiments as follows: "John Roorda had left a legacy of elegant experimental and comparative theoretical work on the nature of imperfections"). Time does not permit a detailed discussion of the many careful experiments on shell buckling, carried out in recent years in many institutes... It would seem that a reasonable measure of agreement between theory and experiment has now been achieved, at least to the extent that theorists and experimentalists can safely attend the same meeting...

> In an extensive series of experiments on the buckling of cylindrical shells under axial compression at the California Institute of Technology, in which the geometric imperfections of the shells were measured quite carefully before the test, it was found that asymmetric imperfections with a long axial wave length dominate.

Moreover,

> most calculations of imperfection-sensitivity have been carried out for *simple shapes* [author's emphasis] of the imperfection distribution, selected for the *convenience of analysis* [author's emphasis]. It is generally realized, of course, that actual imperfections are unlikely to follow this regular pattern.

Probably the most striking example of the early shell imperfection measurements was provided by Arbocz and Williams (1977). They presented an extensive imperfection survey on a 10-feet-diameter integrally stiffened cylindrical shell. Impressive recent test includes 20-feet-tall, 27-feet-wide cylinder simulating a rocket's shell and made of aluminum and lithium (Grant, 2011; Morring, 2011; Roop, 2011).

Professor Josef Singer (1999) testifies on the influence that Caltech group had on him and on the subsequent research at the Technion — Israel Institute of Technology:

> My . . . sabbatical at Caltech in the late '60s was an even more effective two-way technology transfer. I worked with Chuck Babcock and Johann Arbocz on axially compressed, stiffened cylindrical shells. I brought to Pasadena our accumulated Technion experience on the fabrication and testing of closely-stiffened cylindrical shells, while they introduced me to their imperfection measurement and data reduction methods, previously developed for isotropic shells. The result was a joint program on prebuckling, buckling and postbuckling of stringer — and ring — stiffened cylinders (Singer *et al.*, 1971) with careful recording of initial imperfections and their growth. These tests were closely correlated with nonlinear theory for imperfect shells and produced experimental data that was extensively used in the verification of nonlinear codes for stiffened cylindrical shells during these decades. This joint work also initiated the activities of the International Imperfection Data Bank (with branches at Delft Technical University and Technion, Haifa, see Singer and Abramovich (1995) or Arbocz and Hol (1995)).
>
> Upon my return to Haifa, we built an imperfection measurement system that was a development of the one used at Caltech, and since then all the shells tested at Technion were fully imperfection scanned. A decade later we developed a similar and more automated system for larger specimens.

Good correlation between theoretical and experimental results was a major achievement in shell stability analysis. Still, as Fraser and Budiansky (1969) stress:

> an analysis based on a detailed knowledge of the geometrical imperfection of a particular structure is often too difficult [but realizable], and in any case it would be *too specific* [author's emphasis] for the results to be widely applicable to similar structures with slightly different imperfection shapes.

Therefore, "A more useful approach is to attempt to correlate the buckling strengths of imperfections sensitive structures to an appropriate statistical description of their initial imperfections."

This idea was adopted also by the researchers at Caltech. The PhD thesis of Rena Scher Fersht (1968a) and her papers (1968b; 1974) were devoted to the probabilistic analysis of initial imperfections. Apparently following the work by Fraser (1965) she states, "Considering the problem of long cylindrical shell, a particular class of random imperfections, which is of practical significance, is the stationary state of imperfections with respect to the axial variable."

Fersht notes: "Further simplification is obtained by assuming that the random variables satisfy the ergodic property." The author also notes: "After completion of the present report the author's attention was drawn to a study carried out at Harvard (Amazigo, 1967) on the same subject using different techniques. At the present time no comparison of results has been made."

Numerical and experimented results of the Caltech group culminated with the dissertation of Arboucz (1968), who continued this line of work also after obtaining the degree, first as the Senior von Humboldt Fellow at the DFVLR in West Germany, and then at the Delft University of Technology, The Netherlands since 1 September 1976. In 1975 he visited the Technion and presented three lectures (see the copy of the announcement as Fig. 1.5). The lectures were titled as follows:

(1) "Prediction of buckling loads on experimentally measured initial imperfections"
(2) "Some problems associated with data handling when measuring initial imperfections"
(3) "Accurate numerical methods for computing the buckling load of axially compressed imperfect cylindrical shells."

As can be seen, these lectures did not deal with a general methodology for allocating scientifically justified knockdown factors. The general idea propagated in these lectures, namely the "prediction of buckling loads on experimentally measured initial imperfections" appeared to me as an impractical goal. Indeed, one cannot visualize that every shell that will ever be constructed will be amenable to measurement of its initial imperfections. Perhaps Fraser and Budiansky (1969) were implicitly criticizing Caltech's approach to the problem, when noting that "... an analysis based on a detailed knowledge of the geometrical imperfection of a particular structure is often too difficult, and in any case it would be too specific for the results to be widely applicable to similar structures with lightly different imperfection shapes."

הטכניון - מכון טכנולוגי לישראל
הפקולטה להנדסה אוירונוטית

Technion — Israel Institute of Technology
Department of Aeronautical Engineering

סמינריון

PROF. DR. J. ARBOCZ
SENIOR VON HUMBOLDT FELLOW
DFVLR — WEST GERMANY

יתן סדרת הרצאות בנושאים ובימים הבאים:

ביום רביעי 23.4.75 בשעה 15.30 בחדר 216 בבנין אוירונוטיקה

על הנושא: " PREDICTION OF BUCKLING LOADS BASED ON EXPERIMENTALLY MEASURED INITIAL IMPERFECTIONS "

ביום שני 28.4.75 בשעה 16.00 בחדר 216 בבנין אוירונוטיקה

על הנושא: " SOME PROBLEMS ASSOCIATED WITH DATA HANDLING WHEN MEASURING INITIAL IMPERFECTIONS "

ביום רביעי 30.4.75 בשעה 17.00 בחדר 240 בבנין אוירונוטיקה

על הנושא: " ACCURATE NUMERICAL METHODS FOR COMPUTING THE BUCKLING LOAD OF AXIALLY COMPRESSED IMPERFECT CYLINDRICAL SHELLS "

ההרצאות תנתנה בשפה האנגלית.

כבוד קל יוגש לפני כל הרצאה.

Fig. 1.5. Announcement on lectures by Professor J. Arbocz at the Technion.

Chapter 2

Probabilistic Resolution

Particularly in stability problems the variability of the imperfections plays an essential part... To obtain a clear picture of the reliability of such structural elements it becomes imperative to take into account of the uncertainties that enter into any real application... It is realized that imperfections are by nature random quantities...

J. Roorda (1980)

This chapter deals with the probabilistic resolution of the paradox created by the work of Fraser and Budiansky (1965) and other studies performed at Harvard. Bolotin's (1962) contribution is described in some detail. In it Bolotin utilized a single-term Bubnov–Galerkin method to derive reliability of the compressed structure. It is shown that multimode analysis was needed to find realistic reliability of structures.

1. Bolotin's Pioneering Work

The reliability approach is practically the only way to logically based structural design in situations when various and just contradicting requirements to the structure and many loading conditions must be taken into account.

V.V. Bolotin (1979)

Some people hate the very name statistics, but I find them full of beauty and interest.

F. Galton (1889)

In his pioneering paper, Bolotin (1958, 1962) deals with buckling of elastic shells. He notes that "the actual critical values" of the load, "which are determined by experiment... are scattered... with a considerable dispersion that depends on the thoroughness with which the experiment is carried out." He poses the following question: "What value of the load should be considered hazardous, and thus be used as a basis for engineering calculations?" Referring the design load as the "hazardous load," Bolotin replies to his question by stating that

> a correct solution of the problem of hazardous loads can be obtained only on the basis of statistical methods. Actually, the substantial scattering of the experimental values for critical loads is explained by the fact that such values are extremely sensitive to initial deformations, defects in manufacture and fixing, irregularities in loading, etc.

Fig. 2.1. Professor V.V. Bolotin lecturing at a conference.

> These in turn are random factors and are subject to certain statistical laws. Knowing the distribution of these random factors, one can establish the distribution laws of the parameters that characterize the deformed state of the shell. Then the problem of the engineering design of a shell is reduced to finding the values of the loads for which the probability of buckling (or of a hazardous condition defined in some other sense) does not exceed a certain pre-established value for this particular design.

Bolotin (1958, 1962) suggested that one has to first establish the functional dependence

$$w_k = \varphi_k(q_1, q_2, \ldots, q_n), \ (k = 1, 2, \ldots, m) \tag{2.1}$$

where w_k are parameters characterizing the deformed state of the shell, whereas q_k are some determining parameters of the initial imperfections, by using theory of shells.

Bolotin (1958, 1962) further postulates that

> ... without loss of generality, it can be assumed that all of the defining parameters $q_1, \ldots, q_n$ are random quantities (if part of them are determined [quantities], this may be taken into account by introducing singularities into the distribution laws)

with corresponding joint probability density $p(q_1, \ldots, q_n)$. Due to supposed randomness of q_j, the quantities $w_1, \ldots, w_n$) turn out to be random variables too. Then an assumption is made that a hazardous state of the shell, specifically, in Bolotin's terminology, "buckling, reaching dangerous deflections or dangerous stresses" occurs when the condition

$$\Psi^*(q_1, \ldots, q_2) > 0 \tag{2.2}$$

is fulfilled.

Then the "probability of encountering a hazardous condition," or probability of failure $P(*)$ is obtained as

$$P(*) = \int \cdots \int p(q_1, \ldots, q_n) \mathrm{d}q_1, \ldots, \mathrm{d}q_n \\ \psi^*(q_1, \ldots, q_n) > 0 \tag{2.3}$$

The critical load q_* is expressed by Bolotin as

$$q_* = q_* = (q_1, \ldots, q_n) \tag{2.4}$$

requiring solution of the corresponding stability problem. Bolotin (1958, 1962) stresses:

> ...from the experimental point of view it is of the greatest interest to consider the following problem: knowing the distribution of the initial deflection of a shell (and also the distribution of its elastic properties, reinforcements, etc.) to find the law of distribution for critical loads, and in particular to calculate their mathematical expectation and variance.

The first attempt of realization of Bolotin's methodology was provided by then his student, Makarov (1962), who noted (in a free translation) that,

> "[a]t present there are no trustworthy information on character of [probability] distribution of initial imperfections in shells. Hence it makes much sense to pose an inverse problem — by the experimentally found distributions of critical loads to obtain the distribution law of parameters that characterize initial imperfections."

Although the first rigorous methodology of probabilistic methods in stability was proposed by Bolotin (my Grand Teacher for 9 years during 1962–1971 and then forever) in 1958, it was apparently first advocated by Arnold Sergeevitch Volmir in 1956:

> One has to establish in which cases the stability analysis can be conducted based on the linear theory, with sufficient accuracy. It is especially important for nonlinear stability analysis... to conduct theoretical and experimental investigations of the initial imperfections' influence and [that of] initial stresses on the buckling process with the use of statistical methods.

Harris *et al.* (1957) also invoked probabilistic methodology to the data of buckling loads. Specifically, they developed design curves based on a 90% probability criterion.

It should be noted that one of the earliest applications of probabilistic methodology in shell buckling belongs to Mann-Nachbar (1965). They mention that "the chessboard buckle patterns in the solution of the linearized Donnell equations for buckling of a thin, cylindrical shell under axial compression is so sensitive to uncertainties in shell dimensions that the number of circumferential waves and the aspect ratio of the buckles is indeterminate." The authors treated the problem statistically

Fig. 2.2. Professor A.S. Volmir lecturing at the conference on plates and shells.

by treating shell dimensions as random variables with probability density functions dependent on both nominal values and the manufacturing tolerances. The probability densities of the aspect ratio and number of circumferential waves were found by the Monte Carlo method. The authors concluded that there is always a preferred buckle mode.

Almroth and Rankin (1983) also resorted to the probabilistic methodology, using 20 independent nonlinear analyses with determination of the critical load with 99% probability of success.

2. Studies by the University of Waterloo Group

To pinpoint the exact form and magnitude of the initial imperfection in a given shell, and how this may vary with shell geometry and from shell to shell, seems a practical impossibility.

J. Roorda (1972)

Bolotin's (1958, 1962) methodology was also utilized and further extended by Roorda (1969, 1972) and Thompson (1967). Specifically, Roorda (1980) stresses:

> Bolotin (1962, 1969) has taken an approach to this problem which is basically very simple. He proposes the existence of a functional relationship between the critical load

> and various random variables which describe the initial imperfections. This functional relationship can be used as a transfer function to obtain a probabilistic description of the critical load. These ideas may be extended to the concept of *failure* (achievement of a limiting state such as a critical load or a specified magnitude), and to the concept of reliability. Other researchers, which follow this kind of approach, have been published by Roorda (1969, 1972, 1975), Roorda and Hansen (1972), Hansen and Roorda (1974a, 1974b) and Johns (1976).

The difference between Bolotin's (1958, 1962) and Roorda's (1972) works lies in the fact that Roorda uses Koiter's (1945, 1963) asymptotic relationships to derive probabilistic analyses. Specifically, the initial imperfection of the shell is postulated to be axisymmetric one:

$$w_0(x) = \mu h \sin p_0 x, \tag{2.5}$$

where μ denotes the imperfection magnitude as a fractional value of the shell thickness h.

The parameter p_0 is chosen to minimize the critical load, and x is the axial coordinate. Koiter's (1945) relationship reads:

$$2(1-\lambda)^2 - 3c\mu\lambda = 0, \tag{2.6}$$

with $\lambda = 3(1-v^2)sR/(Eh) = csR/(Eh)$ being the normalized critical load; s, v, E, and R are the axial compressive stress, Poisson's ratio, Young's modulus, and shell radius, respectively. Expressing λ as a function of μ yields

$$\lambda(|\xi|) = 1 + \frac{3}{4}|\xi| \pm \frac{1}{2}\left(6|\xi| + \frac{9}{4}|\xi|^2\right)^{1/2} \tag{2.7}$$

with $\xi = c\mu$. Roorda (1972) suggests that the absolute value of ξ is used in the above equation because for shells of infinite length the sign of ξ is of no practical importance.

Roorda (1972) then considers ξ to have the normal probability density function

$$f_\xi(\xi) = \frac{1}{\sqrt{2\pi\sigma^2}} \exp\left[-\frac{(\xi - \bar{\xi})^2}{2\sigma^2}\right] \tag{2.8}$$

where $\bar{\xi}$ is the mean value of the equivalent imperfection, σ^2 represents the variance. In another paper, Hansen and Roorda (1974b) noted: "Perry (1966) has shown by an extensive process of accurate measurements that for certain simple structures the initial imperfections are distributed as Gaussian variables."

Roorda (1972) then states:

> The function $\lambda(|\xi|)$ given in Eq. (2.7) is independent of the sign of ξ and thus equal positive and negative imperfections will cause equal reductions in the critical stress. The question of discontinuity in Eq. (2.7) may therefore be eliminated by expressing $f_\xi(\xi)$ as the sum of two normal probability density functions with equal means of opposite signs, and retaining only the part for which $\xi \geq 0$. Also, $\bar{\xi}$ may be assumed positive without loss of generality.

He derives shell reliability as follows:

$$R(\xi_s) = \frac{1}{\sqrt{2\pi}\sigma}\int_0^{\xi_s}\left\{\exp\left[-\frac{(\xi-\bar{\xi})^2}{2\sigma^2}\right] + \exp\left[\frac{(\xi-\bar{\xi})^2}{2\sigma^2}\right]\right\} d\xi \tag{2.9}$$

with load λ_s corresponding to $|\xi| = \xi_s$ in Eq. ().

$$\lambda_s = \lambda(\xi_s) = 1 + \frac{3}{4}\xi_s - \frac{1}{2}\left(6\xi_s + \frac{9}{4}\xi_s^2\right)^{1/2} \tag{2.10}$$

Roorda (1972) expresses ξ_s in terms of the mean and standard deviation of ξ:

$$\xi_s = \bar{\xi} + N_s\sigma \tag{2.11}$$

where N_s represents the number of standard deviations ξ_s is from the mean.

The author postulates that both the mean and variance of the equivalent axisymmetric imperfection are, to a first approximation, linear functions of R/h and can be written as

$$\bar{\xi} = \beta R/h, \quad \sigma^2 = \alpha R/h. \tag{2.12}$$

According to Roorda (1972) "the constants, β and α, are to be evaluated on the basis of experimental results." The author derived his summarizing equation

$$\lambda_s\left(\frac{R}{h}\right) = 1 + \frac{3}{4}\left(\beta\frac{R}{h} + N_s\sqrt{\alpha\frac{R}{h}}\right) - \frac{1}{2}\left[6\left(\beta\frac{R}{h} + N_s\sqrt{\alpha\frac{R}{h}}\right) + \frac{9}{4}\left(\beta\frac{R}{h} + N_s\sqrt{\alpha\frac{R}{h}}\right)^2\right]^{\frac{1}{2}} \tag{2.13}$$

expressing the stress level at which the shell system has a reliability $R(\xi_s)$ in terms of the shell radius to thickness ratio, the factor N_s being dependent on R/h as well as the prescribed reliability R.

The similarity between Roorda's (1972) work to Bolotin's (1958, 1962) studies lies in the fact that both utilized a single-term approximations to express buckling loads as a function of initial imperfection parameter. Bolotin (1958, 1962) utilized the single-term Bubnov–Galerkin approximation, whereas Roorda (1972) employed Koiter's (1963) asymptotic theory. Both lacked the introduction of the real data into analysis. Roorda (1972) recognized limitations of his analysis: "On the other hand the linear approximations in Eq. (2.12) may not be very good for larger R/h values. In fact an additional nonlinear term in R/h included in the expressions for $\bar{\xi}$ and σ^2 might bend the curves enough to follow the experimental data more closely. It should, however, be reiterated here that Koiter's formula is strictly valid only for small values of ξ and thus the inclusion of additional terms is probably not justified."

In the subsequent publication, Roorda and Hansen (1972) utilized a skewed probability density for the initial imperfection amplitude, with the property that when the skewness parameter approaches zero the distribution approaches normality. They again utilized the shell reliability as their goal, noting: "In the actual design for stability of a cylindrical shell the objective is to arrive at a situation in which the shell has a specified reliability in its performance." Their calculations showed, among others, the following: "A long cylindrical shell, selected from a population of such shells which have axisymmetric deviations with mean amplitude at only one tenth the shell thickness and an uncertainty in amplitude (as measured by the standard deviation) of the same magnitude, will have a 95% reliability at a low load of 0.45 times the ideal buckling load." The authors concluded: "It is perhaps possible, on the basis of a statistical analysis of the wealth of experimental results now available, to come to some conclusions regarding the variability of imperfections in shells of different geometry and support conditions, and to introduce the concept of equivalent axisymmetric imperfections in such shells. Via this route it seems possible to use the present theory to develop safe design curves covering a wide range of shell geometrics and end conditions."

Naturally, it is expected that in the general case shell will possess *nonsymmetric* imperfections. Therefore, it is not clear how the general nonsymmetric imperfections could be equivalent to the axisymmetric imperfections.

Hansen and Roorda (1974a) studied, *inter alia* buckling of an axially loaded prismatic finite beam on a linear elastic foundation. In another paper, Hansen and Roorda (1974b) considered the case when the applied axial load is also a random variable, with applied load and the magnitude of the initial imperfections being independent Gaussian random variables.

Studies performed at the University of Waterloo by Roorda and Hansen were summarized in the monograph by Roorda (1980). He emphasized that he followed Bolotin's (1958, 1962) approach: "In the following sections the approach by Bolotin will be employed to gain some insight into the application of statistical methods in the theory of structural stability."

3. Studies by the Harvard Group (Stochastic Subgroup)

We postulate then that one of the basic responsibilities of the mathematician is to examine the structure of the problems of society and to provide mathematical formulations which are easily susceptible to numerical solution in terms of the current technology. This means not only an examination of the computational aspects, but also of the experimental aspects. What information is required for which formulation, what sensing devices are available, what accuracy do they possess, what should be measured when, and so on?

R. Bellman (1962)

Rice (2008) writes: "Some very well-known contributions, made in part with his Ph.D. student and later faculty colleague John W. Hutchinson, were on the sensitivity to initial imperfections (1966). These could greatly reduce the buckling load of a real structure relative to that of, say, a perfect cylinder or spherical shell."

This is what Hutchinson (2002), the ASME Timoshenko Medal recipient, had to say:

> When pressed to state what I regard as the most remarkable single contribution of an individual in solid mechanics in my lifetime, I am inclined to say that it was Warner Koiter's Ph.D. thesis, *"On the stability of elastic equilibrium,"* published in Amsterdam in 1945. The thesis developed the theory of elastic buckling and post-buckling behavior, the effect of initial geometric imperfections on buckling, and applied this theory to columns, plates and shells. But, that was not all, most of Koiter's subsequent seminal contributions to shells, both linear and nonlinear, had their beginnings in this thesis, and many aspects were already well developed there. I take pride in the fact that Bernie Budiansky and I were among the first to discover Koiter's thesis, and that was not until 1963. Incidentally, the thesis work was carried out during the war in occupied Holland. Koiter later told me he did much of the work in a closet by the light of a candle — he may have been exaggerating. The thesis was published in Dutch. Budiansky and I relied on our astrophysics colleague, Max Krook, who knew Afrikaans and, therefore, a little Dutch to provide us translations of critical sections. Some years later, after Koiter's approach was widely appreciated, I naively asked Koiter why he never published his work on stability. He looked at me down his long nose and informed me it has been published! In Dutch, as his thesis! Shell buckling was one of the hot areas in the 60's, motivated by rockets and other aerospace structures. The perplexing aspect everyone was trying to come in terms with at the time was the notorious discrepancy between the collapse load of actual shells and what was predicted theoretically for buckling of a perfect shell. Thin cylindrical shells under compression were observed to collapse at loads as small as 20% of the theoretical prediction in contrast to columns and plate structures which showed good agreement between experiment and theory for the perfect structures. The key to understanding the discrepancy was the highly nonlinear post-buckling behavior and the extreme sensitivity to imperfections which were related and clarified by Koiter's thesis. Skeptics at the time thought that the basic theory for the perfect shell was intrinsically flawed, but it wasn't. In fact, in the late 60's, Rod Tennyson at the University of Toronto succeeded in making shells so nearly perfect that the buckled within 95% of the prediction of the perfect shell. All this is now history. Buckling problems of all kinds arise continually in many areas of technology. Sometimes I wonder where the expertise on buckling will reside when all of us aging bucklers cross the bar. ABAQUS can solve buckling problems, but it can't pose or understand them. I'm afraid it would not take long to count the number of courses on buckling now taught in this country.

The above statement can be supported by quoting Guran and Lebedav (2011):

> . . . in engineering practice we see that actual critical values can differ significantly from predicted values. Why does this happen? First, conditions in practice are quite far from

> those of precise experimentation. What does this mean? In short, it could be said that the discrepancy in values is produced by inappropriate modeling for studying these effects.

Also, the "difference in the correlation between theory and experiments . . . " led Niedenfuhr (1963), to speculate "on other [than imperfection-sensitivity] possible explanations of the wide scatter of experimental values." In his words, "one idea that appears to hold some promise is the following: a *mechanical systems that is loaded by non-conservative generalized forces may buckle dynamically as well as statically*. . . Yet another effect that ought to be considered in certain cases is the time history of the applied pressure in a test . . ."

In their response, Herrmann and Bungay (1964) noted on the possible presence of the nonconservative force component in the loading: ". . . such type of loading . . . could hardly be held responsible for the scatter of shell buckling loads. As pointed out by Bolotin (1963), not a single experiment has ever been carried out in which buckling would have been produced by a non-conservative static force. The fact of the matter is that such forces are quite easily introduced into the analytical treatment of a model by means of arrows, but their realizability in test present great difficulties . . ." Although Niedenfuhr (1964) "having studied the writers' argument . . ." remained "unconvinced of their validity," his suggestion was not taken by the research community.

In words of Bažant (2000b), "Despite persistent efforts, the discrepancy remained unexplained until von Kármán and Tsien (1941) found the answer in the extreme imperfection sensitivity of non-linear postcritical behavior which causes bifurcation to be asymmetric [their celebrated result was later found to fit the broader content of Koiter's (1945) postcritical theory . . .]." Moreover, "That the critical load can be closely approached was experimentally demonstrated only when Almroth, Holmes and Brush (1964) and Tennyson (1969) succeeded in fabricating cylindrical shells with extraordinarily small imperfections."

Having discovered Koiter's thesis (1945), Harvard group went on to attack the problem in two distinct ways. On one hand, Budiansky and Hutchinson either together, or separately, with their respective collaborators, dealt with deterministic imperfection-sensitivity problems. On the other hand, Budiansky and his collaborators, either jointly or separately, dealt with stochastic imperfections.

The first appearance of the stochastic imperfections in the Harvard group starts with the survey by Budiansky and Hutchinson (1966). Analyzing Koiter's (1945) work and works performed following Koiter they comment: "The kinds of investigations just discussed serve to demonstrate whether or not a given configuration is imperfection sensitive but indicate only qualitatively the degree of such sensitivity; they cannot be used to predict the actual buckling load of a given structure that is imperfection sensitive."

Fig. 2.3. Professor Bernard Budiansky (1925–1999) of Harvard University recognized the importance of probabilistic approaches to the imperfection-sensitivity of structures.

As is seen, Harvard group recognized the limitation of works that specify initial imperfections and then find the loads that such a structure will support. They were after prediction of "actual buckling load of a given structure that is imperfection sensitive."

Budiansky and Hutchinson (1966) continue their discussion on how this noble goal can be accomplished. They criticize one possibility of approaching the problem:

"On the other hand, it does not seem very sensible to attempt to develop methods of analyses based upon a very detailed knowledge of the imperfection under consideration."

We will come back to this idea. Let us now give the podium to Budiansky and Hutchinson (1966): "A more useful goal might be to attempt to correlate the buckling strengths of imperfect structures with appropriate statistical descriptions of their initial imperfections."

Referring to the Ph.D. thesis by Fraser (1965), Budiansky and Hutchinson describe, in some detail, a "pilot problem" of infinitely long column that rests "on a non-linear 'softening' foundation and is supposed to have a initial displacement which is assumed to be a stationary random function of position along the length of the beam."

Apparently referring to Koiter-type analysis, Budiansky and Hutchinson (1966) note: "It is evident that not only will initial imperfections in the shape of the critical buckling mode influence the actual static buckling load of the imperfect structure, but so will, to some extent, imperfections having any other shape."

We will stop quoting further from this paper as pertaining to the stochastic imperfection sensitivity but turn to Fraser's (1965) thesis itself as well as the article by Fraser and Budiansky (1969) that resulted from the thesis, under supervision by Budiansky.

The work by Fraser and Budiansky (1969) has been generalized by Amazigo and his co-authors in several publications. As Amazigo (1976) notes in his review paper:

"Based on the assumption of ergodicity *deterministic* buckling loads have been obtained for numerous structures using approximate methods such as equivalent linearization, truncated hierarchy, and perturbations."

Specifically, linearization technique was utilized by Amazigo (1969); truncated hierarchy method was applied by Amazigo (1969) and Amazigo *et al.* (1971); Poincare perturbation method was used by Amazigo (1971, 1974a, 1974b).

Amazigo (1976) tries to explain the results obtained by the "stochastic" part of Harvard group, which embraced the ergodicity assumption, as follows:

"The method discussed here which is based on the ergodicity hypothesis leads to the conclusion that the structure will buckle statically or dynamically at the corresponding load... with probability 1. *This result may appear paradoxical* [author's emphasis]. However, to dispel the apparent contradiction we note that no matter how the origin of an infinitely long column is defined the buckling load for such columns with imperfections $\bar{w}(x) = \sin(x + \varphi)$ is independent of φ and hence independent of any probabilistic distribution we may assign to φ."

Unfortunately, Videc (1974) and Videc and Sanders (1976) were in agreement with above works on deterministic buckling loads possessed by the stochastic

Fig. 2.4. Professors A.N. Guz (standing), V.V. Bolotin, Yu. N. Rabotnov, and B. Budiansky (seated, from left to right).

structures. Lockhart and Amazigo (1975) and Amazigo and Frank (1975) utilized the ergodicity assumption for some dynamic instability problems of stochastically imperfect shells.

Tvergaard (1976) surveys the work done on stochastic treatment of initial imperfections:

> While most of the investigations mentioned previously consider deterministic imperfections in the shape of the critical buckling mode, some results have been directed towards the *realistic situation* [author's emphasis] where imperfections are known as stochastic rather than deterministic properties. One approach is taken by Amazigo, Budiansky and others, who consider the initial imperfection to be a sample function from an ensemble of ergodic, zero-mean, stationary Gaussian random functions with known autocorrelation. The analyses based on methods of stochastic differential equations lead to asymptotic estimates of the buckling load corresponding to a given imperfection amplitude, and *due to ergodicity hypothesis* [author's emphasis], this buckling load is given with probability. Shell buckling results have been obtained by this method for axially compressed cylinders with random axisymmetric imperfections (Amazigo and Budiansky, 1972; Fersht, 1974) and for externally pressurized cylinders with imperfections that vary randomly in the circumferential direction (Amazigo, 1974).

Moreover,

> Another approach takes an imperfection of a given shape with random amplitude, or the sum of a finite number of given imperfections shapes with random amplitudes. The applied load can also be taken as a random parameter. Then, using the deterministic relations between imperfection parameters and buckling load, the probability of failure occurs can be calculated provided the joint probability density function of these random parameters is known. Recent treatment of stochastic stability problems from this point of view, for an axially compressed cylinder and for other structures, have been given by Roorda and Hansen (1972), Hansen and Roorda (1974)], Augusti and Baratta (1976) and Johns (1976). Amazigo (1976) has diseased the two approaches and used the latter on an externally pressurized cylindrical shell.

Roorda (1980, p. 96), describing the work by the Harvard group, writes:

> A completely different approach [than that by Bolotin] is to use the methods of stochastic differential equations. In this approach the view is taken that imperfections (such as initial geometrical deviations from perfection) are random in shape as well as magnitude. The measure used for the intensity of the initial imperfections takes the form of a power spectral density or the root-mean-square deviation, and results are obtained which relate the critical load to such intensity indicators. Researchers that have followed this approach are Amazigo (1969, 1976), Fraser and Budiansky (1969), van Slooten and Soong (1972), Fersht (1968) and Amazigo *et al.* (1970). These authors generally obtain *the surprising result* [author's emphasis] that the critical load is deterministic. This can be attributed to their assumption (explicit or implicit) that the imperfection is an ergodic process. Such an assumption *essentially eliminates the randomness* [author's emphasis] in the problem since it implies that almost every sample has identical spatial averages.

The work by van Slooten and Soong (1972) is closely related to the Harvard group's studies on stochastically imperfect structures. They utilized

> a perturbation technique developed by Dym and Hoff (1968) . . . to reduce the nonlinear stochastic differential equations to an infinite set of linear deterministic equations with random forcing functions. The perturbation parameter used is the standard deviation of the initial imperfection. The mean and variance of the critical load are determined by solving the first few equations in this set. In the important case when the initial imperfection is small, a simple asymptotic formula is obtained for the critical load.

Their asymptotic formula reads, for the mean buckling load:

$$E(\lambda) = 1 - 1.207[\pi\phi^2 S_{gg}(1)]^{2/5} c^{4/5} \tag{2.14}$$

where $\lambda = \sigma/\sigma_{CL}$, $\phi = 12(1 - v^2)$, and S_{gg} is the spectral density. The authors arrived at the following conclusion: "since the random stress parameter λ is equal to its expected value with probability one, we may treat the relationship between λ and χ [dimensionless end–shortening parameter] as deterministic."

They also concluded, in the abatract of their work: ". . . the maximum load is seen to be equal to its mean value with probability one."

Thus, they arrived at the same paradoxical conclusion as the stochastic members — if such a term can be used — of the Harvard group.

Naturally the above result is an artifact of the assumption of "an infinite cylindrical shell" that was employed "in order to obtain a mathematically tractable problem." The authors rightfully noted: "For a finite shell, the initial imperfection cannot be treated as a stationary random process because of boundary conditions imposed at the ends. Therefore, the spectral analysis developed here for an infinite shell is not applicable for a finite shell."

van Slooten and Soong (1972) also noted: ". . . it is known from deterministic analysis (Dym and Hoff, 1968) that for long cylinders the effect of shell length on the buckling load is small. We, therefore, use the infinite model with some assurance that the critical loads calculated are realistic."

It appears that the latter statement cannot be justified. When dealing with infinite shell, some stochastic information is lost; specifically, only the mean buckling load can be determined, whereas the finite shell possesses the variability of buckling loads.

In the discussion of the paper by van Slooten and Soong (1972), Amazigo and Budiansky (1973) noted that "unfortunately, the authors' conclusion is incorrect. Within the framework of the Kármán–Donnell theory these can be no limit-point (load maximum) in any cylinder having axisymmetric initial imperfections." van Slooten and Soong (1973) appreciated "the comments advanced by Professors Amazigo and Budiansky regarding the paper" but disagreed with the criticism. The invoking of the assumptions of infinitely long structures on one hand, and ergodicity of initial imperfections on the other, invalidates, in our humble opinion, the above efforts.

4. Ergodicity May Induce Large Errors

> *You have probably heard the famous saying (attributed by Mark Twain to Benjamin Disraeli) that there are three kinds of lies: lies, damned lies, and statistics. To add insult to injury, Darrel Huff wrote a famous book in 1954 called* How to Lie with Statistics *that described how statistics can be used in deceptive ways.*
>
> P. Oloffson (2007)

As mentioned above the papers utilizing stochastic treatment of initial imperfections, written at Harvard, heavily utilized the assumption of ergodicity of initial imperfection. Softpedia answers the question "What is ergodicity?" as follows, in laymen's terms:

> Why are election polls inaccurate?... Why are your assumptions often mistaken? The answers to all these questions and to many others have a lot to do with non-ergodicity of human ensembles...
>
> Suppose you are concerned with determining what the most visited parks in a city are. One idea is to take momentary snapshot: to see how many people are this moment in park A, how many are in park B and so on. Another idea is to look at one individual (or few of them) and follow him for a certain period of time, e.g. a year. Then, you observe how often the individual is going to park A, how often he is going to park B and so on.
>
> Thus, you obtain two different results: one statistical analysis over the entire ensemble of people at a certain moment of time, and one statistical analysis for one person over a certain period of time. The first may not be representative for a longer period of time, while the second one may not be representative for all the people.
>
> The idea is that an ensemble is ergodic if two types of statistics give the same result. Many ensembles, like the human population, are not ergodic.

According to Lebowitz and Penrose (1973), ergodicity "means, roughly, whether the system, if left to itself for long enough, will pass close to nearly all the dynamical states compatible with conservation of energy."

In this section we consider two examples that demonstrate that this assumption may introduce large errors.

Following Scheurkogel *et al.* (1981), we consider the following stochastic differential equation:

$$\frac{\mathrm{d}^2x}{\mathrm{d}t^2}+\frac{x}{\xi^2}=\sqrt{4-pA^2-pB^2},\quad t>0,\quad 0<p\leq 4 \tag{2.15}$$

with initial conditions

$$\begin{aligned} x(0)&=\xi A+\xi^2\sqrt{4-pA^2-pB^2}\\ x'(0)&=B \end{aligned} \tag{2.16}$$

Fig. 2.5. Professor W.T. Koiter, Ir. A. Scheurkogel, and the author (from left to right).

where parameter is defined as follows:

$$\xi^2 = \lim_{T\to\infty} \int_0^T x^2(t)\mathrm{d}t, \quad \xi > 0. \tag{2.17}$$

In Eqs. (2.15) and (2.16), p is a control parameter. The random variables A and B are jointly uniformly distributed and the unit circle $A^2 + B^2 \leq 1$. Our objective is to find mathematical expectations $E[x(t)]$ and $E[x^2(t)]$, where $E[\ldots]$ denotes mathematical expectation. It should be noted that Eq. (2.15) is not associated with any physical problem. It is specially devised solely to illustrate some subtle points associated with making probabilistic assumptions.

We first derive approximate solution based on the ergodicity assumption. If is assumed to be mean-square ergodic, the mean-square value follows from Eqs. (2.16) and (2.17):

$$\begin{aligned}\xi^2 &= E[x^2(t)] = E[x^2(0)] = E\left[\xi A + \xi^2\sqrt{4 - pA^2 - pB^2}\right]^2 \\ &= \frac{1}{4}\xi^2 + \left(4 + \frac{1}{2}p\right)\xi^4\end{aligned} \tag{2.18}$$

Thus

$$\xi^2 = E[x^2(t)] = \frac{3}{2(8-p)}. \tag{2.19}$$

In view of this formula, taking expectation of Eqs. (2.15)–(2.17) we find, after solving for $E[x(t)]$:

$$E[x(t)] = \frac{3}{2(8-p)} E\left[\sqrt{4 - pA^2 - pB}\right] \tag{2.20}$$

Introducing the random variable

$$Z = A^2 + B^2 \tag{2.21}$$

uniformly distributed on the interval [0, 1], we obtain

$$\begin{aligned} E[x(t)] &= \frac{3}{2(p-8)} E\left[\sqrt{4-pZ}\right] \\ &= \frac{3}{2(p-8)} \int_0^1 \sqrt{4-pz}\mathrm{d}z = \frac{8-(4-p)^{3/2}}{p(8-p)} \end{aligned} \tag{2.22}$$

As for the exact solution, we note that for each realization of the solution $x(t)$, ξ is independent of t. Consequently the solution of Eqs. (2.15)–(2.17) is

$$x(t) = a\xi \cos\frac{1}{\xi} + b\xi \sin\frac{t}{\xi} + \xi^2(4 - pa^2 - pb^2)^{1/2} \tag{2.23}$$

where the value of ξ is obtained from the substitution of (2.23) into (2.17):

$$\xi^2 = \frac{1-(a^2+b^2)/2}{4-p(a^2+b^2)}. \tag{2.24}$$

One immediately observes from Eq. (2.24) that ξ depends, in general, on the particular realization of the random variable a and b, which implies that $x(t)$ is not mean-square ergodic for arbitrary p. However, as Scheurkogel *et al.* (1981) showed, $x(t)$ is wide-sense stationary, that is, $E[x(t)]$ is constant and the autocorrelation function $E[x(t)x(t+\tau)]$ depends only on the time lag τ only but not t. For details, one can consult the paper by Scheurkogel *et al.* (1981). Specifically, they derived the following exact solution:

$$E[x(t)] = \frac{1}{3p^2}\left[12p - 16 - (5p-8)(4-p)^{1/2}\right], \tag{2.25}$$

$$E[x^2(t)] = \frac{1}{2p}\left[1 - \frac{2(p-2)}{p}\log\left(1-\frac{p}{4}\right)\right]. \tag{2.26}$$

In accordance with Eq. (2.24), the necessary condition for mean-square ergodicity is that $p = 2$ since otherwise ξ would depend on the particular realization of a and b. From Eq. (2.26) we obtain $E[x^2(t)] = 1/4$ for $p = 2$. On the other hand, from Eqs. (2.17) and (2.24) we have

$$\xi^2 = \lim_{\tau\to\infty} \int_0^T x^2(t)\mathrm{d}t = \frac{1}{4} \tag{2.27}$$

so that $\xi^2 = E[x^2(t)]$. This implies the mean-square ergodicity of $x(t)$ iff $p = 2$.

It is intriguing, that for $p = 0$, the solution based on the ergodicity assumption also coincides with the exact solution, although for this particular value of p process is not ergodic in the mean-square sense. This is an example of a "good (or beneficial) error": in this case the wrong assumption leads to correct results.

The mathematical expectation $E[x(t)]$ and the mean square $E[x^2(t)]$ are shown in Figs. 2.6 and 2.7, respectively. It is remarkable that both the exact and approximate

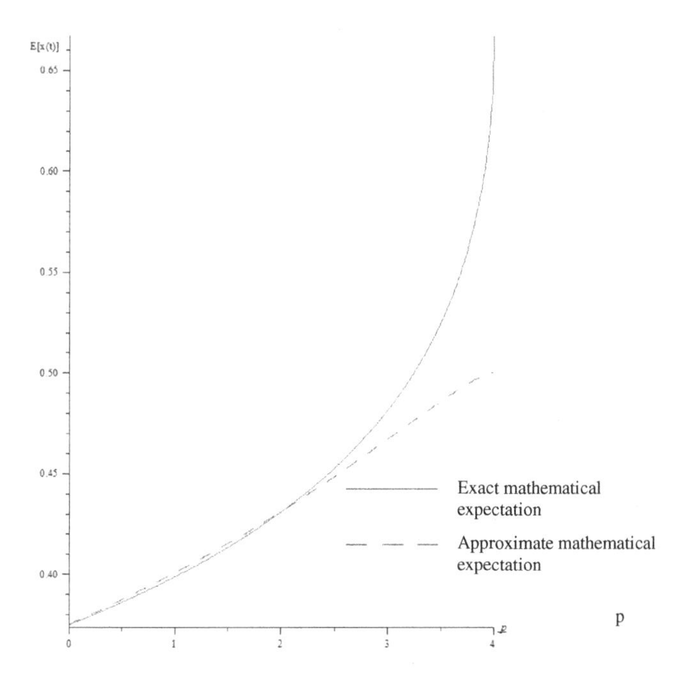

Fig. 2.6. Mathematical expectation as a function of control parameter p.

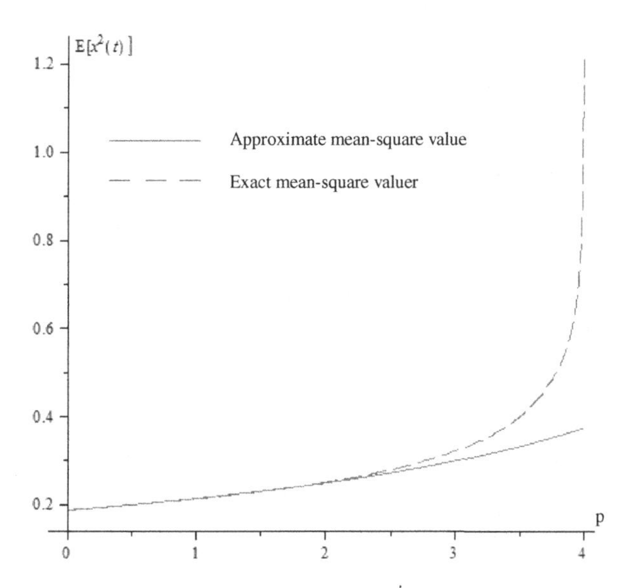

Fig. 2.7. Mean-square value as a function of p; at $p = 0$ and $p = 2$, the exact and approximate solutions coincide.

solutions are very close in the range $0 \leq p \leq 2$, coinciding at the ends of the interval. The percentage error relative to the exact value induced by the ergodicity assumption is of the order of -0.5% in this range.

The foregoing error increases rapidly with p and reaches its maximum at $p = 4$. For the mathematical expectation this error is 25%, whereas for the mean-square value the error approaches 100%: The exact mean-square value $E[x^2(t)]$ tends to infinity, while the approximate one remains finite.

As Softpedia stressed above, "Many ensembles, like the human populations, are not ergodic." In some studies that appeared well after our work ergodicity assumptions validity is doubted too. Grime and Langley (2008) note: "The ergodicity assumption is more likely valid for long T_L." Der Kiureghian (2005), describing his work, notes:

> A framework formula for performance-based earthquake engineering, advocated and used by researchers at the Pacific Earthquake Engineering (PEER) Center, is closely examined. The formula was originally intended for computing the mean annual rate of a performance measure exceeding a specified threshold. However, it has also been used for computing the probability that a performance measure will exceed a specified threshold during a given period of time. It is shown that the use of the formula to compute such probabilities could lead to errors when non-ergodic variables (aleatory or epistemic) are present.

It appears instructive to quote also Earman and Rédei (1996), who stress that "ever since Boltzmann (1871) introduced the concept of ergodicity, its role in statistical mechanics has been controversial."

5. Bolotin's Problem and Ergodicity Assumption

> *Probabilistic methods are "often regarded as a kind of 'magic wand' which produces information out of a void. This is a fallacy; the theory of probability only enables information to be transformed, and conclusions on inaccessible phenomena to be drawn from data on observable ones."*
>
> E.C. Wentzel (1980)

In the previous section we have shown that the ergodicity assumption, widely adopted by the Harvard group, may yield either good or poor approximations, depending on circumstances. It should be stressed that the example presented in the previous section outside the mechanics context there will consider mechanics example. Bolotin (1971) utilized this assumption to derive the second moment of the solution for the case of the white-noise input, and further showed that the first few terms of a perturbation solution for which he again, albeit implicitly, assumed ergodicity, were in agreement with his original result. Thus, Bolotin (1971) did not derive an exact solution to demonstrate the accuracy of the ergodicity assumption.

Here, following Scheurkogel and Elishakoff (1985) an exact solution is given for the Bolotin's problem for particular loading conditions. It will be demonstrated that in this applied mechanics problem the assumption of ergodicity of the output may yield a large error even for a mean-square ergodic input.

Consider the following random differential equation:

$$Dw'''' - (Eh/2)M^2w'' + (Eh/R^2)w = P(x) \tag{2.28}$$

where

$$M^2 = \lim_{L\to\infty} \frac{1}{L}\int_{-L/2}^{L/2} [w'(x)]^2 \mathrm{d}x, \quad D = Eh^3/12(1-v^2) \tag{2.29}$$

where E is the modulus of elasticity, h is the thickness, R is the radius, L is the length, D is the flexural rigidity, and is the Poisson ratio.

Equation (2.28) describes the deformation of a cylindrical shell under a load $P(x)$. The latter is assumed to be a random field that is wide-sense homogenous with zero mean and specified autocorrelation function. For derivation of Eq. (2.28) the reader may consult Bolotin's monograph (1971, 1974), with the following notation:

$$\alpha = Eh/4D, \quad \beta = (Eh/DR^2)^{1/2} \quad Q(x) = P(x)D. \tag{2.30}$$

Equation (2.28) can be recast in the form

$$w'''' - 2\alpha M^2 w'' + \beta^2 w = Q(x). \tag{2.31}$$

We seek the mean-square value denoting mathematical expectation.

Mathematically this is a quasilinear differential equation with random inhomogeneous term and, through M, with random coefficients. We seek the mean square value $E(M^2)$, with E denoting mathematical expectation.

Since Bolotin's (1971, 1974) book is not uniformly available, we first reproduce his solution to the above problem. Bolotin (1971, 1974) postulated mean-square ergodicity of $Y'(x)$. In this case, we have from Eq. (2.29)

$$M^2 = \lim_{L\to\infty}(1/l)\int_{-L/2}^{L/2} [Y'(x)]^2 dx = E\{[Y'(x)]^2\} \tag{2.32}$$

with probability 1. Denoting the second moment of M by σ^2, it follows from this equation that

$$M^2 = E[M^2] = \sigma^2 = E\{[Y'(x)]^2\} \tag{2.33}$$

with probability 1. This implies that the random variable M can be treated as deterministic constant and may be replaced by σ in Eq. (2.31), which thus becomes

$$Y'''' - 2\alpha\sigma^2 Y'' + \beta^2 Y = Q(x). \tag{2.34}$$

With σ a given constant — to be determined eventually from Eq. (2.35) — we now have a differential equation with deterministic coefficients and random input. We consider the spatial spectral density of functions Q and Y, representing the Fourier transforms of the corresponding autocorrelation functions K_{QQ} and K_{YY}, respectively:

$$S_{QQ}(\lambda) = \frac{1}{2\pi}\int_{-\infty}^{\infty} K_{QQ}(u)\mathrm{e}^{-\mathrm{i}\lambda u}\mathrm{d}u \tag{2.35}$$

$$S_{YY}(\lambda) = \frac{1}{2\pi}\int_{-\infty}^{\infty} K_{YY}(u)\mathrm{e}^{-\mathrm{i}\lambda u}\mathrm{d}u. \tag{2.36}$$

The following relation is then valid (Elishakoff 1983, 1999) between the spectral densities of the input and output of Eq. (2.34):

$$S_{YY}(\lambda) = \frac{S_{QQ}(\lambda)}{(\lambda^4 + 2\alpha\sigma^2\lambda^2 + \beta^2)^2}. \tag{2.37}$$

The second moment of $Y'(x)$ is then

$$E\{[Y'(x)]^2\} = \int_{-\infty}^{\infty} \lambda^2 S_{YY}(\lambda)\mathrm{d}\lambda. \tag{2.38}$$

Finally, substitution of Eqs. (2.33) and (2.37) into Eq. (2.38) leads to

$$\sigma^2 = \int_{-\infty}^{\infty} \frac{\lambda^2 S_{QQ}(\lambda)\mathrm{d}\lambda}{(\lambda^4 + 2\alpha\sigma^2\lambda^2 + \beta^2)^2}. \tag{2.39}$$

This equation coincides with Eq. (19) of Bolotin (1971, 1974).

Scheurkogel and Elishakoff (1985) worked out a special example when the loading on the cylindrical shell is

$$Q(x) = a\sqrt{2}\sin\Theta\cos(\omega x - \Phi) + a\cos\Theta \tag{2.40}$$

where

$$a = \left[\frac{6(\omega^4 + \beta^2)^3}{\alpha\omega^4}\right]^{1/2} \tag{2.41}$$

with Φ and Θ being independent random variables, the former uniformly distributed on $[0, 2\pi]$, and the latter having the following probability density function:

$$P_\Theta(\theta) = \begin{cases} \frac{1}{2}\cos\theta,\ |\theta|,\ |\theta| \le \frac{\pi}{2} \\ 0,\ |\theta| \le \frac{\pi}{2}. \end{cases} \tag{2.42}$$

Scheurkogel and Elishakoff (1985) showed that the random function $Q(x)$ is wide-sense homogenous and mean square ergodic.

The spectral density of $Q(x)$ reads:

$$S_{QQ}(\lambda) = \frac{a^2}{6}\delta(\lambda - \omega) + \frac{a^2}{6}\delta(\lambda - \omega) + \frac{2a^2}{3}\delta(\lambda) \tag{2.43}$$

where $\delta(x)$ is the Dirac's delta function. Substitution of Eq. (2.43) into Eq. (2.39) yields

$$\sigma^2 = \frac{a^2\omega^2}{3(\omega^4 + 2\alpha\sigma^2\omega^2 + \beta^2)^2}, \tag{2.44}$$

which, after substitution of the expression for a namely, Eq. (2.41) can be simplified to

$$\sigma^3 + 3\gamma^2\sigma - 6\sqrt{3}\gamma^3 = 0 \tag{2.45}$$

where

$$\gamma = \left(\frac{\omega^4 + \beta^2}{6\alpha\omega^2}\right)^{1/2}. \tag{2.46}$$

From Eq. (2.45) we get the only real root for σ and the associated value

$$\sigma^2 = 3\gamma^2. \tag{2.47}$$

The exact solution (see for details Scheurkogel and Elishakoff, 1985) reads:

$$E(M^2) = \left(\frac{p^5 - p^{-5}}{30} + \frac{p - p^{-1}}{6} - 2\right)\gamma^2 \tag{2.48}$$

where

$$p(9 + \sqrt{82})^{1/3} \tag{2.49}$$

or, finally,

$$E(M^2) = 2.51557\gamma^2. \tag{2.50}$$

The percentagewise error of the approximate solution given in Eq. (2.47), relative to exact solution, constitutes 19.26%. In other words, assumption of the mean-square ergodicity for the shell slope may lead to a considerable error, even if the loading is mean-square ergodic.

One can construct other not less vivid examples in which ergodicity assumption leads to even greater error, though sometimes it may constitute a useful approximation.

6. Simulation of Initial Imperfections

The idea for what was later called the Monte Carlo method occurred to me when I was playing solitaire during [an] illness. I noticed that it may be much more practical to get an idea of the probability of the successful outcome ... by ... experimenting with the process and merely noticing what proportion comes out successfully ... This is intellectually surprising, and if not exactly humiliating, it gives one a feeling of modesty about the limits of rational or traditional thinking.

S. Ulam (1976)

Already in 1977, during his sabbatical at the Technion, I informed Professor Bernard Budiansky that I intended to conduct large-scale numerical experiments on checking the feasibility of the conclusion made in the paper by Fraser and Budiansky (1969) and Amazigo (1971) on determinicity of the buckling load in nonlinear beams with stochastic initial imperfections. In past two sections we showed that the ergodicity assumption of the initial imperfections may lead to large errors. Still, it was necessary to tackle the specific problem studied in above papers: imperfect column on a nonlinear elastic foundation.

Professor Budiansky responded that the Ph.D. dissertation of Wynstone Barrie Fraser already utilized the Monte Carlo method. However, the inspection of the thesis showed that it employed the Monte Carlo method for a single-term Bubnov–Galerkin approximation. However, multiterm analysis appeared to be needed.

It appeared necessary first to develop a simulation procedure of space-random fields. This task was accomplished in 1979 (see Elishakoff, 1979a) although the manuscript was submitted over two years earlier. The random field $\bar{w}(x)$ was represented by the following series in terms of the eigenfunctions $f_j(x)$:

$$\bar{w}(x) = \sum_{j=1}^{\infty} a_j f_j(x) \tag{2.51}$$

where a_j is a random variable for every fixed j. The series ( ) is truncated to some n (number of eigenfunctions taking into consideration) deriving from the requirement that the probabilistic characteristics of the described random field $\bar{w}(x)$ can be determined with sufficient accuracy. Equation (2.51) is replaced by

$$\bar{w}(x) = \sum_{j=1}^{n} a_j f_j(x). \tag{2.52}$$

The mean values of the coefficients a_j are determined by

$$E(a_j) = \frac{\{E[w(x)], f_j(x)\}}{v_j^2}, \quad v_j^2 = \{f_j, f_j\}; \quad (j = 1, \ldots, n) \tag{2.53}$$

where $\{\phi, \psi\}$ is the inner product of $\varphi(x)$ and $\psi(x)$:

$$\{\varphi, \psi\} = \int_0^1 \phi(x)\psi(x)dx. \tag{2.54}$$

The autocorrelation function of $\bar{w}(x)$

$$E\langle\{\bar{w}(x_1) - E[\bar{w}(x_1)]\}\{\bar{w}(x_2) - E[\bar{w}(x_2)]\}\rangle = K_{\bar{w}}(x_1, x_2) \tag{2.55}$$

is given by

$$K_{\bar{w}}(x_1, x_2) = \sum_{j=1}^{n}\sum_{k=1}^{n} \sigma_{jk} f_j(x_1) f_k(x_2) \tag{2.56}$$

where

$$\sigma_{jk} = \frac{1}{v_j^2 v_k^2}\int_0^L\int_0^L K_{\bar{w}}(x_1, x_2) f_j(x_1) f_k(x_2) \mathrm{d}x_1 \mathrm{d}x_2. \tag{2.57}$$

In deriving Eqs. (2.53) and (2.57) we utilized the property of orthogonality of the eigenfunctions $f_j(x)$:

$$\{f_j(x), f_k(x)\}, \quad j \neq k \tag{2.58}$$

The problem is now reduced to simulation of the random vector $[a]^{\mathrm{T}} = [a_1, a_2, \ldots, a_n]$ with mean function (2.53) and variance covariance matrix $\Sigma = \{\sigma_{jk}\}_{n\times n}$ defined in Eq. (2.57). We assume that the mean value coincides with the zero vector. This entails no loss of generality because if a vector $[a]$ is simulated as above, the vector $[a + \mu]$ has the mean $[\mu]$ and the same variance–covariance matrix $[\Sigma]$. Thus we consider the simulation of a random vector $[a]$ that is normally distributed $N([0], [\Sigma])$, specifically the multivariate normal distribution with zero mean and variance–covariance matrix $[\Sigma]$. The problem of simulation of normal random vectors was considered by Scheur and Stoller (1962). The method uses the following fact: Let $[b]$ be distributed $N([0], [I_n])$ and let

$$[a] = [C][b] \tag{2.59}$$

where $[I_n]$ is a unit matrix of size n. Then the vector a is distributed $N([0], [C][C]^{\mathrm{T}})$. In the present case we want $[C][C^{\mathrm{T}}]$ to equal $[\Sigma]$:

$$[\Sigma] = [C][C^{\mathrm{T}}]. \tag{2.60}$$

The matrix $[\Sigma]$ is positive definite and in uniquely decomposable in the form ( ), where $[C]$ is a lower triangular matrix with positive diagonal elements. It can be obtained by using the Cholesky's procedure for factoring a positive definite matrix.

Elishakoff's (1979a) paper gives further details of simulation when $\bar{w}(x)$ is part of a *homogenous* random field autocorrelation function $K_{\bar{w}}(x_1, x_2) = K_{\bar{w}}(\xi)$, $\xi = x_1 - x_2$, where ξ is a separation distance. Specifically, the following expression was obtained for the element σ_{jk}:

$$\sigma_{jk} = \frac{1}{v_j^2 v_k^2} \int_0^1 K(\xi) M_{jk}(\xi) \mathrm{d}\xi \tag{2.61}$$

$$M_{jk} = P_{jk}(\xi) + P_{kj}(\xi) \tag{2.62}$$

$$P_{jk} = \int_{\xi/2}^{1-\xi/2} f_j \left(\eta - \frac{1}{2}\xi \right) f_k \left(\eta + \frac{1}{2}\xi \right) \mathrm{d}\eta \tag{2.63}$$

For simply supported beams

$$f_j(x) = \sin(j\pi\xi), \quad \xi = x/L \tag{2.64}$$

and

$$\sigma_{jj} = 4 \int_0^1 K(\zeta)[(1-\zeta)\cos(j\pi\zeta) + (1/j\pi)\sin(j\pi\zeta)]\mathrm{d}\zeta \tag{2.65}$$

$$\sigma_{jk} = \frac{4[1-(-1)^{j+k+1}]}{\pi(k^2 - j^2)} \int_0^1 K(\zeta)[k\sin(j\pi\zeta) - j\sin(k\pi\zeta)]\mathrm{d}\zeta, \quad j \neq k \tag{2.66}$$

Elishakoff (1979a) studied the numerical example for the space-random field with zero mean and the autocorrelation function

$$K(\xi) = e^{-A|\xi|} \cos B\xi \tag{2.67}$$

where A and B are positive nondimensional quantities. This autocorrelation function was utilized by Fraser and Budiansky (1969) and Amazigo (1969) to describe initial imperfections. For further details the reader may also consult textbook by Elishakoff (1983, 1999). This simulation procedure was utilized also in papers by Elishakoff (1978a, 1978b, 1980a) in a dynamic setting. It must be noted that there are other spectral decomposition methods such as the spectral representation (Shinozuka and Deodatis, 1991) or the Kahrunen–Loève expansion that are widely used in the literature for simulation of initial imperfections (see, e.g., Schenk and Schuëller, 2005 and references cited there).

7. Resolution of Fraser–Budiansky–Amazigo Paradox for Stochastically Imperfect Columns on Nonlinear Foundation

The general statistical theory should moreover be extended to include systems involving more than two independent parameters and to embrace the more complex points of bifurcation arising in certain problems of shell stability.

J.M.T. Thompson (1967)

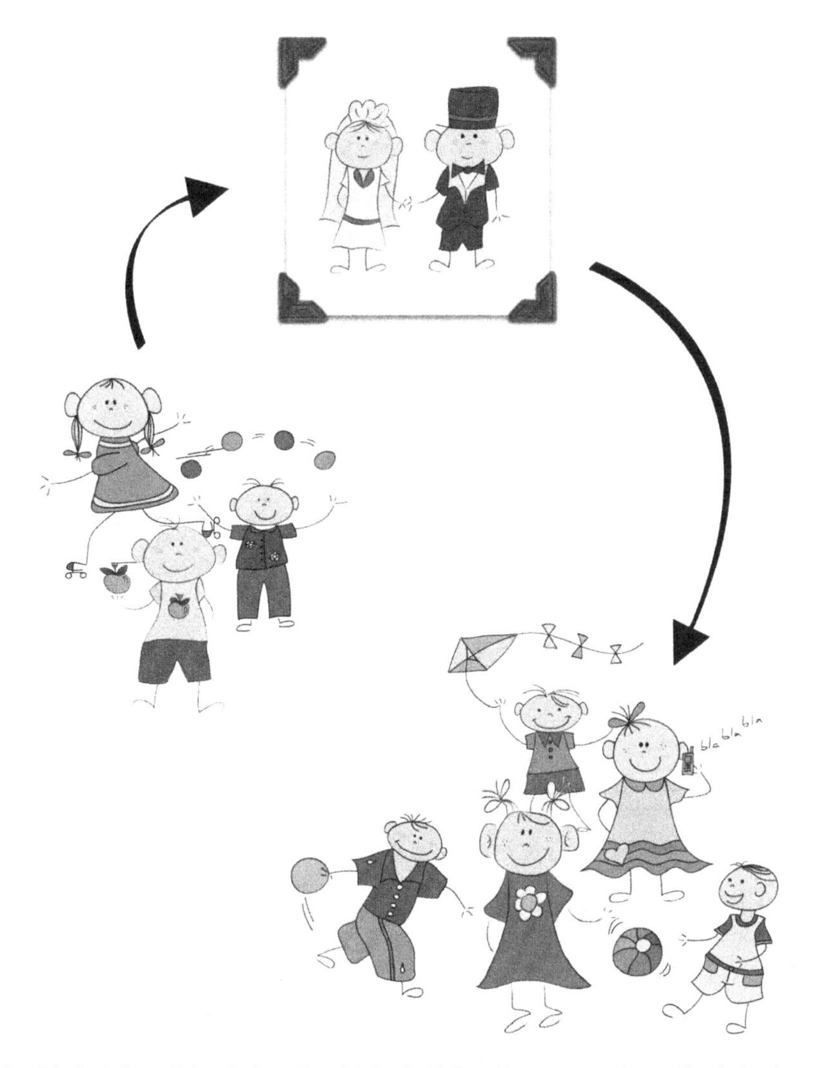

Fig. 2.8. Methodolgy of simulation of multiple shells based upon experimentally derived mean and autocorolation functions of initial imperfection (the "kids" on the left corner represent experimental data pertaining to measured structures; these allow us to construct "parents", namely the mean function, and the auto-correlation function; these in turn allow us to "create" (or simulate) new "kids", or new multiple realizations of the structure that are probabilistic "cousins" of the original structures).

Having described the simulation procedure, which is the cornerstone of the Monte Carlo method, let us turn now the resolution of the Fraser–Budiansky–Amazigo paradox as was accomplished in the paper by Elishakoff (1976b). Consider the equation for the deflection on an imperfect column on a nonlinear "softening" foundation (Fig. 2.1):

$$EI\frac{\mathrm{d}^4 w}{\mathrm{d}x^4} + P\frac{\mathrm{d}^2(w+\bar{w})}{\mathrm{d}x^2} + k_1 w - k_3 w^3 = 0 \tag{2.68}$$

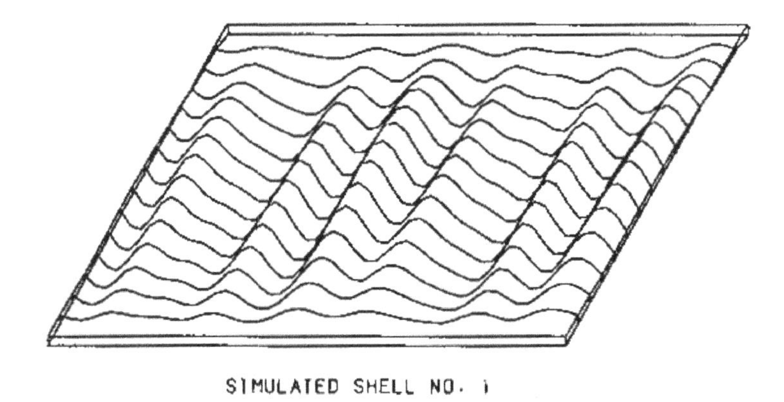

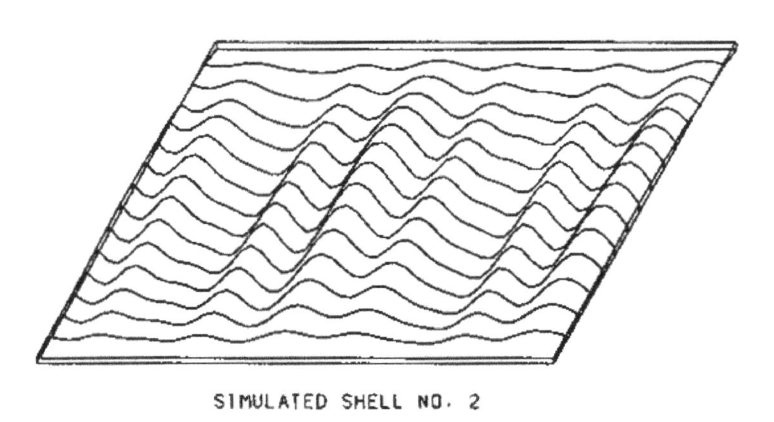

Fig. 2.9. Profiles of simulated initial imperfection based upon experimentally derived mean and autocorrelation functions of initial imperfections.

where E is the Young's modulus, I is the section moment of inertia, $\bar{w}(x)$ is the initial imperfection function, $w(x)$ is the additional deflection due to the axial load P, k_1 is the linear "spring" constant of the foundation, k_3 is the nonlinear "spring" constant of the foundation, and x is the axial coordinate. We introduce nondimensional quantities

$$\begin{gathered} u = w/\Delta, \quad \bar{u} = \bar{w}/\Delta, \quad \eta = x/L \\ \alpha = P/P_{\mathrm{cl}}, \quad \gamma(\kappa_1) = n_*^2\pi^2 + \kappa_1/n_*^2\pi^2 \\ \kappa_1 = k_1 L^4/EI, \quad \kappa_3 = k_3 L^4 \Delta^2/EI \\ P_{\mathrm{cl}}/P_{\mathrm{E}} = \gamma(\kappa_1)/\pi^2, \quad P_{\mathrm{E}} = \pi^2 EI/L^2 \end{gathered} \tag{2.69}$$

where $\Delta = \sqrt{I/A}$ is the radius of inertia, A is cross-sectional area, P_{E} is the critical load of the beam without elastic foundation, P_{cl} is the critical buckling load of a column on a linear elastic foundation, and n_* denotes the number of half-waves during buckling of the linear structure. Unlike the case of columns without elastic

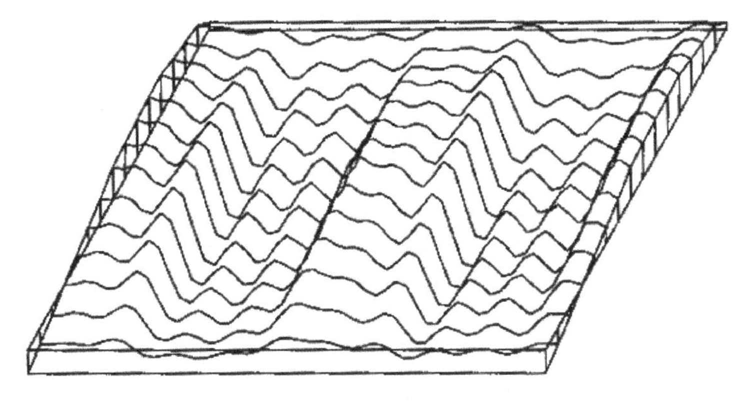

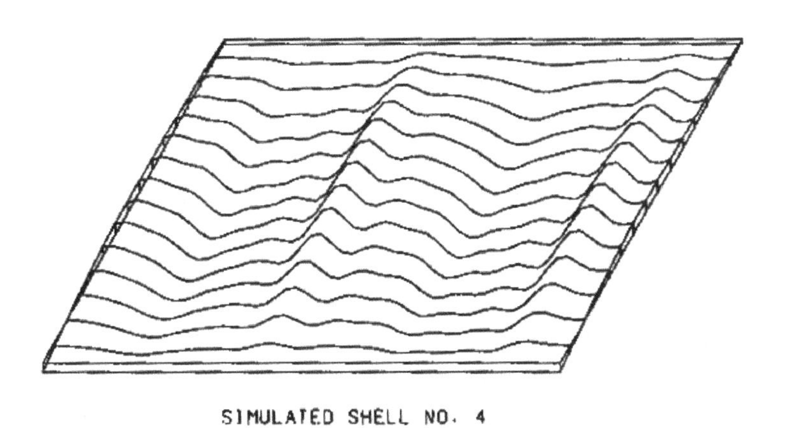

Fig. 2.9. *(Continued)*

foundation, n_* does not necessarily equal unity and is determined from the equation $n_*^2(n_*+1)^2 = \kappa_1/\pi^4$. As κ_1 increases so does the number of half-waves at buckling, and for $\kappa_1 \gg 4\pi^4$, n_* is determined from the condition $n_*^4 = \kappa_1/\pi^4$, leading to the well-known formula $P_{\rm cl}/P_{\rm E} = 2\sqrt{\kappa_1}/\pi^2$, which coincides with the buckling load of an infinite beam.

One reduces Eq. (2.68) to the following form:

$$\frac{d^4u}{d\eta^4} + \alpha\gamma(\kappa_1)\frac{d^2u}{d\eta^2} + \kappa_1 u + \kappa_3 u^3 = -\alpha\gamma(\kappa_1)\frac{d^2\bar{u}}{d\eta^2}. \tag{2.70}$$

The column is simply supported so that the boundary conditions are

$$u = \frac{d^2u}{d\eta^2} = 0, \quad \text{at } \eta = 0 \text{ and } \eta = 1. \tag{2.71}$$

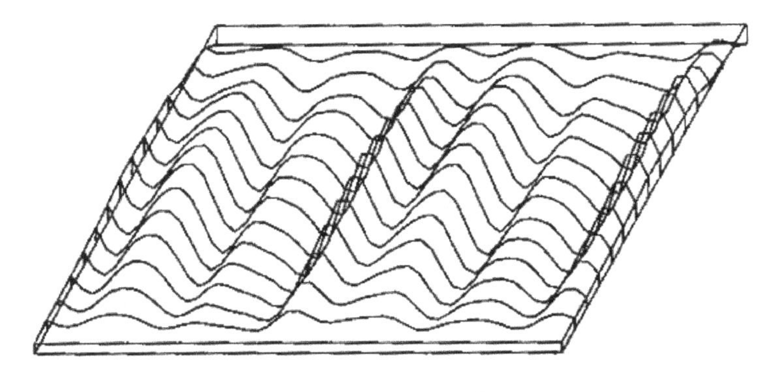

SIMULATED SHELL NO. 5

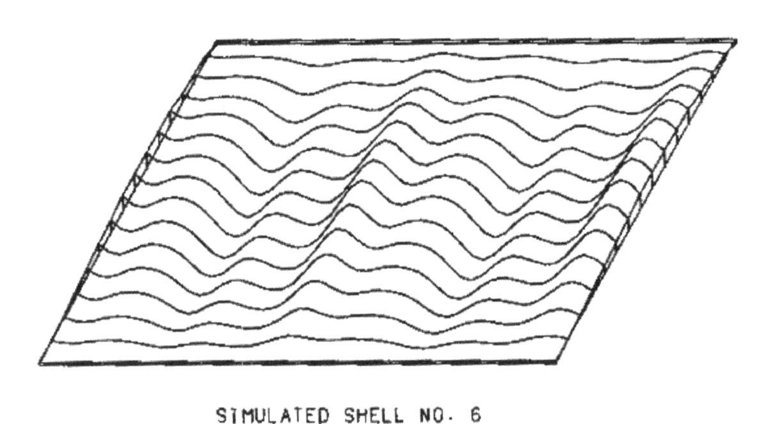

SIMULATED SHELL NO. 6

Fig. 2.9. (*Continued*)

The problem is formulated as follows: Given the probabilistic characteristics of $\bar{u}(\eta)$, find those of the buckling load α^* (for specific values of κ_1 and κ_3).

For a given realization of the imperfection function $\bar{u}(\eta)$, the buckling load is defined from the requirement

$$\frac{\mathrm{d}\alpha}{\mathrm{d}F} = 0, \qquad \alpha = \alpha^* \tag{2.72}$$

where $F = F(u, \bar{u})$ is some functional of the nondimensional displacements increasing with u. The one we use is the end shortening of the bar (the distance of the bar ends moving together under load):

$$\mathrm{d} = \int_0^L \left[\frac{1}{2}\left(\frac{\mathrm{d}w}{\mathrm{d}x}\right)^2 + \left(\frac{\mathrm{d}w}{\mathrm{d}x}\right)\left(\frac{\mathrm{d}\bar{w}}{\mathrm{d}x}\right)\right]\mathrm{d}x \tag{2.73}$$

In terms of the non-dimensional quantities, the equation reads:

$$F(u, \bar{u}) = \frac{\mathrm{d}L}{\Delta^2} = \int_0^1 \left[\frac{1}{2} \left(\frac{\mathrm{d}u}{\mathrm{d}\eta} \right)^2 + \left(\frac{\mathrm{d}u}{\mathrm{d}\eta} \right) \left(\frac{\mathrm{d}\bar{u}}{\mathrm{d}\eta} \right) \right] \mathrm{d}\eta. \tag{2.74}$$

An approximate solution is obtained from Eqs. (2.70)–(2.74) by expanding u and $\bar{u}$ in series in terms of the modes of stability loss of the associated linear structure:

$$\bar{u} = \sum_{k=1}^{\infty} \bar{\xi}_k \sin(k\pi\eta) \tag{2.75}$$

$$u = \sum_{k=1}^{\infty} \xi_k \sin(k\pi\eta). \tag{2.76}$$

Substituting Eqs. (2.75) and (2.76) into Eq. (2.70), we apply the approximate technique by Bubnov–Galerkin. We multiply both sides of the resulting equation by $\sin(m\pi\eta)$, $m = 1, 2, \ldots, N$ (N being the number of retained terms) and integrate over the span, to obtain the following set of coupled nonlinear algebraic equations for ξ_m:

$$\alpha_m \xi_m - \alpha(\xi_m + \bar{\xi}_m) - s\left(\frac{n_*}{m}\right)^2 \sum_{p=1}^{N} \sum_{q=1}^{N} \sum_{r=1}^{N} \xi_p \xi_q \xi_r A_{pqrm} = 0 \quad (m = 1, 2, \ldots, N) \tag{2.77}$$

where

$$\alpha_m = \frac{(m\pi)^2 + \kappa_1 (m\pi)^{-2}}{(n_*\pi)^2 + \kappa_1 (n_*\pi)^{-2}}, \qquad s = \frac{2\kappa_3}{n_*^4 \pi^4 + \kappa_1} \tag{2.78}$$

$$\begin{aligned} A_{pqrm} &= \int_0^1 \sin(p\pi\eta)\sin(q\pi\eta)\sin(r\pi\eta)\sin(m\pi\eta)\mathrm{d}\eta \\ &= \frac{1}{8}[\delta_{p+q,r+m} - \delta_{|p-q|,r+m} - \delta_{p+q,|r-m|} + \delta_{|p-q|,|r-m|} + \delta_{p,q}\delta_{r,m}] \end{aligned} \tag{2.79}$$

and $\delta_{i,j}$ is the Kronecker delta. Equation (2.77) can be rewritten as

$$\alpha_m \xi_m - \alpha(\xi_m + \bar{\xi}_m) + \frac{s}{8}\left(\frac{n_*}{m}\right)^2 I_m = 0, \quad (m = 1, 2, \ldots, N) \tag{2.80}$$

with

$$I_m = 8 \sum_{p=1}^{N} \sum_{q=1}^{N} \sum_{r=1}^{N} \xi_p \xi_q \xi_r A_{pqrm} \tag{2.81}$$

If $\kappa_1 \gg 4\pi^4$, then according to Eq. (2.78), $\kappa_1 \square\, n_*^4\pi^4$, and α_m and s become

$$\alpha_m = \frac{1}{2}\left[\left(\frac{m}{n_*}\right)^2 + \left(\frac{n_*}{m}\right)^2\right], \quad s = \frac{\kappa_3}{\kappa_1} = \frac{k_3\Delta^2}{k_1} \tag{2.82}$$

with $F(u, \bar{u})$ taking the form

$$F(u, \bar{u}) = \frac{1}{4}\sum_{k=1}^{N} \xi_k^2 k^2\pi^2 + \frac{1}{2}\sum_{k=1}^{N} \xi_k\bar{\xi}_k k^2\pi^2. \tag{2.83}$$

Since a closed-form solution is unfeasible, the set of Eqs. (2.80)–(2.83) have to be solved numerically. Yet, one can derive an analytical solution in the case of a single equation, such as the one to which the set (2.80)–(2.83) reduces when only the mth term is retained in series (2.75) and (2.76):

$$\begin{aligned} \alpha &= \frac{\xi_m}{\xi_m + \bar{\xi}_m}\left[\xi_m - \frac{3}{8}\left(\frac{n_*}{m}\right)^2 s\xi_m^2\right] \\ F &= \frac{1}{4}m^2\pi^2\xi_m(\xi_m + 2\bar{\xi}_m) \end{aligned} \tag{2.84}$$

with α_m and s as per Eq. (2.82). Differentiating Eq. (2.84) with respect to F and setting

$$\frac{d\alpha}{dF} = 0, \quad \alpha = \alpha^* \tag{2.85}$$

we obtain, after appropriate algebraic manipulations, the relation between the buckling load α^* and the initial deflection amplitude $\overline{\xi_m}$, corresponding to the Eq. (2.18) of Fraser's (1965) thesis:

$$(\alpha_m - \alpha^*)^3 = \frac{81}{32}\left(\frac{n_*}{m}\right)^2 s\bar{\xi}_m^2(\alpha^*)^2. \tag{2.86}$$

From this equation the buckling load α^* is obtainable given the amplitude $\overline{\xi_m}$ of the initial imperfection.

Note that with Eq. (2.86) available, the single-term approximation can be brought to conclusion to yield an analytic expression for the reliability function $R(\alpha')$, defined as the probability that the random buckling load is not less than a prespecified load. In this case $\bar{\xi}_n$ has a normal distribution $N(\overline{X_n}, \sigma_{nn})$. We are considering the particular case $\overline{X_n} = 0$. As can be seen from Eq. (2.86),

$$R(\alpha') = \mathrm{Prob}(\alpha' < \alpha^* \leq 1) = \mathrm{Prob}(-\overline{\xi_n'} < \overline{\xi_n} < \overline{\xi_n'}). \tag{2.87}$$

In conclusion, we have

$$R(\alpha') = \text{erf}\left[\frac{\overline{\xi'_n}}{2\sigma_{nn}}\right], \quad \overline{\xi'_n} = \frac{4}{9}\sqrt{\frac{2}{s}}\frac{(1-\alpha')^{\frac{3}{2}}}{\alpha}. \tag{2.88}$$

Multiple-mode solution is obtained by transforming the set of Eqs. (2.77)–(2.83) to the initial-value problem as developed by Qiria (1951) and Davidenko (1951). Note that Qiria–Davidenko method was extensively applied by the Rostov-on-Don school of Russian mechanics, headed by Vorovich; the interested reader can consult papers by Vorovich and Zipalova (1965), Zipalova and Nenastieva (1966), Vorovich and Minakova (1967), Vorovich and Shepeleva (1969), and Zipalova and Nenastieva (1971). For numerical simulation of random initial imperfection vectors we utilize the methodology described in Sec. 6 in this chapter.

Now, we develop the method of transformation of an appropriately truncated set of equations

$$B_m \equiv (\alpha_m - \alpha)\xi_m - \alpha\bar{\xi}_m - (s/8)(n/m)^2 I_m = 0, \quad m = 1, 2, \ldots, N \tag{2.89}$$

where F and I_m are determined accordingly by Eqs. (2.83) and (2.81) in their truncated forms to N (N being the number of retained terms) and applied to the initial-value problem as presented by Qiria (1951) and Davidenko (1951). F is now the independent variable, and α and ξ_m are its functions. Differentiating (2.89) with respect to F, we obtain

$$\left(\frac{\partial B_m}{\partial \alpha}\right)\left(\frac{\mathrm{d}\alpha}{\mathrm{d}F}\right) + \sum_{p=1}^{N}\left(\frac{\partial B_m}{\partial \xi_p}\right)\left(\frac{\mathrm{d}\xi_p}{\mathrm{d}F}\right) = 0. \tag{2.90}$$

We also refer to the identity $\mathrm{d}F/\mathrm{d}F - 1 = 0$, in our case in the form

$$\sum_{p=1}^{N}\left(\frac{\partial F}{\partial \xi_p}\right)\left(\frac{\partial \xi_p}{\mathrm{d}F}\right) - 1 = 0. \tag{2.91}$$

Equations (2.90) and (2.91) may be rewritten as

$$-(\xi_m + \overline{\xi_m}\left(\frac{\mathrm{d}\alpha}{\mathrm{d}F}\right) + \sum_{p=1}^{\infty}\left[(\alpha_m - \alpha)\delta_{p,m} - 3\sum_{i=1}^{N}\sum_{j=1}^{N} C_{ijpm}\xi_i\xi_j\right]\left(\frac{\mathrm{d}\xi_p}{\mathrm{d}F}\right) = 0 \tag{2.92}$$

$$\sum_{p=1}^{\infty}(\xi_p + \overline{\xi_p})p^2\pi^2\frac{\mathrm{d}\xi_p/dF}{2} - 1 = 0 \tag{2.93}$$

where

$$C_{ijpm} = s\left(\frac{n}{m}\right)^2 A_{ijpm} \tag{2.94}$$

The result is a set of $N + 1$ differential equations having $N + 1$ variables $\xi_1, \xi_2, \ldots, \xi_N$ and α, subject to initial conditions representing the unstressed state of the column, namely,

$$\alpha = 0, \quad \xi_1 = \xi_2 = \cdots = \xi_N = 0 \text{ at } F = 0. \tag{2.95}$$

Solving (2.92) and (2.93)–(2.95), we have the vector ξ for every α and $\bar{\xi}$, hence also the $\alpha - F$ curve. According to Eq. (2.72) the maximum point on the branch, which originates at zero nondimensional load, represents the nondimensional buckling load α^*.

Note that the reliability function ought to be used as a basis for design of stochastically imperfect structures, the criterion being

$$R(\alpha') \geq r \tag{2.96}$$

where r is the required reliability. The maximum nondimensional load $\alpha^{(R)}$ which satisfies the condition

$$R(\alpha^{(R)}) = r \tag{2.97}$$

is then the design buckling strength for each column in the ensemble.

Fig. 2.10. Assignment of the required reliability r is a societal issue.

Let us demonstrate some numerical results to shed light on the obtaining the design load. The numerical examples were worked out for the autocorrelation function for the initial imperfections as follows:

$$K_{\bar{\xi}}(\tau) = Ae^{-B|\tau|}\cos(C\tau) \tag{2.98}$$

with the product *AB* as their variance and with their mean function as zero. The coefficients should be $A > 0$, $B > 0$ and $C > 0$. Note that in his study, Elishakoff [19] adopted another autocorrelation function. The elements of variance–covariance matrix of the Fourier coefficients σ_{pq} were approximated from Eqs. (2.65)–(2.66) by numerical integration and the elements of matrix C obtained by the Cholesky's decomposition procedure. The initial value problem obtained from Eqs. (2.92)–(2.93) were integrated by numerous "trial" values of the random Fourier coefficients $\overline{\xi_1}, \overline{\xi_2}, \ldots, \overline{\xi_N}$. Continuous load/end-shortening curves and shown in Fig. 2.11 shows the continuous load/end-shortening curves for different values of n_*, i.e., of the "linear" stiffness coefficient chosen as $\kappa_1 = n_*^4\pi^4$ with $A = 0.01$ and $B = 1$ and $C = 20$. The realization of the imperfection vector used in the calculation for Fig. 2.11 was

$$\begin{vmatrix} \bar{\xi}_1 = 0.0057, \bar{\xi}_2 = 0.0226, \bar{\xi}_3 = -0.0368, \bar{\xi}_4 = 0.0271, \bar{\xi}_5 = -0.0196, \\ \bar{\xi}_6 = -0.0717, \bar{\xi}_7 = 0.0265, \bar{\xi}_8 = -0.0018, \bar{\xi}_9 = 0.0906, \bar{\xi}_{10} = 0.0358, \\ \bar{\xi}_{11} = 0.0089, \bar{\xi}_{12} = 0.0333, \bar{\xi}_{13} = 0.0212, \bar{\xi}_{14} = 0.0084, \bar{\xi}_{15} = 0.0178. \end{vmatrix} \tag{2.99}$$

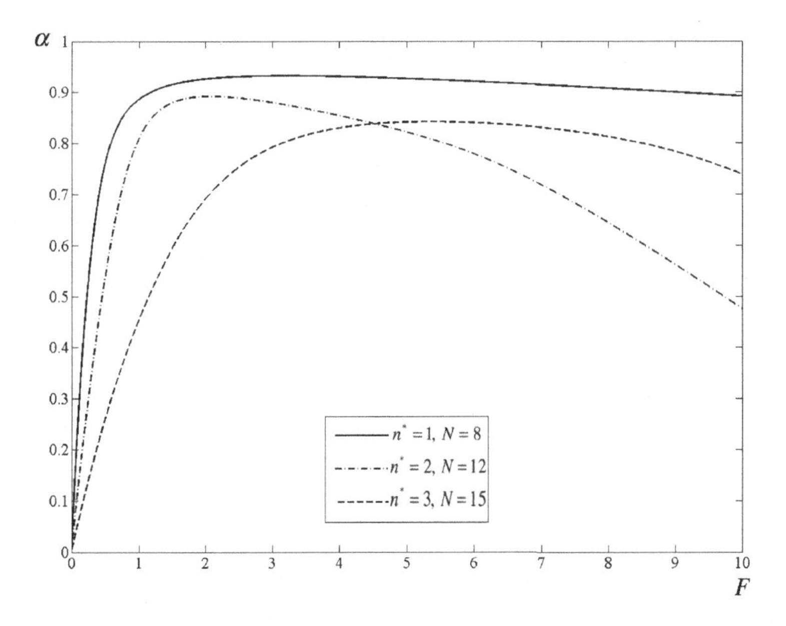

Fig. 2.11. Load/end-shortening curves.

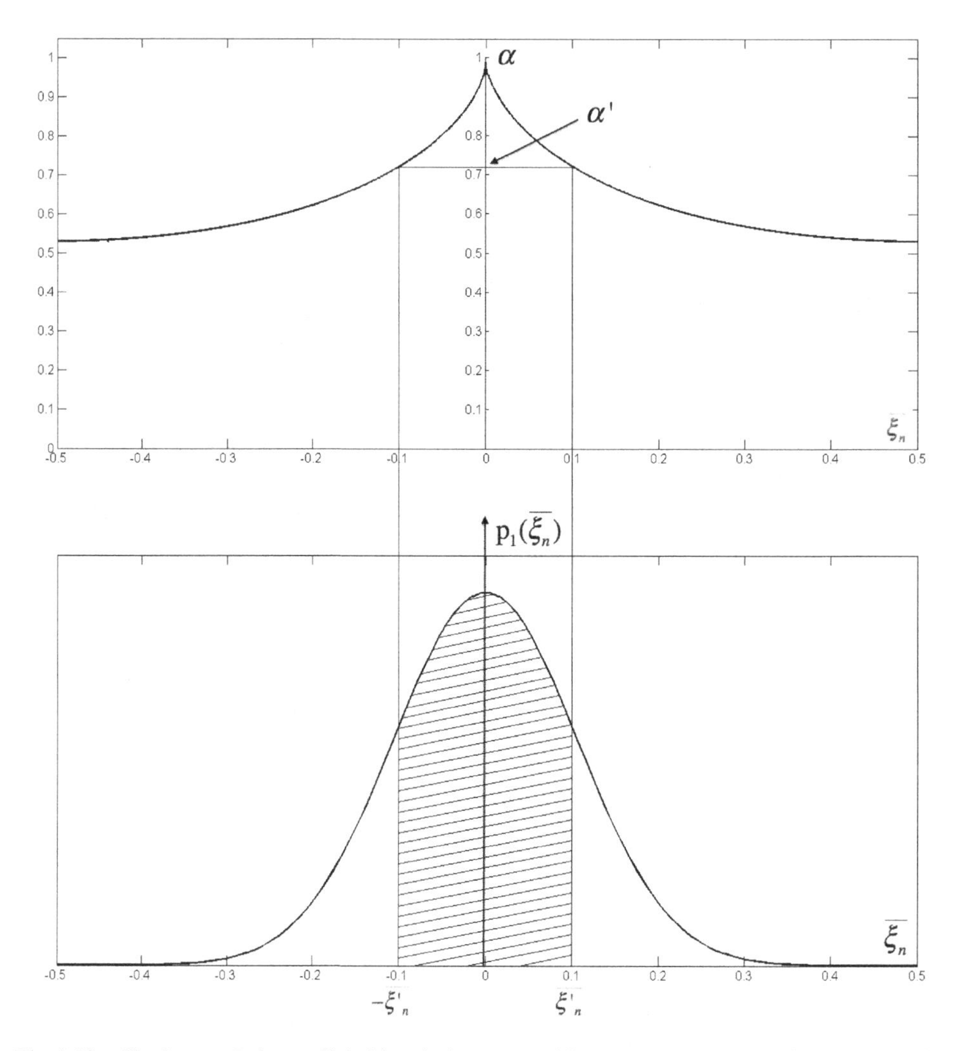

Fig. 2.12. Single-term Bubnov–Galerkin solution: (a) Buckling-load parameter as a function of initial imperfection amplitude and (b) Probability density of initial imperfection amplitude (shaded area equals the reliability of the structure at nondimensional load level α').

For $n_* = 1$, eight modes were needed to be retained in determining the compete load/end-shortening curve for accurate determination of buckling load. It turns out that when the norm of imperfection's vector increases, the value of buckling load decreases. As κ_1 increases, so does the number of modes that must be retained to ensure convergence of the multiple-mode solution: to obtain complete load/end-shortening curve, it is necessary to return at least six modes for $n_* = 2$, eight modes for $n_* = 3$, and 10 modes for $n_* = 4$. When the value B increases, the number of modes to be taken into account generally must increase too. In this paper, the choice was made to use number of terms in excess of the required modes to obtain accurate results.

The evaluation of the buckling loads for different values of the initial imperfections permits to construct the empirical reliability function

$$R''(\alpha') = \frac{1}{M}\mu_M(\alpha') \tag{2.100}$$

where $\mu_M(\alpha')$ is the number of α^* values that exceed the preselected value α' and M is the ensemble size. Brute Monte Carlo method, however, is extremely time-consuming for highly nonlinear problems, especially for large values of n^*. Therefore, a special modification is introduced. The first step is to fit the experimental data obtained for a limited number realizations of the Monte Carlo method theoretically. In order to fit the reliability function, the Kolmogorov–Smirnov test was chosen as the criterion. According to the Kolmogorov–Smirnov test of goodness of fit, and for a sample of 1000 realizations, the critical value of the maximum absolute difference between the theoretical $R(\alpha')$ and the observed empirical reliability function $R^*(\alpha')$ is $1.36/\sqrt{1000} \approx 0.043$ (with a level of significance of 0.05).

The first candidate assumption made was that the buckling load would be normally distributed. Figure 2.13 shows the distribution of the sample and the theoretical associated normal distribution. The maximum distance calculated with Kolmogorov–Smirnov test is 0.0350. This value is lower than the critical value.

However, this assumption cannot be supported for an evident reason: the buckling load cannot be negative whereas normally distributed random variable may take both positive and negative values. Due to this requirement, the log-normal distribution was tested with the sample distribution. Figure 2.14 shows these two distributions.

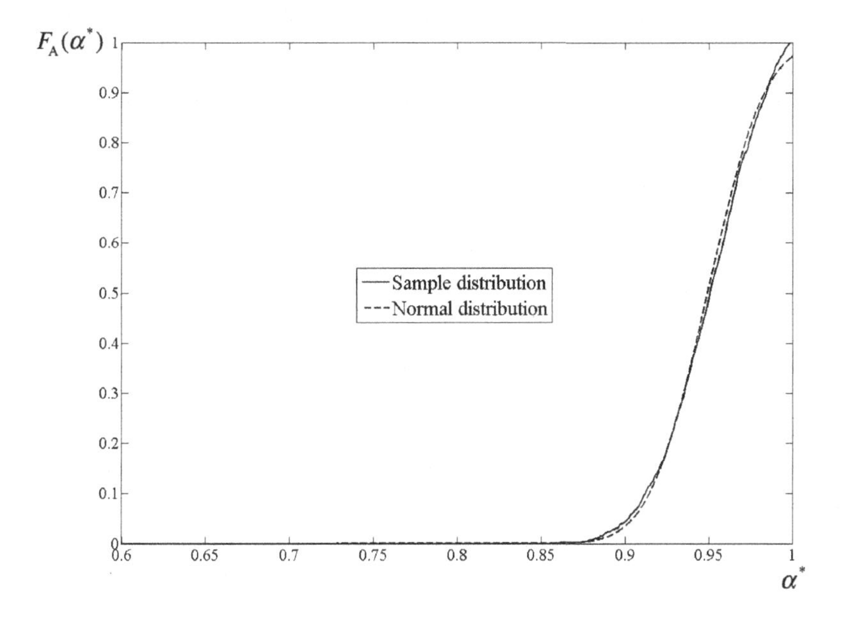

Fig. 2.13. Normal distribution versus sample distribution.

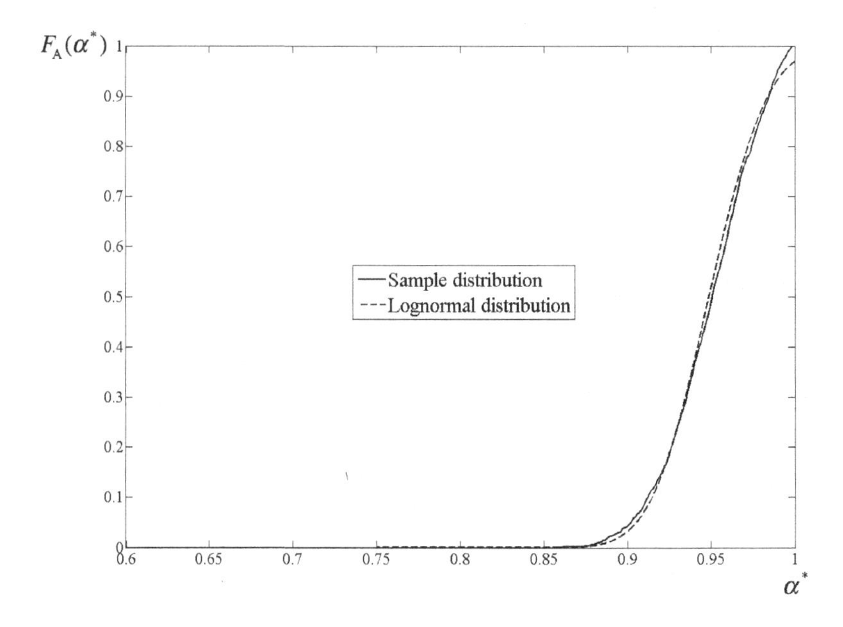

Fig. 2.14. Log-normal distribution versus sample distribution.

Both for normal and log-normal distributions, the parameters of the respective distribution were calculated by the method of moments. For the log-normal case, the distance is 0.0393, which is also lower than the critical value. Because of the maximum distances are relatively close to the critical value for these two widely used distributions, other candidate distributions was also tested.

The generalized extreme value distribution also was tested as a candidate theoretical distribution. This distribution is denoted as GEV. Its probability distribution function is given by the following expression:

$$F_X(x) = \begin{cases} \exp\left\{-\left[1+\xi\left(\frac{x-\mu}{\sigma}\right)\right]^{-1/\xi}\right\} & -\infty < x \leq \mu - \sigma/\xi \text{ for } \xi < 0; \\ & \mu - \sigma/\xi \leq x < \infty \text{ for } \xi > 0; \\ \exp\left\{-\exp\left[-\left(\frac{x-\mu}{\sigma}\right)\right]\right\} & -\infty < x \leq \infty \text{ for } \xi = 0; \end{cases} \tag{2.101}$$

where μ is the location parameter, σ is the scale parameter, and ξ is the shape parameter. Note that Fréchet, Weibull, and Gumbell distributions are particular cases of the GEV distribution. As we can see in the Eq. (2.101), three parameters are used for the GEV distribution. A test of maximum likelihood was used to determine the parameters (Eq. (2.101)) that best fit to the Monte Carlo data. Fisher's maximum likelihood principle applied to our context, as the first step consists in considering the set of values $\alpha^* = \{\alpha_1^*, \alpha_2^*, \ldots, \alpha_n^*\}$ as a set of observations of the buckling load. The random variable α^* can be described by a probability density function $f_A(\alpha^*|\theta)$. $\boldsymbol{\theta}$ is a vector containing the three parameters of the generalized extreme value law, $\theta_1 = \xi$, $\theta_2 = \sigma$ and $\theta_3 = \mu$. One then constructs the likelihood function as the product of probability densities evaluated in observed buckling loads

$$L_A(\theta) = f_A(\alpha_1^*|\theta) f_A(\alpha_2^*|\theta) \ldots f_A(\alpha_n^*|\theta). \tag{2.102}$$

The vector $\boldsymbol{\theta}$ that maximizes the likelihood $L_A(\boldsymbol{\theta})$ is then used as an estimator. This estimator, $\hat{\theta}$, is called the maximum likelihood estimator. In a sense, it is the most likely value for the estimator. The following set of initial parameters was chosen: $A = 0.01$, $B = 1$, $C = 20$, $\kappa_1 = 1$, $\kappa_3 = 1$, $n^* = 1$, and $N = 12$. For this case, the vector $\hat{\theta}$ obtained with the maximum likelihood principle was obtained as follows: $\xi = 0.4137$, $\sigma = 0.0286$, $\mu = 0.9431$. To test generalized extreme value distribution with above parameters against the Monte Carlo simulation, 10,000 samples were generated as sample probability distribution of α^*. The maximal distance evaluated with Kolmogorov–Smirnov test is 0.0267. The maximal distance observed for normal distribution (0.0350) is about 31% above this value; likewise, the distance associated with log-normal distribution is 47% more than its counterpart to GEV. Due to this fact, we conclude that the preferable distribution for buckling load is the generalized extreme value distribution. Figure 2.15 both shows the generalized extreme value distribution and the distribution of the sample.

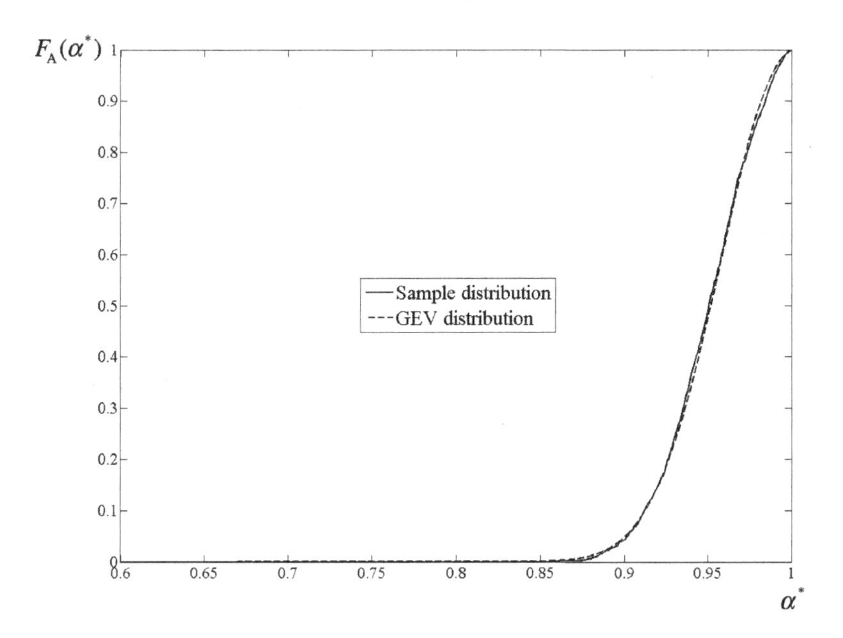

Fig. 2.15. Generalized extreme value (GEV) distribution versus sample distribution.

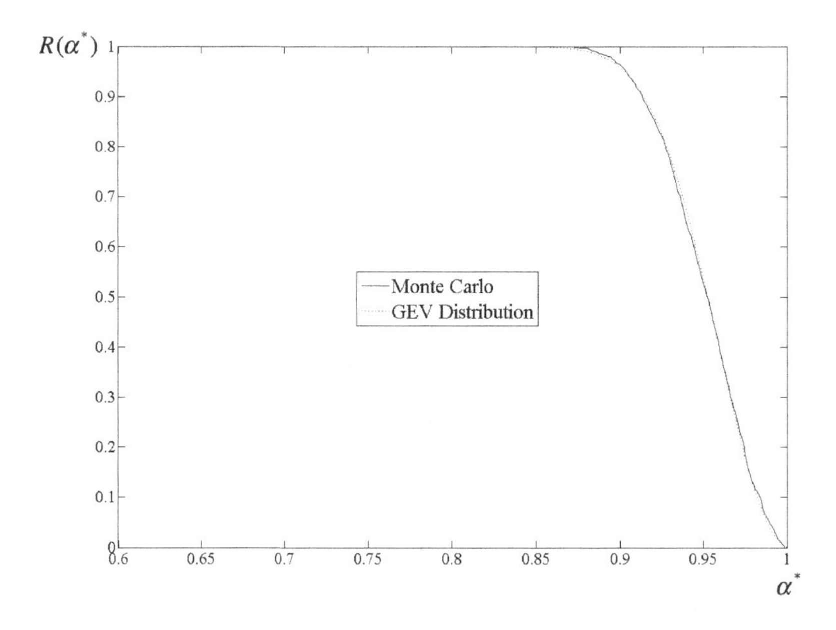

Fig. 2.16. Reliability function from Eq. (2.33) versus GEV cumulative distribution.

Figure 2.16 shows the reliability function evaluated with Eq. (2.34) and the cumulative distributive function calculated with Eq. (2.37).

Figure 2.17 shows the influence of the coefficient B on $R(\alpha')$, which is seen to decrease as B increases for $A = 0.01$. In this way, at load level 0.95, the empirical reliability equals 0.54 (i.e., about 54% of the columns buckle above that level) for $B = 1$, and only 0.33 for $B = 3$. This is understandable, as a larger B signifies a higher variance of the initial imperfections.

In Fig. 2.18 the reliability function is plotted for different values of A at $n_* = 2$. As expected, the reliability decreases as A increases. For example, at load level 0.95 reliability equals 0.52 for $A = 0.01$, and $R = 0.20$ for $A = 0.03$.

In the case $n_* = 1$ the nondimensional buckling loads are plotted as a histogram (Fig. 2.19). Most columns buckle in the load-level interval [0.92;1]. Another way to use the buckling load vesus reliabilty figure with different parameters is to choose a value of required reliability r that the structure must reach, and then to determine the corresponding design load. Figure 2.20 shows this probabilistic method of design. For different values of A, we are seeking the design loads associated with the required reliability $R = 95\%$. For $A = 0.01$, the design load is 0.90 whereas for $A = 0.03$, the load which satisfies the reliability criterion is about 0.85.

Table 2.1 lists the values of the variance calculated from samples of different sizes with the following formula:

$$\mathrm{Var}(\alpha^*) = \frac{\sigma^2}{\xi^2}[\Gamma(1-2\xi) - \Gamma^2(1-\xi)]. \tag{2.103}$$

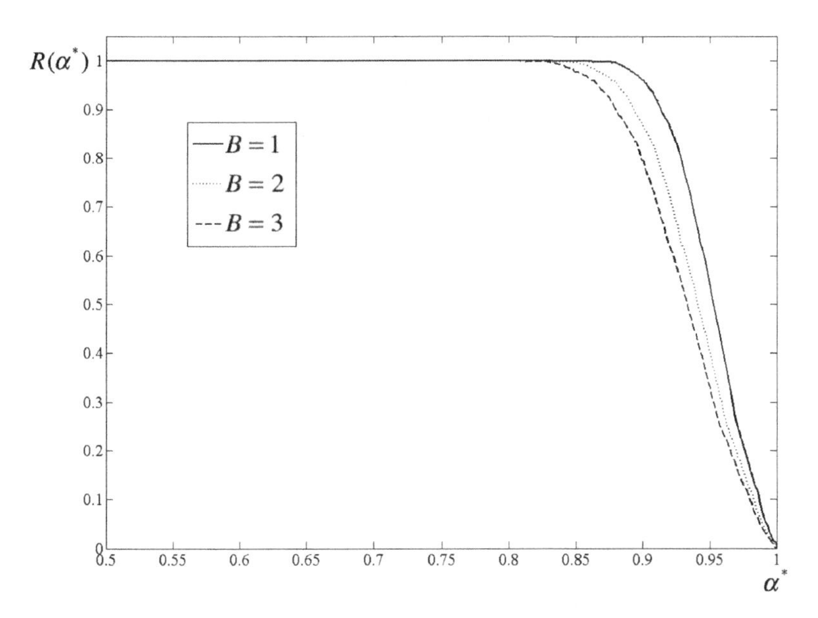

Fig. 2.17. Empirical reliability function $R^*(\alpha')$ versus nondimensional actual load α' ($A = 0.01$, $\kappa_1 = \kappa_3 = \pi^4, n^* = 1, N = 12$).

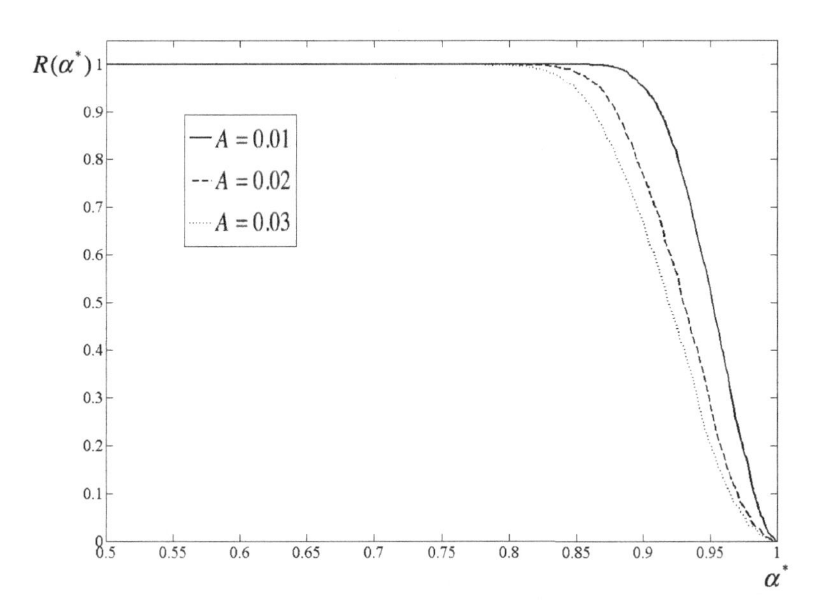

Fig. 2.18. Empirical reliability function $R^*(\alpha')$ versus nondimensional actual load α' ($B = 1$, $\kappa_1 = \kappa_3 = \pi^4, n^* = 2, N = 8$).

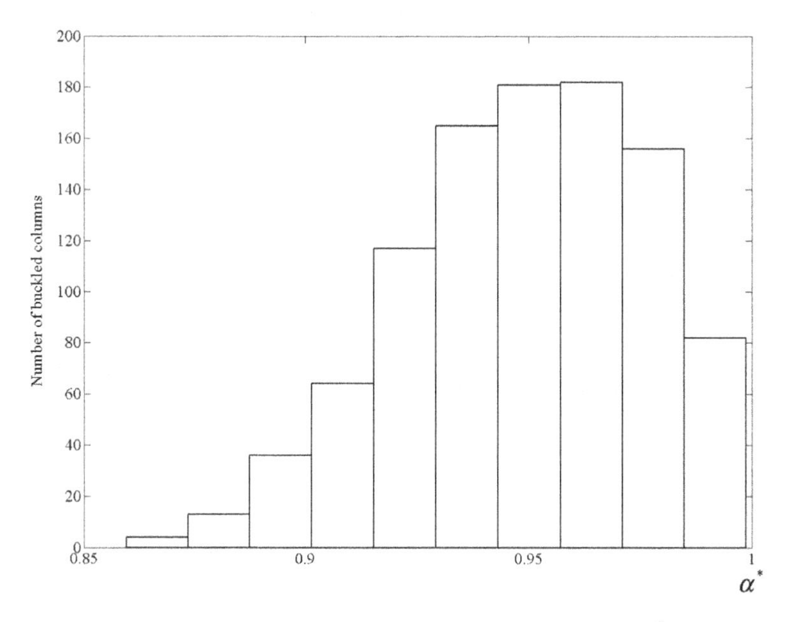

Fig. 2.19. Histogram of nondimensional buckling loads α^* ($\kappa_1 = \kappa_3 = \pi^4$, $A = 0.01$, $B = 1$, $n^* = 1, N = 8$).

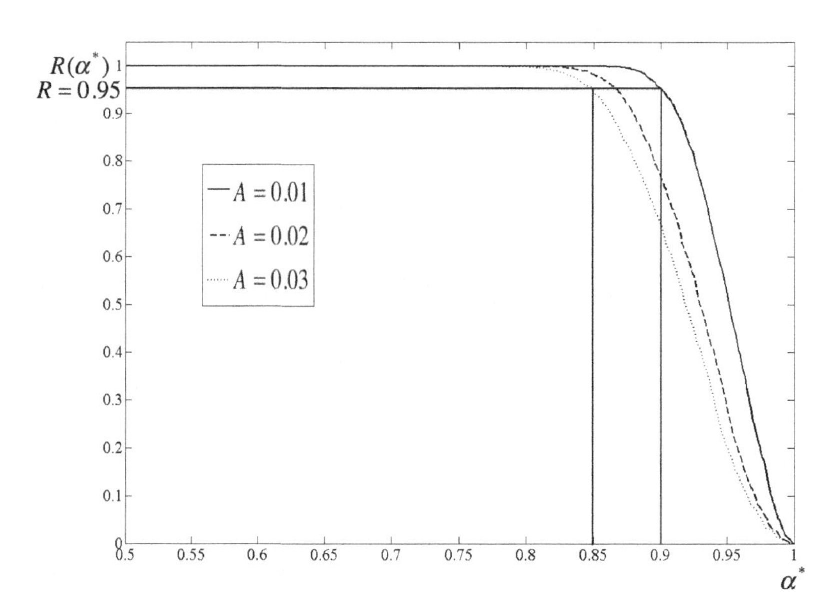

Fig. 2.20. Determination of the nondimensional load P/P_{cl} at 95% reliability for different values of A.

Table 2.1. Variance for different parameters, $\kappa_1 = \kappa_3 = 1$, $N = 12$.

N	Variance (Sample)	Variance (Eq. 2.37)	Percentagewise error e_1	Percentagewise error e_2	CPU Time (s)
500	7.5698e-4	7.5696e-4	2.67%	2.66%	1750
250	7.3835e-4	7.3585e-4	0.14%	0.20%	880
100	7.1826e-4	7.2050e-4	2.59%	2.28%	377
50	6.6195e-4	6.8801e-4	10.22%	6.69%	182

ξ and σ in Eq. (2.37) are estimated with the maximum likelihood principle, Γ is the gamma function:

$$\Gamma(n) = (n-1)! \tag{2.104}$$

The first column lists the size in the Monte Carlo simulations. The second column represents the variance calculated from the realizations of Monte Carlo simulation. The third column is the variance evaluated from the Eq. (2.103). The fourth and fifth columns list the errors: e_1 is the error in comparison with the 10,000 Monte Carlo simulation results (7.3744e-4), e_2 is the error between the variance estimated with Eq. (2.103) versus the sample variance evaluated by Monte Carlo simulation. In the last column the CPU times are listed.

The results in the Table 2.1 show that an acceptable compromise between computation time and a good sufficient of the variance is obtained for a size $N = 100$ columns for the Monte Carlo simulation. Naturally, the error is smaller with a sample of 250 simulations. The problem, in this case, may be the computation time. For high κ_1, the computation time is more than a day for $N = 100$.

Finally, Fig. 2.21 shows the variance of the buckling load versus κ_1 for $\kappa_3 =$ const. The variance of the buckling load decreases in the interval $n_* \leq 4$. This decrease of the variance with increasing κ_1 and constant κ_3 could be expected, and as the structure tends closer to linearity in this process, the mean value of the buckling load approaches unity. The following comment is made in: "Decrease of the variance for the constant ratio κ_3/κ_1 may support the conclusion of Fraser and Budiansky that for an infinite column (to which $\kappa_1 \to \infty$ is equivalent), the buckling load is a deterministic quantity." This conclusion is supported by current investigation. Remarkably, we go beyond the study, since it was able to evaluate the variance up to $n^* = 3$, whereas the modified Monte Carlo method allows us to do the calculation for nearly arbitrary value of n^*.

Now, in retrospect, instead of claiming that the "decrease of the variance . . . may support the conclusion of Fraser and Budiansky (1969)," I had to write that the above decrease did not contradict the conclusion by the Fraser and Budiansky on the determinicity of the buckling load in the infinite column with spacewise

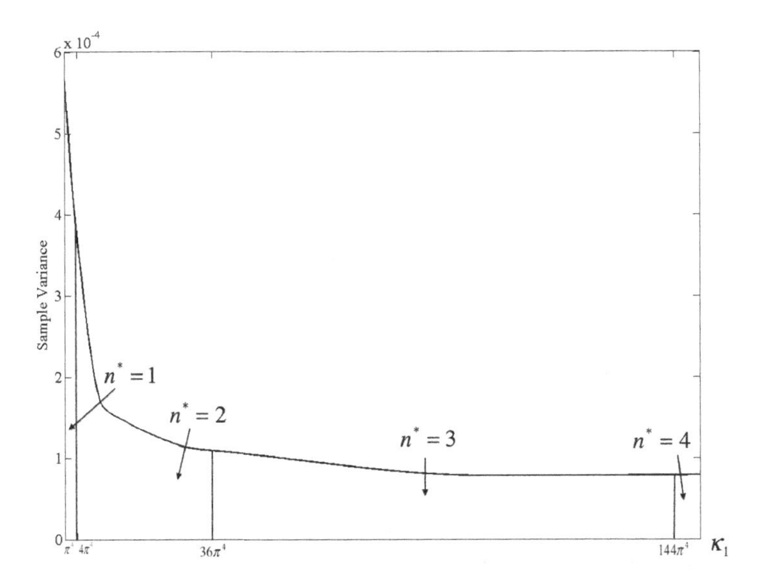

Fig. 2.21. Sample variance of the nondimensional buckling load versus κ_1 (κ_3 = constant).

homogeneous and ergodic random imperfections. Elishakoff (1979) further noted:

> . . . it should be expected, however, for a finite column the buckling load depends on the particular realization of the initial imperfection function, which in turn depends on the probabilistic measures (mean and autocorrelation function) of the initial imperfections.

The conclusion of the above paper read:

> It is hoped that the method outlined in this work can be extended to shell structures. The first and most difficult step is to compile extensive experimental information on imperfections, classified according to the manufacturing process, with a view to finding statistical measures for them. On this basis, available analytic facilities permit strength prediction and determination of the reliability associated with a given manufacturing process. By the means, the imperfection-sensitive concept can be introduced in design.

This paper appears to have been the first one that had its goal to be extremely ambitious: nothing less but the introduction of imperfection-sensitivity concept into design.

I was very pleased when I got an acceptance letter from the Associate Editor of *ASME Journal of Applied Mechanics*, Professor John Hutchinson; this was actually a nice postcard sent out from Denmark.

The paper was positively accepted by the community. As its result, Dr. Fraser of the University of Sydney, Australia, was able to pay a visit to the Technion — Israel Institute of Technology, in 1985. Professor Budiansky organized a lovely reception for us at his spacious home in Lexington, Massachusetts, in September 1980 when I served as a Visiting Scientist at M.I.T.; among many others, Professors Hutchinson and Amazigo were in attendance.

Professors Nicholas Hoff and Wilhelm Flügge attended my seminar at Stanford University, kindly organized by Professor Chuck Steele. There, during the dinner, Flügge surprised us all by making statements of political nature that led Hoffs nearly to leave the dinner. Hoffs also kindly entertained us during dinner at their home at Stanford when I was spending a sabbatical at the Naval Postgraduate School in Monterey, California. This was the longest time that we ever drove (nearly 1.5 h each way) for the dinner, but we also heard the maximum number of jokes and maxims too during that time. Here is one of these: "Once one is recognized three times, many other recognitions follow, somehow!"

Thirty years a flew publication of another paper on this topic Thompson (2013) mentioned "... a significant early paper by Isaac Elishakoff (1983) which pointed the way towards a probabilistic theory of imperfection sensitivity. The appearance of a subsequent book on this topic was most welcome (Elishakoff, Li and Starnes, 2001)."

Professor Koiter advised me to generalize this problem to include quadratic–cubic nonlinear foundation. This task was accomplished by 1985. It was natural to dedicate that paper (Elishakoff, 1985) to Koiter's 70th birth anniversary (Fig. 2.2). Dynamic buckling case for the quadratic–cubic foundation case was accomplished earlier (Elishakoff, 1980b) generalizing Budiansky and Hutchinson's (1964) theoretical and numerical results for the single degree of freedom model. Professor Koiter not only endorsed my approach, but was looking for any predicament it might be associated with. For example, during one of the lectures in Delft Professor Koiter raised a question of possible high sensitivity of structural reliability to small effects. I took this question as an advice to investigate this topic, and addressed it in my 1983 paper; likewise in 1989 Professor Yakov Ben-Haim and I proposed a nonprobabilistic way of introducing theoretically justified knockdown factor (see also Ben-Haim and Elishakoff, 1990; Elishakoff *et al.*, 2001). Thus, if two alternative ways of the conundrum's resolution was not enough, we lucked out with the third, nonprobabilistic way. Description of the earlier papers is beyond the scope of this study. Interested reader can consult our monograph (Elishakoff, Li and Starnes, 2001). Chapters 4 and 6 are fully dedicated to the non-probabilistic resolution of the conundrum.

Since in this section Monte Carlo method was advanced, for the column on the nonlinear elastic foundation, it appears instructive to quote from Hunt's (2002) review of the above book: "The book delivers a clear attempt to come to terms with this unpredictability by probabilistic, rather than deterministic, means. Large simulations, run many times under the Monte Carlo method are the order of the day." In this respect some explanations appear to be in order. Shinozuka and Deodatis (1991) write:

> ... Monte Carlo method is the only method available to solve a large number of stochastic problems involving nonlinearity, system stochasticity, stochastic stability, parametric excitations, etc.

Spanos and Zeldin (1998) elucidate:

> The Monte Carlo simulation is a statistical sampling experiment . . . it involves a repeated generation of random variates and a subsequent determination of the system response for each realization of the random variates by using deterministic methods. Proper statistical processing provides estimates of useful response statistics.

Key element of the above explanation is the "determination of the system response . . . by using deterministic methods." Thus, from this point of view Monte Carlo method and deterministic analysis are not antagonistic; in fact, Monte Carlo method contains in itself deterministic analysis. Hence, contrast made in the above review, when mentioning "probabilistic, rather than deterministic, means" appears to be not precise. One more quote appears to be instructive in this regard. It is due to Shinozuka and Yamazaki (1988):

> Monte Carlo simulation methods can provide useful alternative for the problems to which the perturbation method does not apply very well. . .

For the general notes on philosophy of the Monte Carlo method the interested reader can consult with present writer's paper (Elishakoff, 2003).

8. Studies by the Group of the University of Toronto

> *A central difficulty in most buckling analyses of imperfection sensitive structures is the choice of an appropriate imperfection form. In particular, there is the difficult option between deterministic and probabilities concepts. Deterministic methods furnish a precise technique in the exploration of the mechanisms involved in imperfection sensitivity, whereas it appears that probabilistic methods provide the key to acceptable design criteria as well as the correct interpretation of experimental data.*
>
> J.S. Hansen (1977)

Tennyson (1980) in his survey paper, after quoting Eq. (1.2) stresses:

> However, buckling experiments over the next 50 years were unable to attain much more than 50% of this value. By careful manufacturing procedures, geometrically "near-perfect" circular cylinders were eventually constructed by the author [R.C. Tennyson] in early 1962, and, using clamped end constrains, buckling loads equal to 0.9 σ_{cl} were obtained. During those years, many theoretical attempts were made to explain the discrepancy between tests and theory. These included the large deflection analysis, postulating external disturbances, energy jumps, effects of different test machine stiffnesses, boundary conditions and shape imperfections. In my paper, I made the following concluding remarks: ". . . when tests are performed an isotropic, elastic cylinders sufficiently free of imperfections, . . . the classical value can be attained and a lower buckling load is not always inevitable.

ACTA MECHANICA

EDITED BY

A. PHILLIPS, NEW HAVEN, CONN.
H. TROGER, WIEN
F. ZIEGLER, WIEN
J. ZIEREP, KARLSRUHE

REPRINT FROM VOL. 55, No. 1—2 (1985)

I. Elishakoff

Reliability Approach to the Random Imperfection Sensitivity of Columns

Dedicated to Prof. Dr. Ir. Warner Koiter on the occasion of his 70th birthday

SPRINGER-VERLAG
WIEN · NEW YORK

Fig. 2.22. Cover page of the paper dedicated to Professor W.T. Koiter's 70th birthday; paper was inspired by Koiter's suggestion.

Fig. 2.23. Professors W.T. Koiter and J. Singer at the Israel's Annual Conference on Aeronautics in 1984.

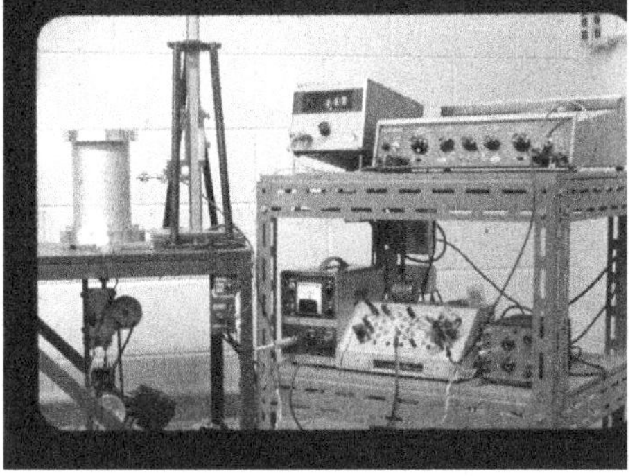

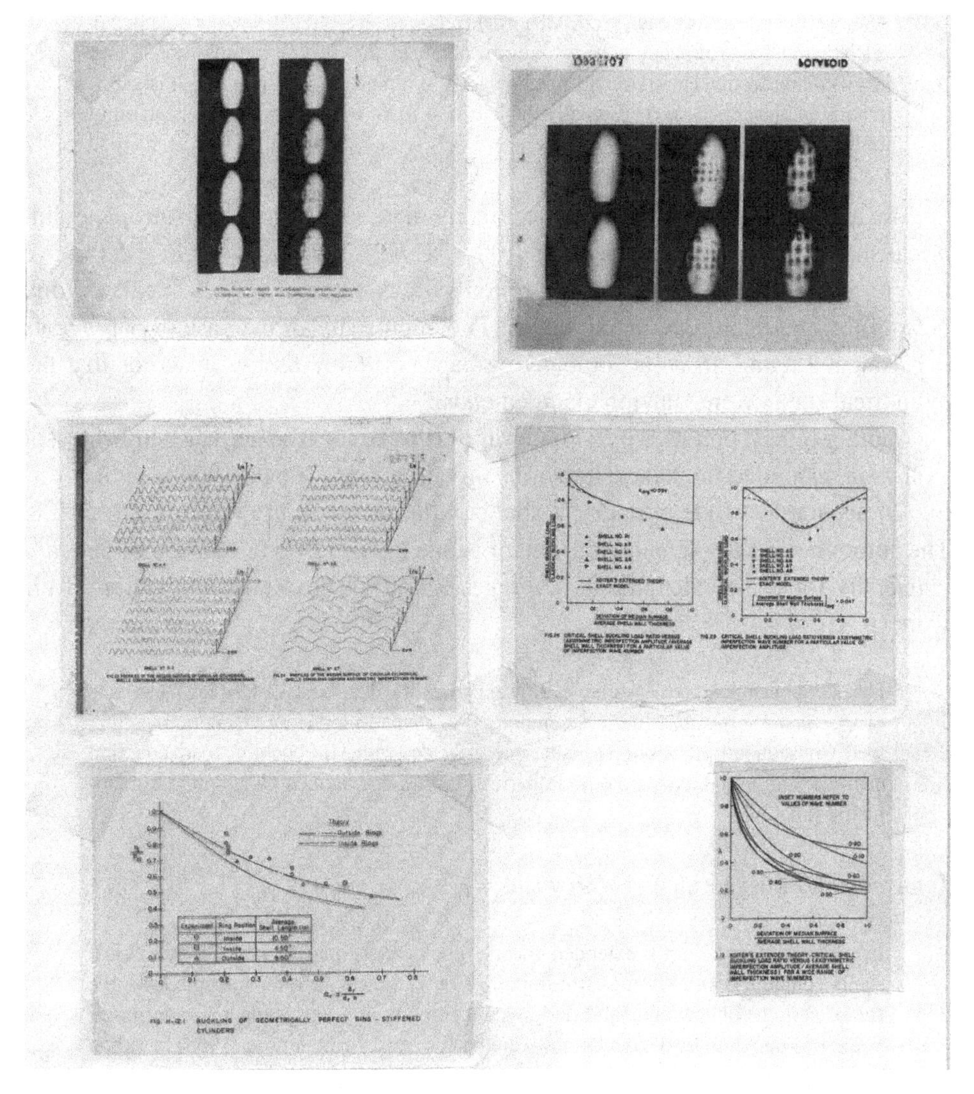

Fig. 2.24. Professor Rod Tennyson of University of Toronto is one of the most active contributors to both theoretical and experimental analyses of shell buckling.

Tennyson (1964, p. 1) explains his breakthrough approach to experimental side of imperfection sensitivity:

> This investigation was started in 1961 with a view to developing a manufacturing technique to produce "near-perfect" circular cylindrical shells such that they would buckle completely elastically near the classical value. To ensure elastic buckling and repeatable results, shells were made from a photoelastic plastic.

Furthermore (p. 9):

> It must be noted that because of the elastic behavior of the shells, as many as 20 to 30 tests were conducted on each shell. All the data reported on each shell was not obtained in a single run, but is an accumulation of results obtained after several runs. By comparing axial end-deflection curves versus machine load, strain gauge data, pictures of isochromatics and isoclinics, and eventually the buckling loads for each run, the repeatability of the tests was established.

The author concluded that "five accurately made, homogenous, isotropic cylindrical shells of circular cross-section ($100 \leq R/h \leq 180$, $2.0 \leq L/R \leq 5.0$) have been tested in axial compression and found to buckle completely elastically within 10% of the classically predicted values. This represents an increase over previous experimental data." In an accompanying paper (1964b) Tennyson writes that the cylindrical shells were "having clamped ends."

Another report, prepared by Muggeridge (1969) under the guidance of Professor Tennyson, extended the latter's previous investigations (1964a, 1964b) by measuring 26 accurately made cylindrical shells with $100 \leq R/h \leq 280$, $L/R \simeq 2.8$. The author notes that "acknowledgement is due to Dr. J.W. Hutchinson of Harvard University for exact calculations." Muggeridge (1969) explains motivation of his work (p. 1):

> Despite substantial theory available, little experimental data exists describing the effects of *specific imperfections* [author's emphasis] in shape in reducing the static buckling load. Consequently, it was of particular interest to determine the buckling load reduction caused by an initial axisymmetric imperfection in shape defined by a simple trigonometric function.

Tennyson and Muggeridge (1969) write:

> In order to test Koiter's extended theory, photoelastic plastic circular cylindrical shells containing an axisymmetric imperfection in shape were manufactured by the spin-casting technique and subjected to axial compression. For comparison purposes, numerical buckling load calculations were performed by Hutchinson (1969) based on an exact analytical model in which the effects of a clamped end constraint and the specific geometrical configuration of the cylinders were taken into account.

Tennyson *et al.* (1971) recognized the need in probabilistic methodology to tackle the drastic differences between theory and practice:

> Clearly, if these imperfection distributions are random in nature, it is exceedingly difficult to formulate a rational design criteria in the absence of a random model theory and a practical method of determining the distribution of imperfections on a shell structure of reasonable size.

At that time two major models for probabilistic analysis were already available, specifically the method of single or finite number of random variables (Bolotin, 1958, 1962; Roorda, 1965a, 1965b, 1969, 1971, 1975, 1980; Thompson, 1967) and the method of random fields by the Harvard group (Fraser and Budiansky, 1969; Amazigo, 1969, 1970, 1971). Tennyson *et al.* (1971a) chose the latter way of describing imperfections. There is nothing wrong with using random fields theory if far-going assumptions of spacewise homogeneity and ergodicity have not been incorporated. The authors expressed some reservations of the approach that they adopted, though:

> At present, theoretical analyses (Fersht, 1968; Amazigo, 1969) are available which describe the effect of random axisymmetric shape imperfections on buckling behavior of circular cylinders. Although it may be argued that existing theory is of limited practical interest, it is first necessary to isolate the major parameters that can be shown to control buckling both theoretically and experimentally.

Tennyson *et al.* (1971a), following Fersht (1968) and Amazigo (1969), postulated spacewise homogeneity of random imperfections. They fabricated the axisymmetric imperfections into the circular cylinders, by using a metal template containing the desired wave forms in conjunction with a hydraulic tracer-tool apparatus. Moreover,

> The profiles were constructed on the templates by means of a circular cam with its center of rotation offset to provide the appropriate amplitudes. The excentric cam was mounted on a motor shaft such that continuous contact was maintained with a spring-loaded cutting tool... In total twenty-three circular cylindrical shells were tested in axial compression... Although the various distributions were not strictly "random" in a statistical sense, it was felt that the profiles were sufficiently general to include a reasonable frequency bandwidth of imperfections.

The authors concluded that "in general, theory and experiment agreed quite well. However, it should be noted that of the nine cylinders tested, only one (AD11) was a constant thickness model as assumed in the theory." It is not clear from the above statement if the comparison was between average value derived by Tennyson *et al.* (1971a) and the asymptotic value obtained by Amazigo (1969). The authors' conclusion that "it would thus appear that the choice of the exponential-cosine autocorrelation peaked at the critical frequency... provides an excellent estimate of the buckling load in terms of the rms [root mean square] amplitude of the imperfection distribution" is questionable. Indeed, authors had to measure the autocorrelation function and check if it indeed was of exponential-cosine type. The subsequent paper by Tennyson *et al.* (1971b) did not address the reliability of the shell but again adopted Amazigo's (1969) asymptotic result.

Hansen (1977) considered effect of general random imperfections in the buckling of axially loaded cylindrical shells based upon his previous deterministic study (Hansen, 1975), which represented an extension of Koiter's axisymmetric solution. In the latter work, he found "the interesting feature . . . that the imperfection parameters associated with the non-axisymmetric modes . . . appear in three and only three distinct summations. Therefore, the behavior of the system depends on the value of these summations and not on the individual imperfections parameters." He also showed "that the inclusion of non-axisymmetric imperfections can lead to intermodal behavior which results in severe reductions of the buckling load. Bifurcation and limit point critical loads have been examined and in this examination it must be concluded that both axisymmetric and non-axisymmetric imperfections play equally important roles." This paper provided a fertile ground for its generalization to include randomness of initial imperfections (Hansen, 1977). The author assumed normality of initial imperfection field that led to the conclusion of the joint Gaussianity of the initial imperfection's Fourier coefficients. Hansen (1977, p. 1253) noted: "In the present case, the simulation proceeded on the assumption that all the modal imperfection amplitudes were *independent and identically distributed with zero mean* [author's emphasis]" Hansen (1977, p. 1255) notes:

> It is the intention here to defend these assumptions rigorously, as it is felt that there is insufficient experimental evidence to make such a strong statement. However, an attempt will be made to provide some justification, as well as to provide an assessment of the influence of these assumptions. The Gaussian assumption is one that has been involved in a number of other analyses (Amazigo, 1969; Roorda and Hansen, 1972). The justification for its use is the central limit theorem. The assumption of zero mean imperfection also has been adopted recently (Amazigo, 1969; Fersht, 1968), even though it may alter the results significantly. The main justification for this assumption is that the perfect structure is the model of the real shell. Thus, it may be anticipated that the randomness in the shell is dispersed about the perfect form. The assumption that the model imperfection amplitudes are statistically independent is perhaps *a far more restrictive assumption* [author's emphasis] than the previous two. It depends entirely on the autocorrelation function . . . of the imperfection process in a set of cylindrical shells. There has been no experimental evaluation of the autocorrelation for a random imperfection in a cylindrical shell."

The evaluation of the autocorrelation function of initial imperfections was performed by Elishakoff and Arbocz (1982, 1985) based on experimental data. It was shown that the normality assumption does not contradict the data (Vermeulen *et al.*, 1984) but the Fourier coefficients are correlated and hence the assumption of independent random variables should be abandoned. Likewise, the assumption, made by Hansen (1977), "that the amplitudes of the modal imperfections are distributed identically" could not be substantiated, based on experimental data. Still, Hansen's papers (1975, 1977) provided perhaps the best possible probabilistic treatment when no real data is involved.

The question arises on how to correlate the conclusions made in the paper by Tennyson *et al.* (1971a, 1971b) with those of Hansen (1977). The latter author notes:

> A comparison between the present results and those of Ref. 8 (Tennyson *et al.*, 1971a) is *not fruitful* [author's emphasis]. This seems basically from the fact that the design curves in Ref. 8 yield "mean" critical loads . . .

because the paper of Tennyson *et al.* (1971) is based on the ergodicity assumption of the initial imperfections and consequent conclusion on deterministic property of the buckling load.

Fig. 2.25. Professor Johann Arbocz and Josef Singer holding a placard "International Imperfection Data Bank" (for the detailed description see the dissertation by Dr. Jan de Vries, 2009).

It appears that the analysis by Hansen (1977) can be usefully revisited by including the available Initial Imperfection Data Banks that would lead to abandonment of strong probabilistic assumptions made, and thus may provide with useful designs. Naturally, the compilation of measurement shells at the Institute of Aerospace Studies at the University of Toronto into a data bank will be a welcome move. It appears also of interest to compare Hansen's (1975) deterministic analysis with combined analytical and numerical techniques by Arbocz, on one hand, and with current finite element codes, on the other.

9. Studies by the Group of Moscow Power Engineering Institute and State University

> *There is a close connection between the concepts of stability and probability. Stable states of equilibrium or motion observed in the natural or engineering systems are the*

> *most probable ones; unstable ones are improbable and even unrealizable. The more stable a state is, the greater is the probability of its realization.*
>
> V.V. Bolotin (1967)

I recall that in 1965 Professor Bolotin traveled for several weeks to the United States. Upon return, he informed Professor Boris Petrovitch Makarov that at Harvard the research on stochastic initial imperfections was going in full swing. He advised Makarov to reengage himself in stochastic imperfection sensitivity (Makarov (1962) dealt with this topic before), to conduct experiments, to measure initial imperfections, to analyze these statistically, and come up with probabilistic design criteria. This topic became the one for the habilitation work of Makarov. Vladimir Mikhailovitch Leizerakh and Nina Ivanovna Sudakova performed their Ph.D. theses on this topic too, all three under direction of Bolotin. In his first work on this topic Makarov (1962) engaged himself in an inverse problem: He notes: "There are no trustworthy data on character of distribution of initial imperfections in shells. Therefore the inverse problem formulation makes a great sense: from the experimental distribution of the buckling loads find the probability distribution of initial imperfections." Makarov used the data reported in the paper by Harris *et al.* (1957).

Fig. 2.26. Professor B.P. Makarov advising students.

In his later paper, Makarov (1970) addressed a direct problem, expanding initial imperfections in trigonometric series. Makarov *et al.* (1968) introduced the assumption of homogeneity in circumferential direction in contrast with the assumption of homogeneity in axial direction that was adopted at Harvard. This assumption is valid for circular shells with no seams. Statistical estimation of the correlation moments of various initial imperfection amplitudes was conducted. This was augmented with experimental measurement of initial imperfections, retaining total of 156 coefficients corresponding to 12 modes in the axial and 13 modes in the circumferential direction. In the paper by Makarov (1970) a series of 50 shells was used, with dimensions: $L = 300$ mm, $R = 100$ mm, $h = 0.18$ mm. He checked the statistical hypothesis on normality of the distribution of Fourier coefficients for the function of the initial imperfections. He concluded (p. 89): "The large-scale computations performed show that the hypothesis of Fourier coefficients' normal distribution is consistent with the experimental data." Shells were made from steel sheets with ends glued to each other with the seams. It appears that the presence of seams should have violated the assumption of homogeneity in circumferential direction. Makarov (1971, p. 246) notes that "deviations from the ideal form appear as a result of imperfection of shell preparation methodology." These papers do not reproduce the buckling load evaluation techniques, though the histograms of buckling loads are reproduced. The tested two series of shells had following mean buckling loads: in the first series the mean buckling load constituted about a half of the theoretical buckling load. The second, less perfect, series had lower experimental buckling loads.

In the subsequent study, Makarov (1969) provides details on the way the critical buckling load was derived, the availability of the relationship between buckling load and imperfection coefficients being of crucial importance for probability evaluations. Mossakovskii (1969), in his review article, comments on Makarov's (1969) paper, in a free translation: "The considered problem when taking into full account the basic factors of moment prebuckling state is very complex. In order to arrive at the final statistical results, Makarov simplified the deterministic solution; specifically he neglected the effect of prebuckling displacements in comparison with forces in the mid-surface of the shell, though the influence of these effects may have the same order of magnitude (Voblykh, 1965). Overall, Makarov's (1969) work appears of the methodic interest." Makarov's work was extended by Leizerakh (1969) and Leizerakh and Makarov (1973).

10. Resolution of Amazigo–Budiansky Paradox for Stochastically Imperfect Cylindrical Shells

When they are not brutalized, but delicately handled by the higher methods, and are warily interpreted their [statistics] power of dealing with complicated phenomena is extraordinary.

F. Galton (1889)

An important contribution of the "stochastic" part of the Harvard group was the realization that the theories of imperfection sensitivity should be combined with the probabilistic analysis of the initial imperfections. The first work in this direction for the most controversial structure — the circular cylindrical shell under axial compression — was pioneered by Amazigo (1969). He treated infinitely long shells with spatially homogeneous, ergodic random axisymmetric imperfections by means of a truncated hierarchy method. The conclusion derived was that the buckling stress is a *deterministic* quantity, depending only on the spectral density of the random axisymmetric imperfections. Moreover, for small values of the standard deviations of the initial imperfections, the deterministic buckling stress depended only on the value of the initial imperfection power spectral density at the spatial frequency corresponding to the classical axisymmetric buckling mode, this dependence being

$$\lambda = 1 - \left[\frac{9\pi^2 c^2}{2\sqrt{2}}\bar{S}_{\bar{\mathrm{w}}}(1)\right]^{2/7}\delta^{4/7}, \quad \lambda = \frac{P_{\mathrm{BIF}}}{P_{\mathrm{cl}}}. \tag{2.105}$$

where P_{BIF} is the buckling load at which the governing nonlinear equations admit an asymmetric solution infinitesimally adjacent to the prebuckling axisymmetric state, P_{cl} is the classical buckling load of the perfect structure, λ is the nondimensional deterministic buckling load (and, as a result, also the *mean* buckling load), $c = \sqrt{3(1-\upsilon^2)}$, υ is Poisson's ratio, $\bar{S}_{\bar{\mathrm{w}}}(1)$ is the normalized critical axisymmetric imperfection power spectral density so that $\int_{-\infty}^{\infty} S_{\bar{\mathrm{w}}}(\kappa)\mathrm{d}\kappa = 1$, κ is the non-dimensional spatial frequency so that $\kappa = 1$ corresponds to the wave number associated with the axisymmetric buckling mode, and δ is the standard deviation of the initial imperfections. Later on, Amazigo and Budiansky (1972) modified Eq. (2.95) to

$$\lambda = 1 - \left[\frac{9\pi c^2}{2\sqrt{2}}\bar{S}_{\bar{\mathrm{w}}}(1)\right]^{2/7}\delta^{4/7}\lambda^{4/7} \tag{2.106}$$

Fig. 2.27. The author with Professor J. Arbocz.

In the works of Amazigo (1969) and Amazigo and Budiansky (1972) the normalized autocorrelation function $\overline{R}_{\bar{w}}(x_1, x_2) = R_{\bar{w}}(x_1, x_2)/\delta^2$ was assumed to be of exponential-cosine type

$$\bar{R}_{\bar{w}}(x_1, x_2) = \mathrm{e}^{-\beta\eta}\cos(\gamma\eta) \tag{2.107}$$

where η is the difference between the nondimensional axial coordinates of the points of observation, and β and γ are some positive constants supposed to depend on the manufacturing process. The normalized spectral density associated with this autocorrelation function is

$$\bar{S}_{\bar{\mathrm{w}}}(\kappa) = \frac{\beta}{\pi}\frac{\kappa^2 + \beta^2 + \gamma^2}{\kappa^4 + 2(\beta^2 - \gamma^2)\kappa^4 + (\beta^2 + \gamma^2)^2} \tag{2.108}$$

so that $\bar{S}_{\bar{\mathrm{w}}}(1)$ entering into Eq. (2.95) by Amazigo (1969) and Eq. (2.96) by Amazigo and Budiansky is

$$\bar{S}_{\bar{\mathrm{w}}}(1) = \frac{\beta}{\pi}\frac{1 + \beta^2 + \gamma^2}{1 + 2(\beta^2 - \gamma^2) + (\beta^2 + \gamma^2)^2} \tag{2.109}$$

At this point I would like to digress and bring an extensive quote from the second edition of my book (1983, 1999):

> [In 1979] I became entitled to a sabbatical leave [from the Technion — Israel Institute of Technology]. This custom of taking an academic "time out" may seem a superfluous luxury at first glance, but, as my experience bears out, it provides a valuable opportunity for renewal and fruitful interaction with other environments and scientists. At any rate, following the Biblical concept of the sabbatical year, and availing myself of the generosity of the Technion, I started to look for an institution that would be interested in my research. Fortunately, several universities in North America [including M.I.T. in the U.S.A. and University of Waterloo in Canada] and Europe offered guest appointments for the 1979/80 academic year. We chose to go to Europe; I had always liked manageable distances, especially since my driving experience was limited to tiny Israel. Of the European institutions that were interested, I chose the Delft University of Technology, mainly because of my desire to conduct specific research on initial imperfection sensitivity, whose deterministic aspects had been so vividly uncovered at Delft some 35 years earlier by Professor Dr. Ir. Warner Tjardus Koiter in his famous Ph.D. dissertation, and then had been pursued analytically, numerically and experimentally — with great success — by Professor Dr. Johann Arbocz. I was dreaming of combining my recently developed probabilistic techniques with the numerical codes developed in Delft. Upon my arrival, I found that Delft was to pose . . . unexpected challenges . . . Professor Arbocz himself told me that he did not believe in probability, though he said I was welcome to try to prove my own belief . . ." (The other challenge in Delft I will write later on in Sec. 13.) I responded that I do not view theory of probability as a "quasi-religion," as some researchers do, according to Kalman (1994); I mentioned that although "probabilities do not exist," in the words of de Finetti (1974), or that, in words of philosopher Yeshaiahu Leibowitz, "the nature did not take a class in probability," still, it may prove to be an extremely useful tool.

Fig. 2.28. Professor J. Arbocz at the Florida Atlantic University between the author (left) and Professor Y.K. Lin (right).

Here it suffices to say that the self-defined task of mine was to try to convince Professor Arbocz and his group in the importance, if not the indispensability, of the probabilistic approach to buckling. I suggested that we deal with numerical checking of the Amazigo's (1969) and Amazigo and Budiansky's (1972) paradox, utilizing previous results on simulation of random fields (1979) and my resolution (1976) of the Fraser and Budiansky's (1969) paradox, associated with buckling of stochastically imperfect columns on nonlinear elastic foundation.

The above task was accomplished in two joint studies (Elishakoff and Arbocz, 1982, 1985).

The former investigation dealt with axisymmetric initial imperfections, whereas the latter one dealt with general nonsymmetric imperfections. The initial imperfections were represented as the following series:

$$\bar{W}(\xi,\theta) = \sum_{i=0}^{N_1} A_i \cos(i\pi\xi) + \sum_{k=1}^{N_2}\sum_{l=1}^{N_3} \times [C_{kl}\sin(k\pi\xi)\cos(l\theta) + D_{kl}\sin(k\pi\xi)\sin(l\theta)] \tag{2.100}$$

where

$$\bar{W}(\xi,\theta) = \frac{\overline{w}(\xi,\theta)}{h}, \quad \xi = \frac{x}{L}, \quad \theta = \frac{y}{R} \tag{2.101}$$

with $\bar{w}(\xi,\theta)$ and $\bar{W}(\xi,\theta)$ being dimensional and nondimensional initial imperfections; h, L, and R represent thickness, length, and radius of the shell, respectively; x is the axial and y is the circumferential coordinate. Notice that in Eq. (2.100) the first sum represents the axisymmetric part of the initial imperfection profile, whereas the second, double sum is associated with its nonsymmetric part. The axisymmetric part is expressed in the half-range cosine series, whereas the nonsymmetric part is represented by half-range sine series, so that the series in Eq. (2.100) sums up to the measured imperfection profile in the range $0 \leq x \leq L, 0 \leq \theta \leq 2\pi$.

The mean value of $W_0(\xi, \theta)$ reads:

$$\langle \bar{W}(\xi, \theta) \rangle = \sum_{i=0}^{N_1} \langle A_i \rangle \cos(i\pi\xi) + \sum_{k=1}^{N_2} \sum_{l=1}^{N_3} \times [\langle C_{k,l} \rangle \sin(k\pi\xi) \cos(l\theta) + \langle D_{kl} \rangle \sin(k\pi\xi) \sin(l\theta)] \quad (2.102)$$

where $\langle \ldots \rangle$ denotes a mathematical expectation. The auto-covariance function

$$C_{\bar{W}}(\xi_1, \theta_1; \xi_2, \theta_2) = \langle [\bar{W}(\xi_1, \theta_1) - \langle \bar{W}(\xi_1, \theta_1) \rangle][\bar{W}(\xi_2, \theta_2) - \langle \bar{W}(\xi_2, \theta_2) \rangle] \rangle \quad (2.103)$$

becomes

$$\begin{aligned} C_{\bar{W}}(\xi_1, \theta_1; \xi_2, \theta_2) = \Bigg\langle \Bigg[& \sum_{i=0}^{N_1} (A_i - \langle A_i \rangle) \cos(i\pi\xi_1) \\ & + \sum_{k=1}^{N_2} \sum_{l=1}^{N_3} (C_{kl} - \langle C_{kl} \rangle) \sin(k\pi\xi_1) \cos(l\theta_1) \\ & + \sum_{k=1}^{N_2} \sum_{l=1}^{N_3} (D_{kl} - \langle D_{kl} \rangle) \sin(k\pi\xi_1) \sin(l\theta_1) \Bigg] \\ & \times \Bigg[\sum_{j=0}^{N_1} (A_j - \langle A_j \rangle) \cos(j\pi\xi_2) \\ & + \sum_{m=1}^{N_2} \sum_{n=1}^{N_3} (C_{mn} - \langle C_{mn} \rangle) \sin(m\pi\xi_2) \cos(n\theta_2) \\ & + \sum_{m=1}^{N_2} \sum_{n=1}^{N_3} (D_{mn} - \langle D_{mn} \rangle \sin(m\pi\xi_2) \sin(n\theta_2) \Bigg] \Bigg\rangle. \end{aligned} \quad (2.104)$$

The autocovariance function becomes

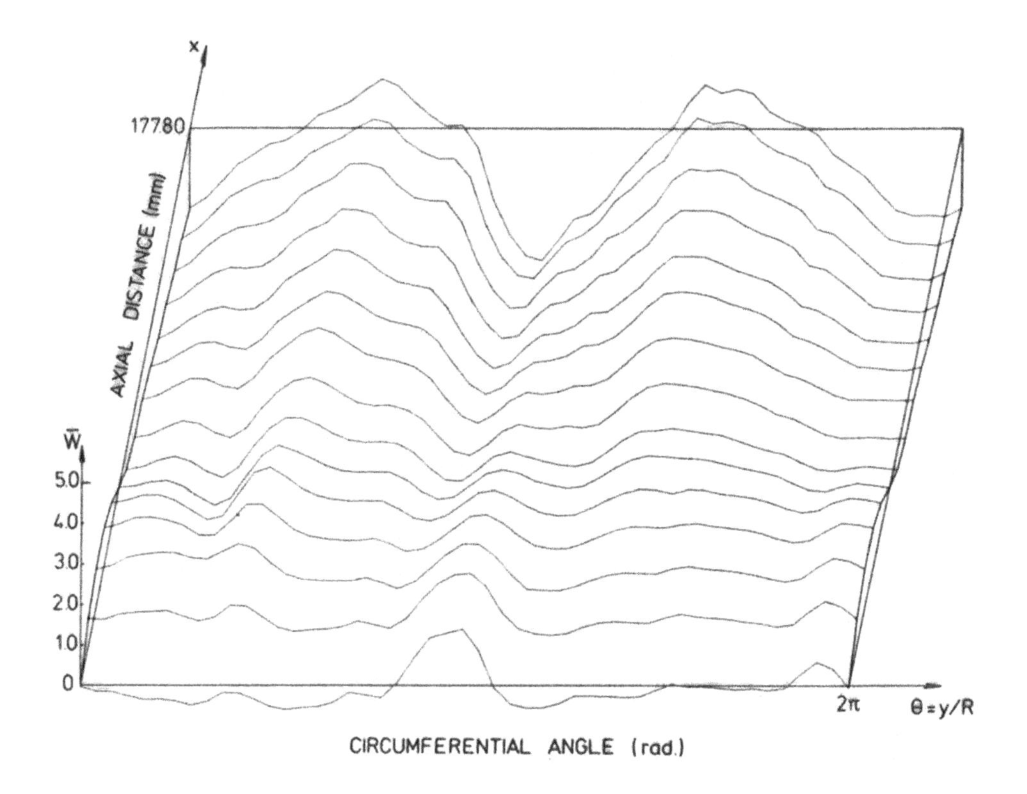

Fig. 2.29. Measured initial shape of the isotropic shell A − 8.

$$
\begin{aligned}
&C_{\bar{W}}(\xi_1, \theta_1; \xi_2, \theta_2) \\
&\quad = \sum_{i=0}^{N_1} \sum_{j=0}^{N_2} K_{A_i A_j} \cos(i\pi\xi_1) \cos(j\pi\xi_2) \\
&\qquad + \sum_{i=0}^{N_1} \sum_{m=1}^{N_2} \sum_{n=1}^{N_3} K_{A_i C_{mn}} \cos(i\pi\xi_1) \sin(m\pi\xi_2) \cos(n\theta_2) \\
&\qquad + \sum_{k=1}^{N_2} \sum_{l=1}^{N_3} \sum_{j=0}^{N_1} K_{C_{kl} A_j} \sin(k\pi\xi_1) \cos(l\theta_1) \cos(j\pi\xi_2) \\
&\qquad + \sum_{k=1}^{N_2} \sum_{l=1}^{N_3} \sum_{m=1}^{N_2} \sum_{n=1}^{N_3} K_{C_{kl} C_{mn}} \sin(k\pi\xi_1) \sin(m\pi\xi_2) \cos(l\theta_1) \cos(n\theta_2) \\
&\qquad + \sum_{i=0}^{N_1} \sum_{m=1}^{N_2} \sum_{n=1}^{N_3} K_{A_i D_{mn}} \cos(i\pi\xi_1) \sin(m\pi\xi_2) \sin(n\theta_2)
\end{aligned} \tag{2.105}
$$

$$+\sum_{k=1}^{N_2}\sum_{l=1}^{N_3}\sum_{j=0}^{N_1} K_{D_{kl}A_j}\sin(k\pi\xi_1)\cos(\pi\xi_2)\sin(l\theta_1)$$

$$+\sum_{k=1}^{N_2}\sum_{l=1}^{N_3}\sum_{m=1}^{N_2}\sum_{n=1}^{N_3} K_{C_{kl}D_{mn}}\sin(k\pi\xi_1)\sin(m\pi\xi_2)\sin(l\theta_1)\sin(n\theta_2)$$

$$+\sum_{k=1}^{N_2}\sum_{l=1}^{N_3}\sum_{m=1}^{N_2}\sum_{n=1}^{N_3} K_{D_{kl}C_{mn}}\sin(k\pi\xi_1)\sin(m\pi\xi_2)\sin(l\theta_1)\sin(n\theta_2)$$

$$+\sum_{k=1}^{N_2}\sum_{l=1}^{N_3}\sum_{m=1}^{N_2}\sum_{n=1}^{N_3} K_{D_{kl}D_{mn}}\sin(k\pi\xi_1)\sin(m\pi\xi_2)\sin(l\theta_1)\sin(n\theta_2)$$

where

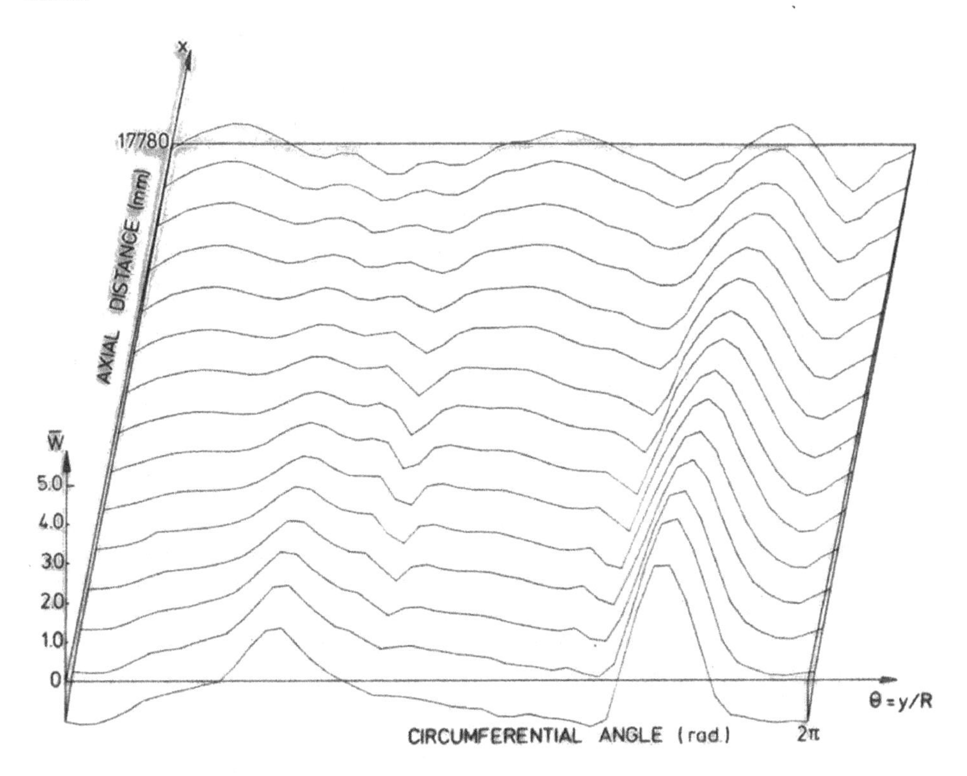

Fig. 2.30. Measured initial shape of the isotropic shell A-7.

$$K_{A_iA_j} = \langle[A_i - \langle A_i\rangle][A_j - \langle A_j\rangle]\rangle$$

$$K_{A_iC_{mn}} = \langle[A_i - \langle A_i\rangle][C_{mn} - \langle C_{mn}\rangle]\rangle$$

$$
\begin{aligned}
K_{C_{kl}A_j} &= \langle[C_{kl} - \langle C_{kl}\rangle][A_j - \langle A_j\rangle]\rangle \\
K_{C_{kl}C_{mn}} &= \langle[C_{kl} - \langle C_{kl}\rangle][C_{mn} - \langle C_{mn}\rangle]\rangle \\
K_{A_iD_{mn}} &= \langle[A_i - \langle A_i\rangle][D_{mn} - \langle D_{mn}\rangle]\rangle \\
K_{D_{kl}A_j} &= \langle[D_{kl} - \langle D_{kl}\rangle][A_j - \langle A_j\rangle]\rangle \\
K_{C_{kl}D_{mn}} &= \langle[C_{kl} - \langle C_{kl}\rangle][D_{mn} - \langle D_{mn}\rangle]\rangle \\
K_{D_{kl}C_{mn}} &= \langle[D_{kl} - \langle D_{kl}\rangle][C_{mn} - \langle C_{mn}\rangle]\rangle \\
K_{D_{kl}D_{mn}} &= \langle[D_{kl} - \langle D_{kl}\rangle][D_{mn} - \langle D_{mn}\rangle]\rangle
\end{aligned} \tag{2.106}
$$

are the elements of associated variance–covariance matrices. In order for the initial imperfections to constitute a homogeneous random field in either axial direction [as was assumed by Fraser and Budiansky (1969) or Amazigo (1969)] or circumferential direction [as was assumed by Makarov (1970)], it is necessary and sufficient the variance–covariance matrices to satisfy certain conditions.

It has been shown by Elishakoff and Arbocz (1982, 1985) that these conditions are violated. Specifically, Fig. 5 in the paper by Elishakoff and Arbocz (1985) portrays the mean and variance functions of the initial imperfections. Neither mean not variance use constant in circumferential direction, implying that random imperfections do not constitute a circumferentially homogeneous random field. Likewise, axial homogeneity property is not present in shells of finite length.

For the detailed account, the interested reader can also be directed to the monograph by Elishakoff *et al.* (2001).

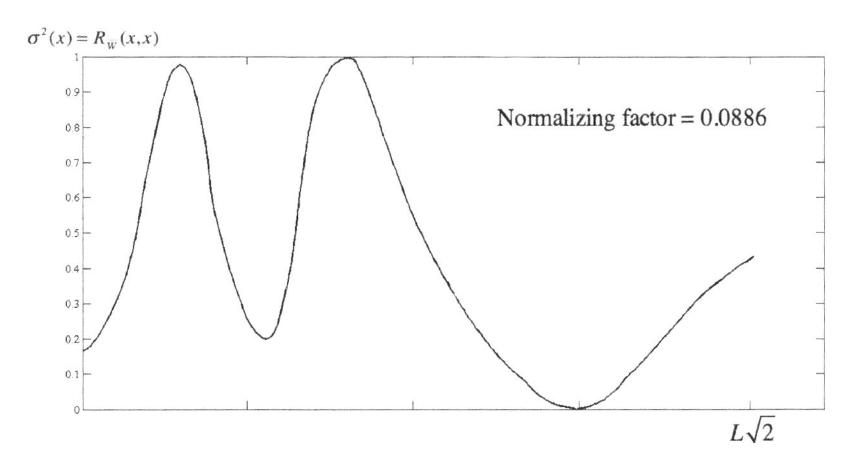

Fig. 2.31. Non-stationarity implies non-ergodicity; variance of the measured initial imperfections (group of 7 A-shells) is not constant.

11. Alternative Resolution of the Conundrum by the First-Order Second-Moment Method

> *. . . a first order second moment mean value reliability index. It is a very long name, but the underlying meaning of each part of the name is very important: First order because we use first order terms in the Taylor series expansion. Second moment because only means and variances are needed. Mean value because the Taylor series expansion is about the mean values.*
>
> A.S. Nowak and K.R. Collins (2000)

Referring to Bolotin's (1958) postulate of availability of a functional dependence, given in Eq. (2.1) (see Sec. 4 in this chapter) between the critical buckling load and the initial imperfection coefficients, Amazigo (1976, p. 175) notes:

> It is however a nontrivial problem to obtain relation on (3) and perform the above [probabilistic] analysis for $n > 2$. It is this difficulty that limits the effectiveness of this method.

A Dutch engineer, Ir. Sipke van Manen, reviewing our Monte Carlo methodology on stochastic imperfection sensitivity suggested that it was not necessary to obtain Eq. (2.1) analytically, and that having a numerical code relating buckling loads with imperfection coefficients would allow a semianalytical treatment of the problem. Such a joint study was conducted by Elishakoff *et al.* (1987), which will be followed hereafter. It was based on the first-order second moment methodology developed by Rzhanitsin (1949) of Moscow ZNIISK (Central Scientific Research Institute of Building Constructions) and apparently independently, some decades later, by Cornell (1969) of the M.I.T. The cornerstone of the method is rewriting Eq. (2.4) in a form

$$Z = Z(q_1, q_2, \ldots, q_n) \tag{2.107}$$

defined in such a manner that the equation

$$Z = 0 \tag{2.108}$$

determines the failure boundary, whereas $Z < 0$ implies failure and $Z > 0$ signifies nonfailure (successful performance). Our goal is the determination of the reliability function

$$R(\lambda) = \text{Prob}(\Lambda_{\text{cr}} > \lambda) \tag{2.109}$$

where Λ_{cr} is the critical buckling load and λ is the (deterministic) level of the load at which one is interested in the reliability. As is seen from Eq. (2.109) we are not interested in achieving an impossible task, namely in preventing buckling. This is due to the fact that due to unavoidable presence of initial imperfections buckling will inevitably occur. In such circumstances, our aim is to make buckling load Λ_{cr} as big as possible. Specifically, we desire this buckling load to exceed any preselected, deterministic load level λ. Thus, reliability is identified with the probability that the random event $\{\Lambda_{cr} > \lambda\}$ takes place.

This leads to the following definition of the function Z:

$$Z(\lambda) = \Lambda_{cr} - \lambda = \varphi(q_1, \ldots, q_n) - \lambda \tag{2.110}$$

Whereas it is impossible to conduct analytical calculation of reliability due to absence of the analytically obtainable function $\varphi_{(q_1,\ldots,q_n)}$ it is possible to conduct numerical analysis to determine the reliability, behaving, as it were, that such an analytical dependence is available. The mean value of the performance function Z is obtained as follows:

$$\begin{aligned} E(Z) &= E(\Lambda_{cr}) - \lambda = E[\varphi(q_1, \ldots, q_n)] - \lambda \\ &\approx \varphi[E(q_1), \ldots, E(q_2)] - \lambda \end{aligned} \tag{2.111}$$

This corresponds to the use of the Laplace approximation (see Elishakoff, 1983, 1999) for the determination of the moments of the nonlinear function. The value of $\varphi[E(q_1), \ldots, E(q_n)]$ should be calculated numerically by any available numerical code. It corresponds to the deterministic buckling load of the structure possessing mean imperfection amplitudes.

The variance of the performance function Z is given by

$$\begin{aligned} \text{Var}(Z) &= \text{Var}(\Lambda_{cr}) \\ &\approx \sum_{j=1}^{n}\sum_{k=1}^{n}\left(\frac{\partial\varphi}{\partial q_j}\right)_{q_j=E(Q_j)}\left(\frac{\partial\varphi}{\partial q_k}\right)_{q_k=E(Q_k)} Cov(Q_j, Q_k) \end{aligned} \tag{2.112}$$

Calculation of the sensitivity derivatives $\partial\varphi/\partial q_j$ is performed numerically by utilizing numerical differentiation formulas. To find sensitivity derivatives it is necessary to carry out n calculations of the buckling problem.

Once the values of $E(Z)$ and $\text{Var}(Z)$ are estimated, the reliability estimate is obtained as follows:

$$R(\lambda) = \text{Prob}(Z > 0) = \Phi(\beta) \tag{2.113}$$

where Φ is the standard normal probability distribution function and β is the reliability index given by

$$\beta = \frac{E(Z)}{\sqrt{\text{Var}(Z)}} \tag{2.114}$$

The above analysis was conducted on the groups of 4 called "B shells." The values of initial imperfection coefficients are in summarized in the paper (Elishakoff *et al.*, 1987). The mean values of the imperfection coefficients were

$$\begin{aligned} &E(Q_1) = -0.0364, \quad E(Q_2) = -0.0050 \\ &E(Q_3) = 0.4436, \quad E(Q_4) = -0.0316 \\ &E(Q_5) = -0.0148, \quad E(Q_6) = -0.0436 \\ &E(Q_7) = -0.0335, \quad E(Q_8) = -0.0034 \end{aligned} \tag{2.115}$$

Elements of the σ_{jk} variance–covariance matrix

$$\sigma_{jk} = \mathrm{Cov}(Q_j, Q_k) \tag{2.116}$$

were estimated as having following values:

$$\begin{aligned} &\sigma_{11} = 0.1319, \quad \sigma_{21} = 0.0566, \quad \sigma_{22} = 0.0402 \\ &\sigma_{31} = -0.7074, \quad \sigma_{32} = -0.1916, \quad \sigma_{33} = 4.686 \\ &\sigma_{41} = -0.0926, \quad \sigma_{42} = -0.0810, \quad \sigma_{43} = -0.0872 \\ &\sigma_{44} = 0.9670, \quad \sigma_{51} = -0.3646, \quad \sigma_{52} = -0.3327 \\ &\sigma_{53} = 0.6154, \quad \sigma_{54} = -0.9956, \quad \sigma_{55} = 3.057 \\ &\sigma_{61} = 0.4771, \quad \sigma_{62} = 0.1986, \quad \sigma_{63} = -2.478 \\ &\sigma_{64} = -0.6529, \quad \sigma_{65} = -1.372, \quad \sigma_{66} = 1.868 \\ &\sigma_{71} = 0.03484, \quad \sigma_{72} = -0.0518, \quad \sigma_{73} = -0.5978 \\ &\sigma_{74} = -0.0833, \quad \sigma_{75} = 0.5647, \quad \sigma_{76} = 0.2623 \\ &\sigma_{77} = 0.3710, \quad \sigma_{81} = 0.0646, \quad \sigma_{82} = 0.0272 \\ &\sigma_{83} = -0.3739, \quad \sigma_{84} = 0.0205, \quad \sigma_{85} = -0.1497 \\ &\sigma_{86} = 0.2068, \quad \sigma_{87} = 0.0022, \quad \sigma_{88} = 0.0369 \end{aligned} \tag{2.117}$$

In order to apply the first-order second-moment method, the mean buckling load has to be calculated first. For this purpose the multimode analysis developed by Arbocz and Babcock (1974) was employed. The result of the calculation of the mean buckling load is given in Fig., $E(\Lambda_s) = 0.746$ (i.e. about 74.6% of the classical buckling load). The mean buckling load calculated via the Monte Carlo method constituted $E(\Lambda_s) = 0.739$, thus, the difference between mean buckling loads, delivered by these two methods, is only 0.007% or 0.95%. Next the sensitivity derivatives in Eq. (2.97) were calculated by the finite difference approximation; for the increment of the Fourier coefficients in Eq. (), 1% of their original values were used, so that $\Delta\xi_1 = 0.01\overline{X_j}$. The calculated derivatives are:

$$
\begin{aligned}
\frac{\partial\psi_s}{\partial X_1} &= 0.09668, \quad \text{for } X_1 \equiv A_{2,0} \\
\frac{\partial\psi_s}{\partial X_2} &= 0.00340, \quad \text{for } X_2 \equiv A_{4,0} \\
\frac{\partial\psi_s}{\partial X_3} &= -0.01854, \quad \text{for } X_3 \equiv C_{1,2} \\
\frac{\partial\psi_s}{\partial X_4} &= -0.05687, \quad \text{for } X_4 \equiv C_{1,6} \\
\frac{\partial\psi_s}{\partial X_5} &= -0.24686, \quad \text{for } X_5 \equiv C_{1,8} \\
\frac{\partial\psi_s}{\partial X_6} &= -0.08183, \quad \text{for } X_6 \equiv C_{1,10} \\
\frac{\partial\psi_s}{\partial X_7} &= -0.01314, \quad \text{for } X_7 \equiv C_{2,3} \\
\frac{\partial\psi_s}{\partial X_8} &= -0.07173, \quad \text{for } X_8 \equiv C_{2,11}
\end{aligned} \tag{2.118}
$$

Elishakoff *et al.* (1987) chose the increments of end-shortening in such a way that the limit loads were found to an accuracy of 0.0001.

The results of the mathematical expectation and variance of Z are $E(Z) = 0.746 - \lambda$ and $\text{Var}(Z) = 0.0175$. The reliability functions calculated with the Monte Carlo method and with the first-order second-moment method are both given by Fig.... These curves appear to be in excellent agreement; however, in the higher reliability region, which is most important for design load derivation, the deviation is more noticeable.

The following conclusions were made in the above study:

(a) The first-order second-moment method can be successfully utilized for determining the reliability function of axially compressed shells.
(b) The number of buckling load calculations necessary for the first-order second-moment method is significantly less than with the Monte Carlo method.
(c) The mean buckling loads due to both methods are in excellent agreement, but still higher than the experimental value. This is caused by the simplified deterministic buckling load analysis. Since the present method does not need as many calculations as the Monte Carlo method, a more advanced and expensive method

(in terms of the computer time) can be employed for the deterministic analysis of buckling load calculations.
(d) The reliability functions delivered by both methods are in good agreement.

Other definitive works on stochastic imperfection sensitivity include papers by Eggwertz and Palmberg (1985), Palassopoulos (1991, 1997), Hussain (1996), Chryssanthopoulos (1998), Baker and Dowling (1991), Chryssanthopoulos and Poggi (1995a, 1995b), Bourinet *et al.* (2000), Gayton *et al.* (2002), and Gayton and Lemaire (2002), Most, Bucher and Schorling (2004), Stefanou and Papadrakakis (2004), Papadopoulos and Papadrakakis (2005), Lagaros and Papadopoulos (2006), Papadopoulos and Iglesis (2007), Dubourg *et al.* (2009), Schenk and Schuëller (2002, 2005), Schenk *et al.* (2000, 2001), Schuëller and Broggi (2011), Kriegesmann *et al.* (2011), Kriegesmann *et al.* (2010, 2011), and many others.

12. Corroboration Project STONIVOKS

Even in the age of the computer, experiments are an essential tool in stability research, and in particular in shell buckling studies.

J. Singer (1999)

During the initial months of my sabbatical stay at Delft University of Technology, Professor Walter Verduyn would often stop by my office, remarking that he was not a scientist but rather an artist who could use his hands very well. It occurred to that his gift could be utilized to validate, or at least corroborate, our probabilistic research on a large scale. Accordingly, I suggested that he develop a test apparatus for measuring the initial imperfection of a large quantity of identically produced shells. If we measure a group of such shells and generate a reliability function based on these measurements, we can later check the predicted reliability values on the subsequent buckling of another, large group.
"Which shells?" he asked.
"Beer cans," was my reply.
Professor Verduyn eagerly accepted my suggestion. He estimated that we need 20,000 Dutch guilders to build the apparatus which the Faculty of Aerospace Engineering kindly agreed to provide (good ol' times!).
I contacted the Heineken beer company in Amsterdam with a request for free beer cans. My limited Dutch and my foreign accent were noticed at once.
"Where are you from?" asked the Heineken manager.
"Israel," was my reply.
"Are the beer cans for peaceful purposes?" he inquired.
"This depends on whether or they are full or empty," I responded.

Accordingly, he kindly sent us 200 empty beer cans to the disappointment of some of my colleagues, who felt that full cans would have been preferable.

In a subsequent paper presented in Haifa at the Seventh International Conference on Experimental Stress Analysis, we wrote (Verduyn and Elishakoff, 1982):

> The main complications inherent in this type of experimental research are twofold. Firstly, the number of test specimens has to be found [Thanks Heineken!]. They have to be identical in dimensions, made out of the same material, by an identical production process. Secondly, all specimens have to be tested under exactly the same conditions. The testing-machine described here was designed and produced at the Aerospace Department [of the Delft University of Technology]. There it is known as "STONIVOKS" which is the acronym for "Statistisch Onderzoek naar de Invloed van Initiële Vormonzuiverheden op de Kniklast van Schaalconstructies" (Statistical Research into the Influence of Initial Imperfections on the Buckling Load of Shells).

The phases of the experimental sequence were defined as follows:

(a) Measurement of the initial imperfections of every specimen at zero axial compressive load
(b) Periodic measurement of the growth of the imperfections under increasing compressive load
(c) Determination of the compressive collapse load of every specimen
(d) Periodic follow-up of the buckling pattern after failure.

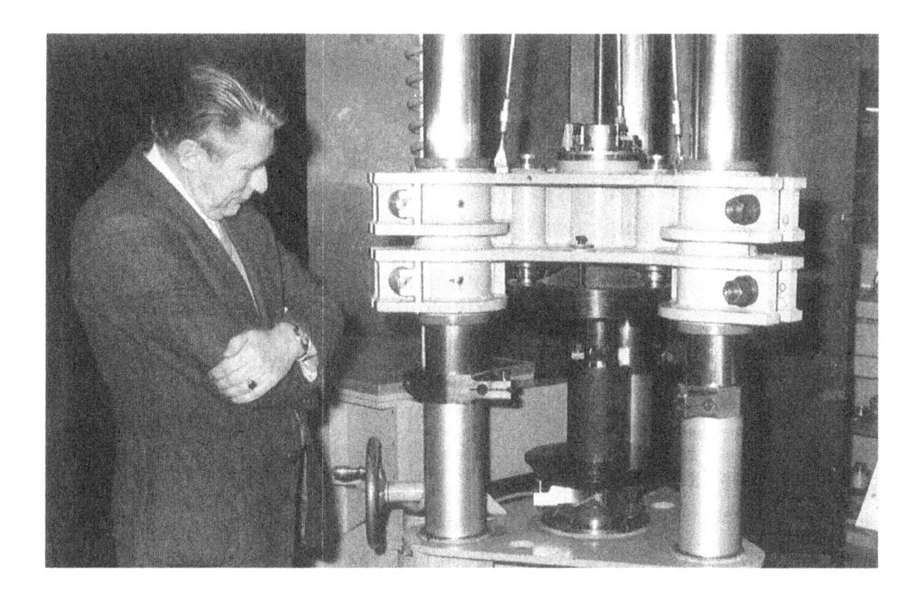

Fig. 2.32. Professor W. D. Verduyn next to the STONIVOKS.

Fig. 2.33. The author next to the STONIVOKS.

Fig. 2.34. STONIVOKS.

Owing to the small diameter of the test specimen (about 66 mm) all measurements had to be taken on its outside surface. In these circumstances, the specimen had to be rotated and the measuring setup stationary in the circumferential direction, moving only in the axial direction of the specimen. The process had to be automated to a high degree.

The spacing of the measuring points in the axial direction was set at 1 mm, and the number of measurements in the circumferential direction at 100, so that the spacing of the measuring points in that direction was at about 2 mm. Since the useful length of the specimen was about 80 mm, the number of measurements per cycle was 8000.

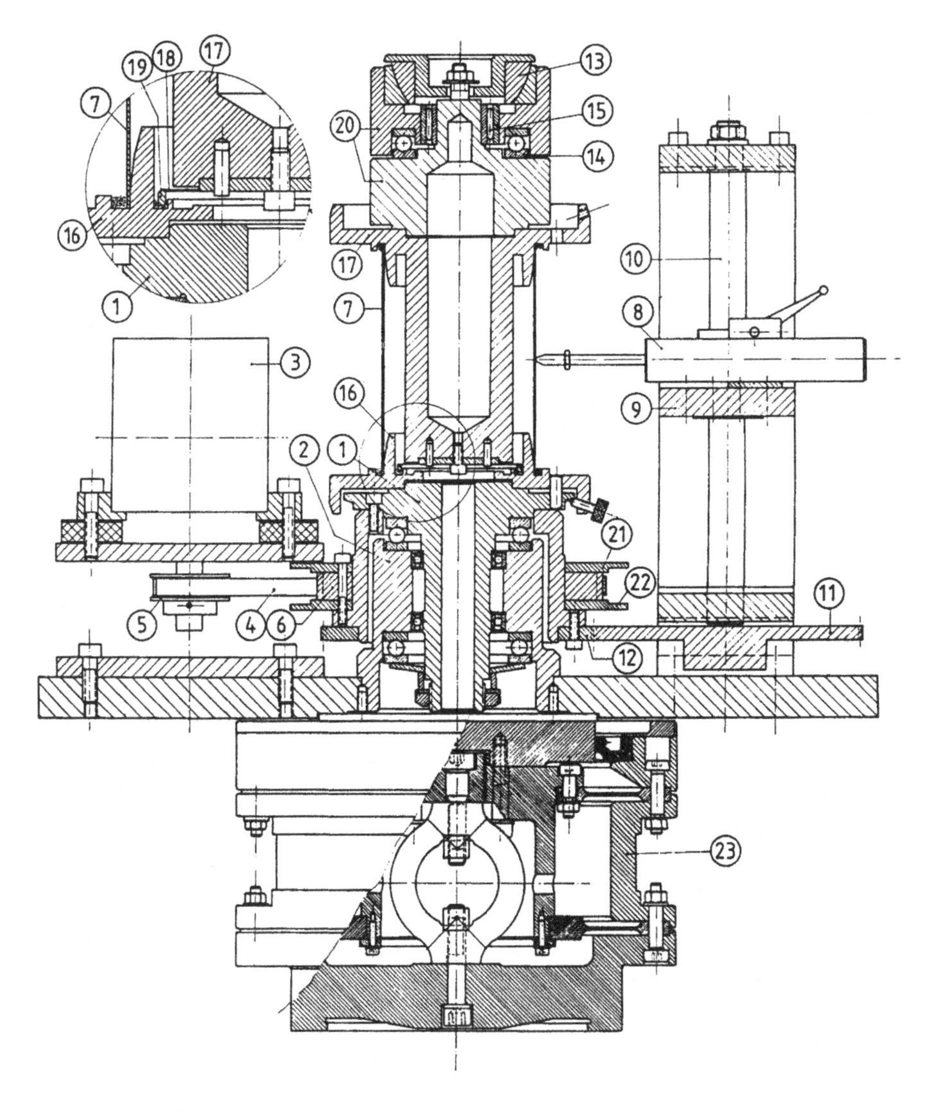

Fig. 2.35. STONIVOKS at the design stage. (courtesy of Dr. Jan de Vries)

Fig. 2.36. STONIVOKS in the assembled form.

Dancy and Jacobs (1988) write:

> There are only a couple of plants in the world producing beer cans in the amount of millions each day. To be more precise, the machine which starts off with a small cup, pushing this cup through 3 or 4 rings spits out 1000 cans each minute. Our beer can is not a unique species, however, it is unique in the sense that a set of 30 was tested 25 years ago in the Faculty of Aerospace Engineering at Delft University of Technology by Dancy and Jacobs.

Buckling of beer cans was also studied theoretically, as a local problem, by de Vries (2006), via resorting to the energy approach.

According to de Vries (2009), "shell IW1-20 is one of over 30 beer cans investigated by Dancy and Jacobs (1988). The thin-walled shell manufactured from steel has a length of 100 mm, a radius of 33 mm, and a thickness of approximately 0.1 mm, yielding

$$R/h = 330, \quad \phi = \sqrt{R/h}/16 = 1.13537 \tag{2.119}$$

Then the knockdown factor is

$$\gamma = 1 - 0.901(1 - \mathrm{e}^{-\phi}) = 0.3885 \tag{2.120}$$

The lower bound buckling load for this shell now becomes

$$P_\gamma = \gamma P_{\mathrm{cl}} = -3102.4N \tag{2.121}$$

as against the experimentally found load

$$P_{\mathrm{exp}} = -3890.0N \tag{2.122}$$

Reliability was evaluated by Arbocz *et al.* (2000), following my suggestions to Professor Arbocz in communications over the years as well as during meetings in Delft or at conferences. Specifically, I suggested to doubling the initial can population and comparing the design loads associated with different population sizes. For the first group of four cans, using the first-order second-moment method (described in Sec. 11 in this chapter), the design load corresponding to the required reliability 0.99999 is 0.49, which constitutes the reliability based knockdown factor. For a sample of 32 cans the reliability-based knockdown factor equals 0.44. It was gratifying to witness the justification of the probabilistic methodologies proposed by the author.

Thus, the inquiry of Mossakovski's (1970)

> The substantial result of the investigations performed up to now is the explanation, in a certain sense, of the mechanism of the effect of initial imperfections...
>
> But the extremely important question remains: Can one bring into correspondence the theoretical and experimental results, obtained in laboratory conditions, if one introduces the real imperfections into the analysis?

was answered in affirmative.

BY APPOINTMENT TO
THE ROYAL DANISH COURT
BY APPOINTMENT TO
THE ROYAL SWEDISH COURT
TUBORG
INTERNATIONAL A/S

Professor Isaac Elishakoff
Israel Institute of Technology
Department of Aeronautical Engineering
Haifa
Israel

9th November, 198
X/LFJ/d4

Dear Sir,

We thank you for your kind letter of 3rd November, 1982 and are pleased to note that you enjoyed your visit to our brewery last summer.

As requested, we have sent you a few samples of our 28 cl, 33 cl and 50 cl cans which we hope will arrive in good condition.

Yours faithfully,
TUBORG INTERNATIONAL A/S

Lars Friis-Jensen

Fig. 2.37. Letter from TUBORG International A/S.

13. Delft's Other Challenge

Science is a team sport.

M.S. Cohen (2012)

In Sec. 8, it is mentioned that (Elishakoff, 1999) "upon my arrival in 1979, I found that Delft posed another challenge," besides the disbelief expressed by my host, Professor Arbocz, in probabilistic methods. I depict the situation in my book *Probabilistic Theory of Structures* (1999), but not in its 1983 edition, as it will be explained later:

> ...to my astonishment, Ir. Dijkshoorn appeared in my office during my first week [in September 1979] there and said: "You will be returning to your home university in a year, but we would like you to leave some pedagogical trace here. Please prepare written lecture notes on probabilistic methods, so that future generations of students and researchers at our university can use them." I was not enthusiastic about such a task. Not only had I been unaware of this traditional Delft requirement, but preparing the lecture notes would disrupt my plan to devote all my time to research. However, thanks to several elements unique to Delft I was able to meet this challenge. These elements included the extremely friendly atmosphere created by John Arbocz, his wife Margot, and the entire "Vakgroep C" of the Aerospace Department. The warm hospitality Arbocz, Koiter, van der Neut, Schijve, van Geer and other families is unforgettable. The extremely fast (rocket like) and skillful typing of Ms. Marijke Schillemans and the beautiful artwork of craftsman Willem Spee were additional blessings. The exceptional situation these elements yielded would eventually produce active scientific cooperation between myself and J. Arbocz, A. Scheurkogel, J. Kalker, W. Verduyn, J. van Geer, T. van Baten, S. van Manen, P. Vermeulen, and in later years with W.T. Koiter.
>
> Another beautiful Delft tradition I had not known about was the distribution of faculty members' memoranda to faculty in other departments. This, too, was to take me in a direction I had not anticipated. In March 1980 a conference organized by W.T. Koiter took place at the Mechanical Engineering Department. Professor Koiter, as its chair, met every participant at the "Centraal Hotel." He asked me — as one who lived in extreme vicinity to this hotel — if I would accompany him home, after all the participants had arrived. So we waited together until nearly midnight, and we had several glasses of Jonge Jenever. At that time, Professor Koiter told me that he had read my memorandum on probabilistic methods, which had been distributed to his department. He complimented me at its explanatory style and contents, and urged me to base a book on these materials. I was very excited. With this blessing from Professor Koiter (who was characterized by Bernard Budiansky — himself one of the champions of American mechanics — as "the sage from Delft"), my energies received a tremendous boost. I decided to devote the remaining months of my sabbatical in advancing this project as far as I could.
>
> Within three years, the manuscript was completed at the Technion Israel Institute of Technology and had been accepted by Wiley. Upon the book's initial publication I refrained from mentioning the fact that Professor Koiter's enthusiastic encouragement was the main driving force behind the book. As I later explained to him, I had felt that I should not use his approval to promote my work, that the work must stand on its own feet.

In fact, I was afraid that should the book have an unfavorable reception, this would reflect on Professor Koiter to such a criticism. Professor Koiter incidentally endorsed my reasoning. Fortunately, the book was well received. I take the liberty of quoting excerpts from some of the reviews that pleased both me and the main "instigators" of the project: Ir. Dijkshoorn and Professor Koiter: (1) "...A treatise on random vibration and buckling..." (*Journal of Applied Mechanics*), (2) "...This extremely well-written text, authored by one of the leaders in the field... excellent graduate course in random vibration and buckling... outstanding instrument..." (*AIAA Journal*), (3) "...Author ties together reliability, random vibration and random buckling... well written... useful book..." (*Shock and Vibration Digest*), (4) "... This is a notable book... a unified well-developed presentation... a good book; a different book" (*Journal of Sound and Vibration*); (5) "... Outstanding..." (*Zeitschrift für Flugwissenschaften und Weltraumforschung*); (6) "...The specially interesting points of the book include the analysis of buckling load for an imperfect column (based on the author's previous work), and both chapters dealing with modeling structural response due to random vibrations" (*Applied Mechanics Reviews*), (7) " Here is a book for structural engineers that proceeds from the concept of a random event to the Monte Carlo method of solving the column-buckling problem without the fortuous detours... a welcome addition to the specialist literature" (*Search*).

There were also personal letters: (1) "...This is an impressive volume..." (Professor W.T. Koiter); (2) "...By far the best book on the market today...," Professor Niels Lind, University of Waterloo; (3) "...This is a beautiful book..." (Professor Stephen H. Crandall, M.I.T.); (4) "...The book certainly satisfies the need that now exists for a readable textbook and reference book..." (Professor Masanobu Shinozuka, Princeton University); (5) "...It seems to me a hard work with great result..." (Professor Hans G. Natke, University of Hanover).

After my lecture at the European Colloquium in Delft in March 1980, the Koiter family invited myself and my wife Esther several times for dinner. Then they asked us to stop by at their home at any weekend, without a formal invitation. We regarded this as a friendly gesture not to be taken advantage of, so you can imagine our surprise when Warner and Luis Koiter appeared in our doorstep 130 Ritveld Straat in the center of Delft. They said that they wanted to show us how easy it was to come without an invitation.

On one occasion, during dinner at their home, Professor Koiter mentioned that he was critical of himself for not having written a book on elastic stability, as he claimed, for fear of imperfect coverage of the topic. Dick H. van Campen (1999) writes in Professor Koiter's obituary:

> Koiter was so critical of his own formulations that, despite his numerous scientific publications and his involvement in many books, he never published a book in English. In 1972 he did publish a textbook in Dutch entitled *Sterkte en Stiffheid 1, Grondslagen*. A new edition appeared in 1985, entitled *Inleiding tot de Leer van Stiffheid en Sterkte*.

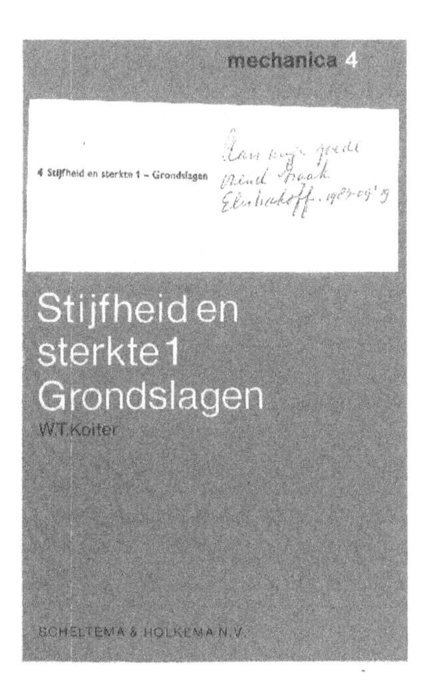

Fig. 2.38. Professor W.T. Koiter's book with a personal inscription.

On two different occasions Koiter presented me these books. The second time, I reminded him that he had already presented me with the first edition. He replied that he wanted me to have the second edition as well. Fig. 2.38 shows the cover page of one of the editions with his inscription in Dutch; the other edition was presented some years later, bearing the same inscription.

In a public lecture (Koiter, 1979, p. 244) he mentions: "I also realize the obligation I have incurred by Van der Heijden's excellent lecture notes of the course I gave this year in Delft!"

Fortunately, van der Heijden (2009) published these lecture notes 12 years after Professor Koiter's passing away. I was greatly honored when Cambridge University Press chose me to serve as reviewer of the manuscript of this book. Koiter's desire to have a book published on elastic stability was thus realized by his former student, disciple, and colleague.

In digressing, I would like to note that unfortunately, not only Koiter but also some other great researchers in elastic stability did not write a book. It is comforting to read the statement on the importance of writing books, recently made Thompson (2013):

> One thing that I should mention, in a wider context, is the importance of (someone) writing a book when there has been a great explosion or breakthrough of research in a field. This is needed to clarify, codify and record the achievement, and it is important

> that one of the key workers should take it upon themselves to summarize the new developments, preferably as a book or monograph. Likily, in my case, I rather like writing books, so there was no problem there. However, when I look back at the advances in shell buckling in the early 1960s, I cannot help wishing that there had been an extensive and clear write-up of the deep theoretical progress that was made in imperfection-sensitivity studies. Unfortunately, key workers, such as Koiter at Delft and Budiansky and Hutchinson at Harvard, did not seem to be the book-writing types (I know that Koiter particularly regretted this).

Thompson (2013) then comments: "There are, of course, a lot of books on shell buckling, as can be seen (for example) on the comprehensive website created by Bushnell (2012)." Researchers also may be interested I the Facebook page created by Hilburger (2011) by NASA Langley Research Center. It announces that "NASA Engineering and Safety Center's Shell Buckling Knockdown Factor Twitter site is now online."

Hilburger (2012) notes: "NASA's Shell Buckling Knockdown Factor (SBKF) project was established in the spring of 2007 by the NASA Engineering and Safety Center (NESC) in collaboration with the Constellation Program and Exploration Systems Mission Directorate.

The SBKF project has the current goal of developing less-conservative, robust shell buckling design factors (a.k.a. Knockdown factors) and design and analysis technologies for light-weight stiffened metallic launch vehicle (LV) structures."

Chapter 3

Hybrid Uncertainty in Imperfections and Axial Loading

Even imperfection itself may have its ideal or perfect state.
T. De Quincey (1881)

This chapter investigates the combined effect of randomness in initial geometric imperfections and the applied loading on the reliability of axially compressed cylindrical shells. In order to gain an insight, we consider simplest possible case when both the initial imperfections and the applied loads are uniformly distributed. It is shown that hybrid randomness may either increase or decrease the reliability of the shell if the latter is treated as experiencing the sole randomness in initial imperfections.

1. Introduction

The randomness in the applied load was first introduced by Roorda (1980), followed by the study of Elishakoff (1983). Later on, Cederbaum and Arbocz (1996) and Li, Elishakoff, Starnes and Shinozuka (1995) studied this topic. The former investigation utilized the first-order second-moment method, whereas the latter work applied the conditional simulation technique.

In this chapter we study the combined effect of random initial imperfections and random applied load. In order to grasp the effect, the simplest possible probability density — the uniform density — is taken in order to describe the random variables involved. Various cases are considered with evaluation of the attendant reliability, that is, the probability that the buckling load exceeds the applied load. It has been demonstrated that the claim appearing in the literature that randomness in the loads reduces structural reliability must be modified. It turns that the load uncertainty can be either detrimental or beneficial.

2. Basic Equations

Koiter in his thesis (1945) and a subsequent paper (1963) provided the following relationship describing the imperfection-sensitivity of a cylindrical shell:

$$(1-\rho)^2 - q\rho|\xi| = 0 \tag{3.1}$$

where

$$\rho = \lambda_s/\lambda_c \tag{3.2}$$

is the ratio between the buckling load λ_s of the imperfection structure and λ_c, which represents the buckling load of the perfect structure, ξ is the nondimensional initial imperfection amplitude, and q is the numerical coefficient. Equation (3.1) in actuality represents the modification of Koiter's original formula, since in a very long shell if is anticipated that initial imperfections either positive or negative will produce the same physical effect. It is seen from Eq. (3.1) that if the structure is perfect, that is, $|\xi| = 0$, then $\rho = 1$ and there is no reduction in the load-carrying capacity. However, for $|\xi| > 0$, the root $\rho < 1$ describes the reduction of the load-carrying capacity because of initial imperfections.

In this chapter we are interested in the case when both the applied load and the initial imperfections are random. Yet, it appears to be instructive to first investigate the case when the initial imperfection alone is treated as a random variable.

Hence, the initial imperfection ξ is treated as a random variable with given probability density $f(\xi)$ or cumulative distribution function F_ξ. We are interested in evaluating the structural reliability or the probability that the cylindrical shell will perform its desired function.

It is not possible to expect the maximalistic task of the shell not buckling altogether, since it is the nature of the shell to undergo buckling, as grasped by Eq. (3.1). However, one can demand that the buckling will not occur prior prespecified non-dimensional load α. Alternatively, we can identify the reliability with the probability that the shell buckles at loads exceeding α:

$$R = \text{Prob}(\rho_s > \alpha) \tag{3.3}$$

In order to elucidate the imperfection-sensitivity in the probabilistic context, we will first deal with the case when initial imperfections are uniformly distributed. Therefore, three cases will be considered: (a) initial imperfections take positive values only, (b) initial imperfections assume negative values solely, and (c) initial imperfections take both positive and negative values. Prior to doing so, some comments appear in need on why uniform density is chosen. Indeed, Vermeulen *et al.* (1984) showed that normality assumption of initial imperfections does not contradict the analyzed data. However in absence of statistical data on axial load, it makes sense to use simplest possible density, for the purpose of consistency.

Since our goal, in this chapter, is to show that the reliability can be either reduced or increased due to hybrid uncertainty, we resort to the form of Ockham's (c. 1287–1347) razor stating: "It is in vain to do more what can be done with fewer." Indeed, "everything should be made as simple as possible, but not simpler" (maxim often attributed to Einstein). Uniform density both for initial imperfection and axial loading is therefore adopted in this chapter.

3. Positive-Valued Uniformly Distributed Imperfections

Assume that initial imperfection's probability density reads:

$$f(\xi) = \begin{cases} 0, & \text{for } \xi < 0,\ \xi < \xi_0 \\ 1/\xi_0, & \text{for } 0 < \xi < \xi_0. \end{cases} \tag{3.4}$$

In order calculate the reliability $R(\alpha)$ we first express from Eq. (3.1) the value of ρ as a function of ξ:

$$\rho_{1,2} = \left(2 + q\xi \pm \sqrt{q\xi(4 + q\xi)}\right)/2. \tag{3.5}$$

As is seen in Fig. 3.1, ρ_1 corresponds to values $\rho > 1$ and hence does not describe the physical phenomenon at hand. Only the curve ρ_2, with attendant values of $\rho < 1$, corresponds to the imperfection sensitivity. Thus, reliability becomes

$$\begin{aligned} R(\alpha) &= \text{Prob}(\rho_s > \alpha) = \text{Prob}(\rho_2 > \alpha) \\ &= \text{Prob}\left[\left(2 + q\xi - \sqrt{q\xi(4 + q\xi)}\right)/2 > \alpha\right]. \end{aligned} \tag{3.6}$$

leading to

$$R(\alpha) = \text{Prob}[\sqrt{q\xi(4 + q\xi)} < 2(1 - \alpha) + q\xi]. \tag{3.7}$$

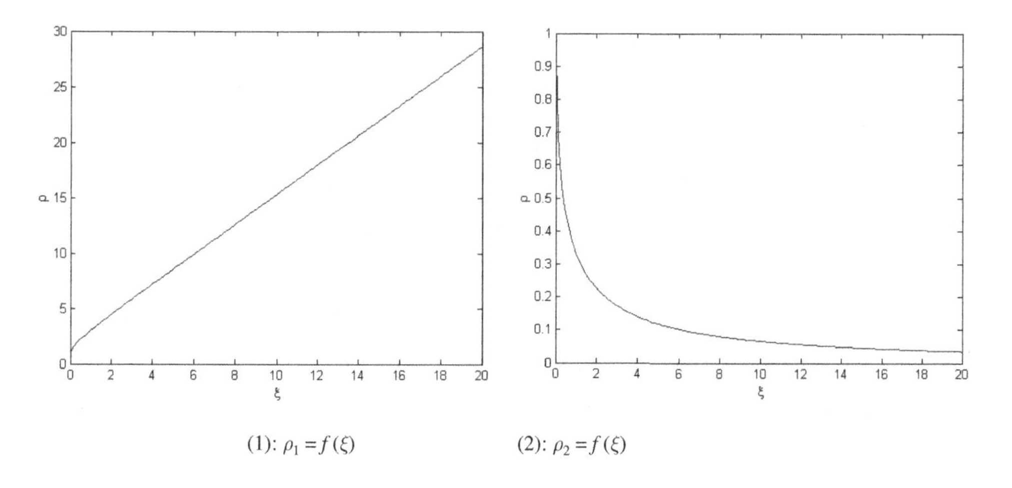

(1): $\rho_1 = f(\xi)$ (2): $\rho_2 = f(\xi)$

Fig. 3.1. Variation of ρ_1 and ρ_2 versus initial imperfection amplitude ξ.

For $\alpha < 1$ and initial imperfection ξ taking positive values, the right-hand side $2(1-\alpha)+q\xi$ is positive. Therefore, the inequality in Eq. (3.7) can be replaced by

$$R(\alpha) = \mathrm{Prob}\big[q\xi(4+q\xi) < (2(1-\alpha)+q\xi)^2\big]. \tag{3.8}$$

After some algebra, we get

$$R = \mathrm{Prob}[\xi < (1-\alpha)^2/q\alpha] \tag{3.9}$$

or

$$R = F_\xi[(1-\alpha)^2/q\alpha] \tag{3.10}$$

where F_ξ is the cumulative probability distribution function of initial imperfection, evaluated at $(1-\alpha)^2/q\alpha$. Therefore, the reliability becomes

$$\begin{aligned} R(\alpha) = 0, \quad & \text{for } (1-\alpha)^2/q\alpha < 0 \\ (1-\alpha)^2/q\alpha\xi_0, \quad & \text{for } 0 < (1-\alpha)^2/q\alpha \le \xi_0 \\ 1, \quad & \text{for} (1-\alpha)^2/q\alpha > \xi_0. \end{aligned} \tag{3.11}$$

The first inequality is invalid. Consider the equality

$$(1-\alpha)^2/q\alpha = \xi_0 \tag{3.12}$$

leading to

$$\alpha_{1,2} = \big(2+q\xi_0 \pm \sqrt{q\xi_0(4+q\xi_0)}\big)/2. \tag{3.13}$$

The branch with the plus sign does not bear a physical sense since it is associated with consequence $\alpha > 1$. Hence only the branch with the minus sign has a physical sense. We denote

$$\alpha^* = \big(2+q\xi_0 - \sqrt{q\xi_0(4+q\xi_0)}\big)/2. \tag{3.14}$$

Hence the requirement $(1-\alpha)^2/q\alpha \le \xi_0$ in Eq. (3.11) is associated with an inequality $\alpha \ge \alpha^*$. The reliability becomes

$$\begin{aligned} R(\alpha) = 1, \quad & \alpha \le \alpha^* \\ (1-\alpha)^2/q\alpha\xi_0, \quad & \alpha \ge \alpha^* \end{aligned} \tag{3.15}$$

Let us deal with the design of the shell, demanding the structural reliability level to be at least r. The value of α corresponding to r is denoted hereinafter as $\alpha_{\mathrm{allowable}}$. It satisfies the following equation:

$$r = \frac{(1-\alpha_{\mathrm{allowable}})^2}{q\alpha_{\mathrm{allowable}}\xi_0}. \tag{3.16}$$

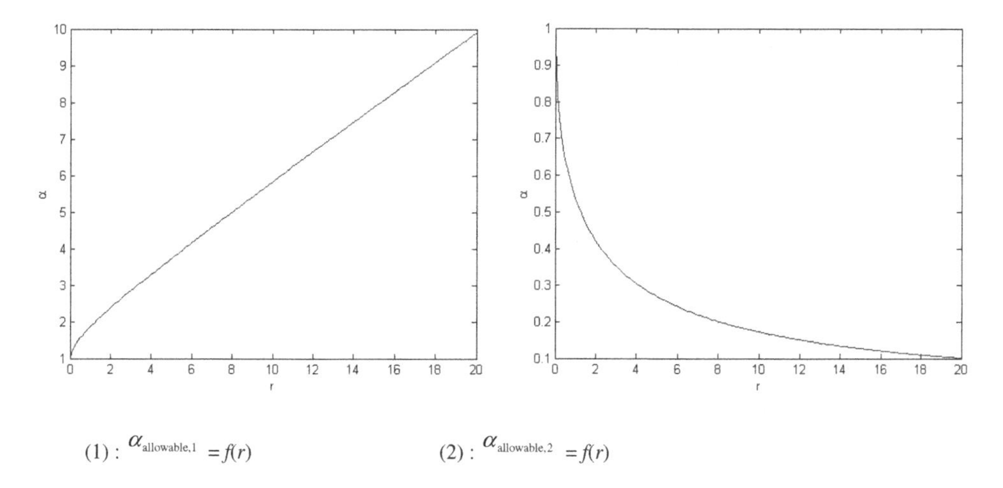

Fig. 3.2. Variation of $\alpha_{\text{allowable},1}$ and $\alpha_{\text{allowable},2}$ as function of r.

For $\alpha_{\text{allowable}}$ we get

$$\alpha_{\text{allowable}1,2} = \left(2 + qr\xi_0 \pm \sqrt{qr\xi_0(4 + qr\xi_0)}\right)/2. \tag{3.17}$$

As is seen in Fig. 3.2 $\alpha_{\text{allowable},1}$ corresponds to values of α in excess of unity and hence does not describe the physical phenomenon at hand. Thus, the final expression for $\alpha_{\text{allowable}}$ becomes

$$\alpha_{\text{allowable}} = \alpha_{\text{allowable},2} = \left(2 + qr\xi_0 - \sqrt{qr\xi_0(4 + qr\xi_0)}\right)/2. \tag{3.18}$$

4. Combined Randomness in Imperfection and Load for Positive Imperfection Values

Let us extend the above considerations to the case when the applied load Λ is a random variable. The simplest possible case is the one when Λ is treated as a uniformly distributed random variable in the interval $[\lambda_1, \lambda_2]$, $\lambda_{j=1,2} \in [0, 1]$, $\lambda_1 < \lambda_2$. Reliability in the new hybrid circumstances is denoted by R and becomes

$$R = \text{Prob}(\rho_s > \Lambda). \tag{3.19}$$

Under the assumption that initial imperfections and load are statistically independent, the reliability is evaluated by the formula

$$R = \int_0^1 [1 - F_{\rho_s}(\alpha)] f_\Lambda(\alpha) \mathrm{d}\alpha \tag{3.20}$$

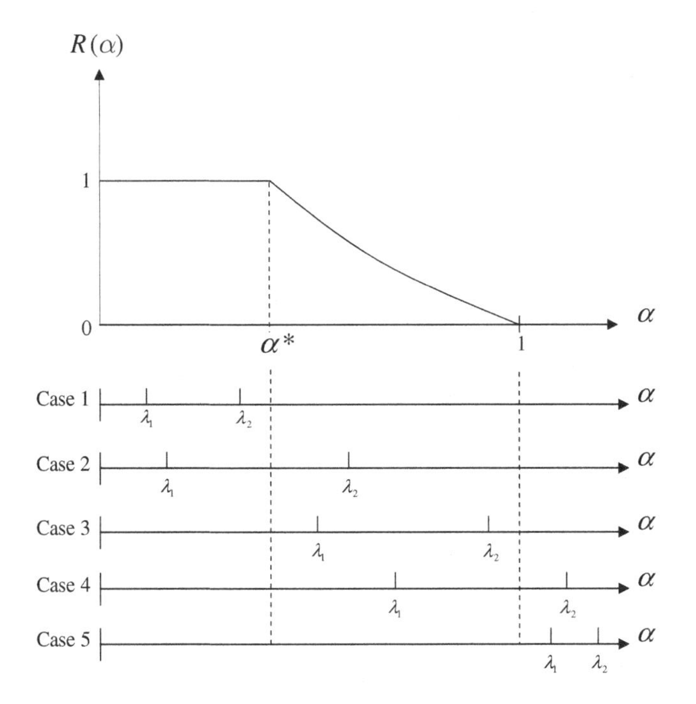

Fig. 3.3. Five possible different locations of load bounds with respect to α^* and unity.

or, since Λ varies in the interval $[\lambda_1, \lambda_2]$,

$$R = \int_{\lambda_1}^{\lambda_2} [1 - F_{\rho_s}(\alpha)]\, f_\Lambda(\alpha)\mathrm{d}\alpha = \frac{1}{\lambda_2 - \lambda_1}\int_{\lambda_1}^{\lambda_2} R(\alpha)\mathrm{d}\alpha. \tag{3.21}$$

The general shape of the reliability curve $R(\alpha)$ is depicted in Fig. 3.3. Five cases arise due to positions of λ_1 and λ_2 regarding α^* value. The reliability $R(\alpha)$ equals unity for $\alpha \le \alpha^*$, and decreases to zero, the latter value being achieved for $\alpha = 1$.

For $\lambda_j \in [0, \alpha^*]$, where α^* is defined here as formally coinciding with Eq. (3.14), we get

$$R = \frac{1}{\lambda_2 - \lambda_1}\int_{\lambda_1}^{\lambda_2} \mathrm{d}\alpha = 1. \tag{3.22}$$

For $\lambda_1 \in [0, \alpha^*]$ and $\lambda_2 \in (\alpha^*, 1]$, we get

$$R = \frac{1}{\lambda_2 - \lambda_1}\left(\int_{\lambda_1}^{\alpha^*} \mathrm{d}\alpha + \int_{\alpha^*}^{\lambda_2} \frac{(1-\alpha)^2}{q\alpha\xi_0}\mathrm{d}\alpha\right) \tag{3.23}$$

or

$$R = \frac{1}{2q\xi_0(\lambda_2-\lambda_1)}\left[2\ln\left(\frac{\lambda_2}{\alpha^*}\right)+\lambda_2(\lambda_2-4)-\alpha^*(\alpha^*-4)\right] + \frac{\alpha^*-\lambda_1}{\lambda_2-\lambda_1}. \tag{3.24}$$

For $\lambda_1 \in (\alpha^*, 1)$, $\lambda_2 \in (\alpha^*, 1]$, we obtain

$$R = \frac{1}{\lambda_2-\lambda_1}\int_{\lambda_1}^{\lambda_2}\frac{(1-\alpha)^2}{q\alpha\xi_0}\mathrm{d}\alpha = \frac{1}{2q\xi_0(\lambda_2-\lambda_1)}\left[2\ln\left(\frac{\lambda_2}{\lambda_1}\right)+\lambda_2(\lambda_2-4)-\lambda_1(\lambda_1-4)\right]. \tag{3.25}$$

For $\lambda_1 \in (\alpha^*, 1]$, $\lambda_2 > 1$, we have

$$R = \frac{1}{\lambda_2-\lambda_1}\int_{\lambda_1}^{1}\frac{(1-\alpha)^2}{q\alpha\xi_0}\mathrm{d}\alpha = \frac{1}{2q\xi_0(\lambda_2-\lambda_1)}\left[2\ln\left(\frac{1}{\lambda_1}\right)-3-\lambda_1(\lambda_1-4)\right]. \tag{3.26}$$

For $\lambda_j > \alpha^*$, we find

$$R = 0. \tag{3.27}$$

To conduct numerical calculation, we set $\xi_0 = 0.3$ and $q = 4/3$ and we get $\alpha^* = 0.5366$.

Then we calculate reliability values for the three cases defined previously:

Table 3.1. Reliability values as depending upon bounds of the applied load.

Case number	1	2	3	4	5
λ_1	0.2	0.2	0.6	0.9	1.5
λ_2	0.3	0.6	0.9	1.5	3
R	1	0.9722	0.2539	0.0015	0

Consider now the case when $\lambda_1 = 0$ and λ_2 numerically coincide with the value $\alpha_{\text{allowable}}$ found in Eq. (3.18). Reliability R, given by the formula (3.20), is evaluated by the formula

$$R = \int_0^{\alpha_{\text{allowable}}}[1-F_{\rho_s}(\alpha)]f_{\Lambda}(\alpha)\mathrm{d}\alpha. \tag{3.28}$$

For $\xi_0 = 0.3$, $q = 4/3$, $r = 0.9$, the calculations yield $\alpha^* = 0.5366$ and $\alpha_{\text{allowable}} = 0.5383$. As is seen, $\alpha_{\text{allowable}}$ is greater than α^*; hence, reliability expression corresponds to case 2 where $\lambda_1 = 0$ and $\lambda_2 > \alpha^*$. Then Eq. (3.24) reduces to

$$R = \frac{1}{2q\xi_0\alpha_{\text{allowable}}}\left[2\ln\left(\frac{\alpha_{\text{allowable}}}{\alpha^*}\right) + \alpha_{\text{allowable}}(\alpha_{\text{allowable}} - 4) - \alpha^*(\alpha^* - 4)\right] + \frac{\alpha^*}{\alpha_{\text{allowable}}}. \tag{3.29}$$

Finally, we get the value of reliability $R = 0.999\,9848$, which signifies that as a result of the load randomness the actual reliability has *increased* above the level $r = 0.99$ demanded for the case when the load was treated as a deterministic quantity. This conclusion on the increase of the structural reliability is not a universal one, however. If the random load takes values beyond the level $\alpha_{\text{allowable}}$, one can anticipate the decrease in structural reliability. It is instructive, therefore, to consider a case when the possible values of Λ are in the following interval: $\lambda_1 \leq \lambda \leq \lambda_2$. For example, $\lambda_1 = \alpha_{\text{allowable}}$, $\lambda_2 > \alpha_{\text{allowable}}$, $\lambda_2 = 1.01\alpha_{\text{allowable}}$ can be chosen. In new circumstances, the reliability becomes

$$R = \int_{\lambda_1}^{\lambda_2} [1 - F_{\rho_s}(\alpha)] f_\Lambda(\alpha) \mathrm{d}\alpha = \frac{1}{\xi_0(\lambda_2 - \lambda_1)} \int_{\lambda_1}^{\lambda_2} \frac{(\alpha - 1)^2}{\alpha q} \mathrm{d}\alpha. \tag{3.30}$$

The final expression for the reliability reads

$$R = \frac{1}{2\xi_0(\lambda_2 - \lambda_1)}\left[2\ln\left(\frac{\lambda_2}{\lambda_1}\right) + \lambda_2(\lambda_2 - 4) - \lambda_1(\lambda_1 - 4)\right]. \tag{3.31}$$

In new circumstances, the actual reliability becomes $R = 0.98357$, which is *less* than $r = 0.99$. We thus conclude that the load randomness can either increase or decrease the shell's reliability for the uniformly distributed initial imperfections and loads.

5. Negative-Valued Uniformly Distributed Imperfections

When initial imperfections take on only negative values, Eq. (3.1) reduces to

$$(1 - \rho)^2 + q\rho_s\xi = 0 \tag{3.32}$$

leading to

$$\rho_{1,2} = \left(2 - q\xi \pm \sqrt{q\xi(q\xi - 4)}\right)/2. \tag{3.33}$$

Only the branch with minus sign has a physical significance. The reliability becomes

$$\begin{aligned} R(\alpha) &= \text{Prob}(\rho_s > \alpha) = \text{Prob}(\rho_2 > \alpha) \\ &= \text{Prob}\big[\big(2 - q\xi - \sqrt{q\xi(q\xi - 4)}\big)/2 > \alpha\big] \end{aligned} \tag{3.34}$$

or

$$R(\alpha) = \text{Prob}\big[\sqrt{q\xi(q\xi - 4)} < 2(1-\alpha) - q\xi\big]. \tag{3.35}$$

For $\alpha < 1$ and initial imperfection ξ taking negative values, the right-hand side $2(1-\alpha) - q\xi$ is positive. Therefore, the inequality in Eq. (3.8) can be replaced by

$$R(\alpha) = \text{Prob}[q\xi(\mathrm{q}\xi - 4) < (2(1-\alpha) - q\xi)^2]. \tag{3.36}$$

After some derivations, we get

$$R(\alpha) = \text{Prob}[\xi < -(1-\alpha)^2/q\alpha] \tag{3.37}$$

or

$$R(\alpha) = F_\xi[-(1-\alpha)^2/q\alpha] \tag{3.38}$$

where F_ξ is the cumulative probability distribution function of initial imperfection, evaluated at $-(1-\alpha)^2/q\alpha$. $F_\xi(\xi)$ reads, in the region $\xi < 0$:

$$\begin{aligned} F_\xi(\xi)0, &\quad \text{for } \xi > 0 \\ \xi/\xi_0, &\quad \text{for } \xi_0 < \xi < 0 \\ 1, &\quad \text{for } \xi < \xi_0. \end{aligned} \tag{3.39}$$

Therefore, the reliability becomes

$$\begin{aligned} R(\alpha) = 0, &\quad \text{for } -(1-\alpha)^2/q\alpha > 0 \\ -(1-\alpha)^2/q\alpha\xi_0, &\quad \text{for } \xi_0 < -(1-\alpha)^2/q\alpha \le 0 \\ 1, &\quad \text{for } -(1-\alpha)^2/q\alpha \le \xi_0. \end{aligned} \tag{3.40}$$

The first inequality $-(1-\alpha)^2/q\alpha > 0$ is invalid. Consider the equality

$$-(1-\alpha)^2/q\alpha = \xi_0 \tag{3.41}$$

resulting in

$$\alpha_{1,2} = \big(2 - q\xi_0 \pm \sqrt{q\xi_0(q\xi_0 - 4)}\big)/2. \tag{3.42}$$

The branch with the plus sign does not bear a physical sense since it is associated with consequence $\alpha > 1$. Hence only the branch with the minus sign has a physical sense. We denote

$$\alpha^* = \big(2 - q\xi_0 - \sqrt{q\xi_0(q\xi_0 - 4)}\big)/2. \tag{3.43}$$

Hence the requirement reliability becomes

$$R(\alpha) = 1, \qquad \alpha \leq \alpha^*$$
$$-(1-\alpha)^2/q\alpha\xi_0, \quad \alpha \geq \alpha^*. \tag{3.44}$$

However, since ξ_0 is negative, then by denoting $\xi_0 = -X$, where X is a positive value, we arrive at

$$R(\alpha) = 1, \qquad \alpha \leq \alpha^*$$
$$(1-\alpha)^2/q\alpha X, \quad \alpha \geq \alpha^*. \tag{3.45}$$

It is remarkable that the above expression for the reliability is exactly the same as Eq. (3.15). This means that whether ξ is positive or negative, the reliability of the structure is the same.

Thus, all results obtained for negative ξ case are exactly the same, as the ones for positive ξ; $\alpha_{\text{allowable}}$ has the same expression as Eq. (3.18); reliability with the load being a random variable has the expressions given by Eqs. (3.22) to (3.27).

6. Uniformly Distributed Imperfections Taking on Either Positive or Negative Values

Let the initial imperfections take on both positive and negative values, $a \leq \xi \leq b$, where $a \leq 0$ and $0 \leq b$. In the general case we should use Eq. (3.1) as a transfer function between imperfections and buckling loads:

$$(1-\rho)^2 - q\rho|\xi| = 0.$$

We deduce the following load expression when ξ can take on either positive or negative values

$$\rho = \left(2 + q|\xi| - \sqrt{q|\xi|(4 + q|\xi|)}\right)/2. \tag{3.46}$$

Equation (3.46) coincides with expressions for ρ_2 in both Eqs. (3.6) and (3.34). Reliability becomes

$$R = \text{Prob}(\rho > \alpha) = \text{Prob}\left[\left(2 + q|\xi| - \sqrt{q|\xi|(4 + q|\xi|)}\right)/2 > \alpha\right] \tag{3.47}$$

or

$$R = \text{Prob}[|\xi| < (1-\alpha)^2/q\alpha] = F_\xi[\xi_0] - F_\xi[-\xi_0] \tag{3.48}$$

where

$$\xi_0 = (1-\alpha)^2/q\alpha. \tag{3.49}$$

Figure 3.3(a) represents the variation of ρ_s with ξ. We fix the level $\rho_s = \alpha$, in order to calculate the probability that $\rho_s > \alpha$. The reliability equals then the probability that the initial imperfections take values between $-\xi_0$ and ξ_0, the values at which the horizontal line $\rho_s = \alpha$ crosses the function $\rho_s = \rho_s(\xi)$. Four different cases may occur:

(I) The initial imperfection interval is fully enclosed in the interval $X = [-\xi_0, \xi_0]$; (II) the lower bound of initial imperfections belongs to the interval X, but the upper bound is outside it, that is, $b > \xi_0$; (III) the lower bound of initial imperfections is outside X, that is, $a < -\xi_0$, whereas the upper bound belongs to X; (IV) both a and b are outside X. These cases are depicted in Fig. 3.4. The hatched areas represent the structural reliability.

Reliability can be summarized as follows:

$$\begin{aligned} R(\alpha) = 1, \quad & \text{for } \xi_0 \geq \max(|a|, b) \\ \frac{\xi_0 - a}{b - a}, \quad & \text{for } |a| \leq \xi_0 < b \\ \frac{b + \xi_0}{b - a}, \quad & \text{for } b \leq \xi_0 < |a| \\ \frac{2\xi_0}{b - a}, \quad & \text{for } \xi_0 < \min(|a|, b). \end{aligned} \tag{3.50}$$

Reliability can be rewritten as a function of α_a and α_b, which are loads corresponding respectively to bounds a and b. There expressions read

$$\alpha_a = \left(2 - qa - \sqrt{qa(qa - 4)}\right)/2 \tag{3.51}$$

and

$$\alpha_b = (2 + qb - \sqrt{qa(qb + 4)})/2. \tag{3.52}$$

First, we consider $|a| < b$. Let us explore different values α_i that α can take for various $i = 1, 2, 3, 4$. The reliability equals the probability that the initial imperfections take values at which the horizontal lines $\rho_s = \alpha_i$ cross the function $\rho_s = \rho_s(\xi)$. Three different cases may occur, as depicted in Fig. 3.4. The first case represents $\alpha_a < \alpha$; the second one shows $\alpha_b < \alpha \leq \alpha_a$; finally, $\alpha < \alpha_b$ is evaluated.

Reliability can be summarized as follows

$$\begin{aligned} R(\alpha) = 1, \quad & \text{for } \alpha \leq \min(\alpha_a, \alpha_b) \\ \frac{\xi_0 - a}{b - a}, \quad & \text{for } \alpha_b < \alpha \leq \alpha_a \\ \frac{2\xi_0}{b - a}, \quad & \text{for } \max(\alpha_a, \alpha_b) < \alpha. \end{aligned} \tag{3.53}$$

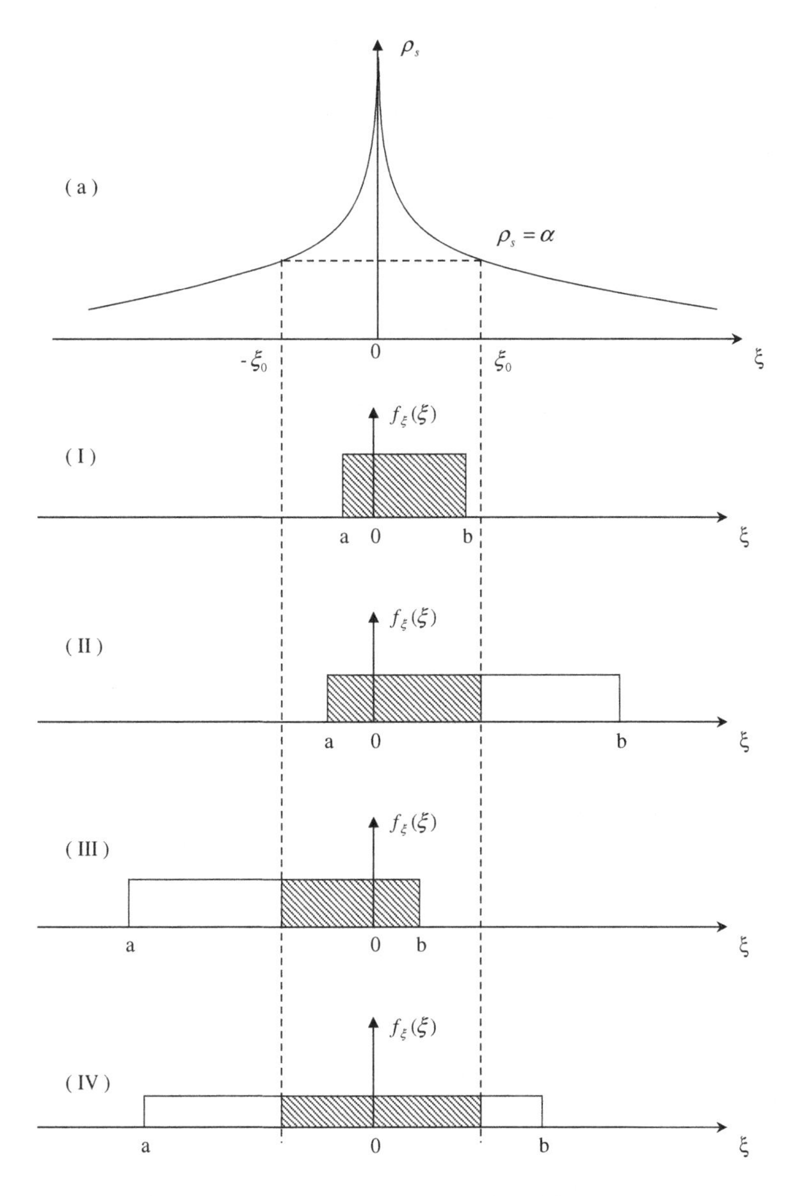

Fig. 3.4. Evaluation of reliability in four different cases.

Consider now the case $|a| > b$. In this case, $\alpha_a < \alpha_b$ and condition $\alpha_b < \alpha \leq \alpha_a$ must be replaced by $\alpha_a < \alpha \leq \alpha_b$. We obtain

$$R(\alpha) = \begin{cases} 1, & \text{for } \alpha \leq \min(\alpha_a, \alpha_b) \\ \dfrac{b+\xi_0}{b-a}, & \text{for } \alpha_a < \alpha \leq \alpha_b \\ \dfrac{2\xi_0}{b-a}, & \text{for } \max(\alpha_a, \alpha_b) < \alpha. \end{cases} \tag{3.54}$$

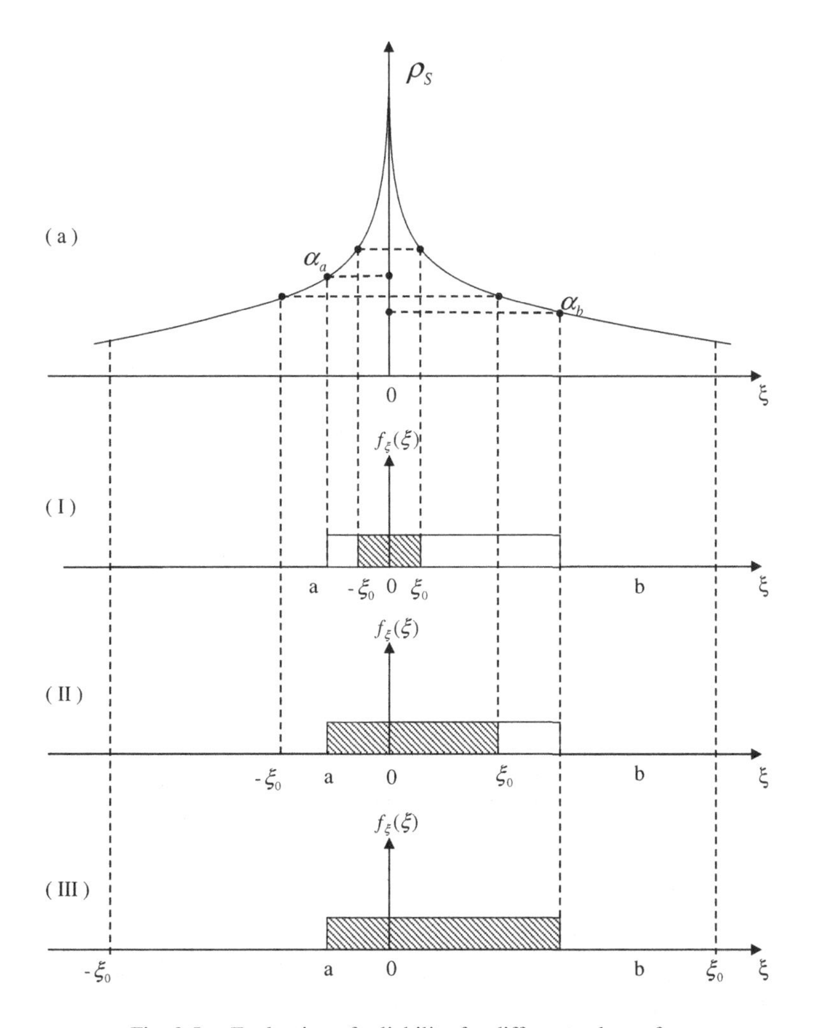

Fig. 3.5. Evaluation of reliability for different values of α.

Combining Eqs. (3.53) and (3.54) results in the following expression:

$$
\begin{aligned}
R(\alpha) = 1, \quad & \text{for } \alpha \le \min(\alpha_a, \alpha_b) \\
\frac{b+\xi_0}{b-a}, \quad & \text{for } \alpha_a < \alpha \le \alpha_b \text{ if } |a| > b \\
\frac{\xi_0 - a}{b-a}, \quad & \text{for } \alpha_b < \alpha \le \alpha_a \text{ if } |a| < b \\
\frac{2\xi_0}{b-a}, \quad & \text{for } \max(\alpha_a, \alpha_b) < \alpha.
\end{aligned}
\tag{3.55}
$$

Note that Eqs. (3.55) and (3.50) coincide, despite the fact that the conditions are written in the different form.

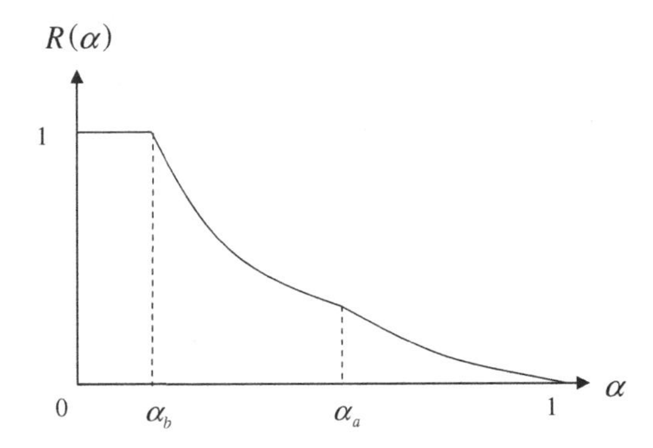

Fig. 3.6. Curve shape of reliability when $\alpha_a < \alpha_b$, $q = 4/3$, $a = -1$, and $b = 5$.

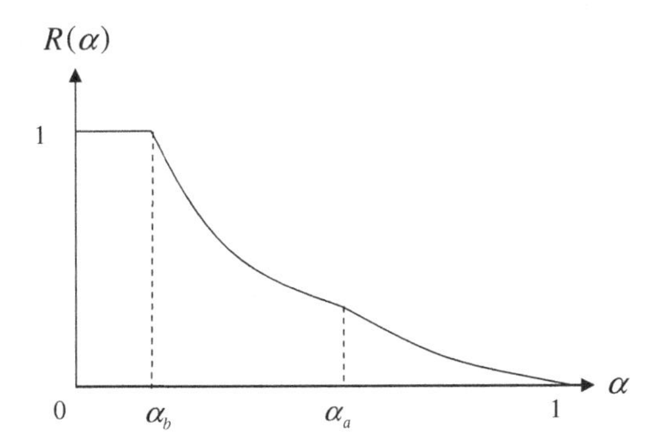

Fig. 3.7. Curve shape of reliability when $\alpha_{\mathrm{a}} > \alpha_{\mathrm{b}}$, $q = 4/3$, $a = -5$, and $b = 1$.

7. Combined Randomness in Imperfection and Load for Either Positive or Negative Imperfection Values

Let us consider combined randomness in imperfection and load. The applied load Λ is defined as random variable in the interval $[\lambda_1, \lambda_2]$, $0 < \lambda_1 < \lambda_2$. Equation (3.20) takes the following form:

$$R = \int_{\lambda_1}^{\lambda_2} R(\alpha) f_\Lambda(\alpha) \mathrm{d}\alpha = \frac{1}{\lambda_2 - \lambda_1} \int_{\lambda_1}^{\lambda_2} R(\alpha) \mathrm{d}\alpha. \tag{3.56}$$

To evaluate this reliability, consider $|a| > b$. We distinguish five different cases.

For $\lambda_j \in [0, \alpha_a]$, $(j = 1, 2)$ where α_a is defined here as formally coinciding with Eq. (3.51), we get

$$R = \frac{1}{\lambda_2 - \lambda_1} \int_{\lambda_1}^{\lambda_2} \mathrm{d}\alpha = 1. \tag{3.57}$$

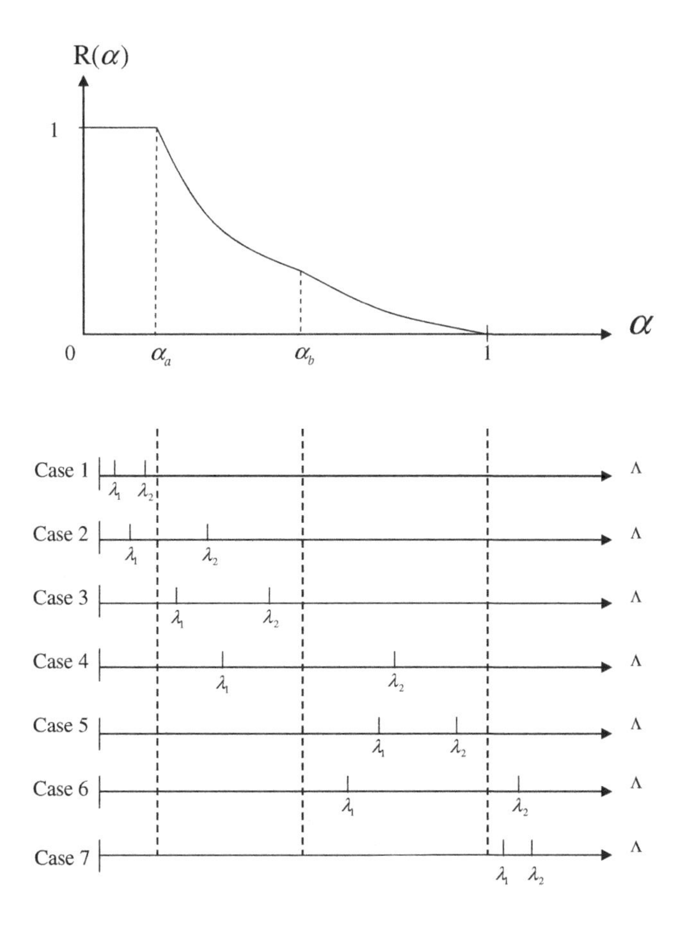

Fig. 3.8. Seven possible different locations of load bounds with respect to α_a, α_b, and unity.

For $\lambda_1 \in [0, \alpha_a]$ and $\lambda_2 \in]\alpha_a, \alpha_b]$, we have

$$R = \frac{1}{\lambda_2 - \lambda_1}\left(\int_{\lambda_1}^{\alpha_a} \mathrm{d}\alpha + \int_{\alpha_a}^{\lambda_2} \frac{\xi_0 + b}{b - a}\mathrm{d}\alpha\right)$$
$$= \frac{\alpha_a - \lambda_1}{\lambda_2 - \lambda_1} + \frac{1}{\lambda_2 - \lambda_1}\int_{\alpha_a}^{\lambda_2}\left(\frac{(1-\alpha)^2}{q\alpha(b-a)} + \frac{b}{b-a}\right)\mathrm{d}\alpha \tag{3.58}$$

or

$$R = \frac{\alpha_a - \lambda_1}{\lambda_2 - \lambda_1} + \frac{1}{2q(b-a)(\lambda_2 - \lambda_1)}\left[2\ln\left(\frac{\lambda_2}{\alpha_a}\right) + \lambda_2(\lambda_2 - 4) - \alpha_a(\alpha_a - 4)\right]$$
$$+ \frac{b(\lambda_2 - \alpha_a)}{(b-a)(\lambda_2 - \lambda_1)}. \tag{3.59}$$

For $\lambda_j \in]\alpha_a, \alpha_b]$, we obtain

$$R = \frac{1}{\lambda_2 - \lambda_1} \int_{\lambda_1}^{\lambda_2} \frac{\xi_0 + b}{b - a} \mathrm{d}\alpha = \frac{1}{\lambda_2 - \lambda_1} \int_{\lambda_1}^{\lambda_2} \left(\frac{(1-\alpha)^2}{q\alpha(b-a)} + \frac{b}{b-a} \right) \mathrm{d}\alpha \tag{3.60}$$

or

$$R = \frac{1}{2q(b-a)(\lambda_2 - \lambda_1)} \left[2\ln\left(\frac{\lambda_2}{\lambda_1}\right) + \lambda_2(\lambda_2 - 4) - \lambda_1(\lambda_1 - 4) \right] + \frac{b}{(b-a)}. \tag{3.61}$$

For $\lambda_1 \in [\alpha_a, \alpha_b]$ and $\lambda_2 > \alpha_b$, we find

$$\begin{aligned} R &= \frac{1}{\lambda_2 - \lambda_1} \left(\int_{\lambda_1}^{\alpha_b} \frac{\xi_0 + b}{b - a} \mathrm{d}\alpha + \int_{\alpha_b}^{\lambda_2} \frac{2\xi_0}{b - a} \mathrm{d}\alpha \right) \\ &= \frac{1}{\lambda_2 - \lambda_1} \left[\int_{\lambda_1}^{\alpha_b} \left(\frac{(1-\alpha)^2}{q\alpha(b-a)} + \frac{b}{b-a} \right) \mathrm{d}\alpha + 2 \int_{\alpha_b}^{\lambda_2} \frac{(1-\alpha)^2}{q\alpha(b-a)} \mathrm{d}\alpha \right] \end{aligned} \tag{3.62}$$

or

$$\begin{aligned} R = \frac{1}{2q(b-a)(\lambda_2 - \lambda_1)} &\left[2\ln\left(\frac{\alpha_b}{\lambda_1}\right) + \alpha_b(\alpha_b - 4) - \lambda_1(\lambda_1 - 4) \right. \\ &\left. + 4\ln\left(\frac{\lambda_2}{\alpha_b}\right) + 2\lambda_2(\lambda_2 - 4) - 2\alpha_b(\alpha_b - 4) \right] + \frac{b(\alpha_b - \lambda_1)}{(b-a)(\lambda_2 - \lambda_1)}. \end{aligned} \tag{3.63}$$

For $\lambda_j \in (\alpha_b, 1]$, we have

$$R = \frac{1}{\lambda_2 - \lambda_1} \int_{\lambda_1}^{\lambda_2} \frac{2\xi_0}{b - a} \mathrm{d}\alpha = \frac{2}{\lambda_2 - \lambda_1} \int_{\lambda_1}^{\lambda_2} \frac{(1-\alpha)^2}{q\alpha(b-a)} \mathrm{d}\alpha \tag{3.64}$$

or

$$R = \frac{1}{q(b-a)(\lambda_2 - \lambda_1)} \left[2\ln\left(\frac{\lambda_2}{\lambda_1}\right) + \lambda_2(\lambda_2 - 4) - \lambda_1(\lambda_1 - 4) \right]. \tag{3.65}$$

For $\lambda_1 \in (\alpha_b, 1]$ and $\lambda_2 > 1$, we get

$$\begin{aligned} R &= \frac{1}{\lambda_2 - \lambda_1} \int_{\lambda_1}^{1} \frac{2\xi_0}{b - a} \mathrm{d}\alpha = \frac{2}{\lambda_2 - \lambda_1} \int_{\lambda_1}^{1} \frac{(1-\alpha)^2}{q\alpha(b-a)} \mathrm{d}\alpha \\ &= \frac{1}{q(b-a)(\lambda_2 - \lambda_1)} \left[2\ln\left(\frac{1}{\lambda_1}\right) - 3 - \lambda_1(\lambda_1 - 4) \right] \end{aligned} \tag{3.66}$$

For $\lambda_j > 1$, we obviously arrive at

$$R = 0. \tag{3.67}$$

Consider now $|a| < b$. We discern also five different cases.

For $\lambda_j \in [0, \alpha_b]$,

$$R = \frac{1}{\lambda_2 - \lambda_1} \int_{\lambda_1}^{\lambda_2} \mathrm{d}\alpha = 1. \tag{3.68}$$

For $\lambda_1 \in [0, \alpha_b]$ and $\lambda_2 \in]\alpha_b, \alpha_a]$, we find

$$\begin{aligned} R &= \frac{1}{\lambda_2 - \lambda_1} \left(\int_{\lambda_1}^{\alpha_b} \mathrm{d}\alpha + \int_{\alpha_b}^{\lambda_2} \frac{\xi_0 - a}{b - a} \mathrm{d}\alpha \right) \\ &= \frac{\alpha_b - \lambda_1}{\lambda_2 - \lambda_1} + \frac{1}{\lambda_2 - \lambda_1} \int_{\alpha_b}^{\lambda_2} \left(\frac{(1-\alpha)^2}{q\alpha(b-a)} - \frac{a}{b-a} \right) \mathrm{d}\alpha \end{aligned} \tag{3.69}$$

or

$$\begin{aligned} R = {} & \frac{\alpha_b - \lambda_1}{\lambda_2 - \lambda_1} + \frac{1}{2q(b-a)(\lambda_2 - \lambda_1)} \left[2 \ln \left(\frac{\lambda_2}{\alpha_b} \right) + \lambda_2(\lambda_2 - 4) - \alpha_b(\alpha_b - 4) \right] \\ & - \frac{a(\lambda_2 - \alpha_b)}{(b-a)(\lambda_2 - \lambda_1)}. \end{aligned} \tag{3.70}$$

For $\lambda_j \in]\alpha_b, \alpha_a]$, evaluation of R reads

$$R = \frac{1}{\lambda_2 - \lambda_1} \int_{\lambda_1}^{\lambda_2} \frac{\xi_0 - a}{b - a} \mathrm{d}\alpha = \frac{1}{\lambda_2 - \lambda_1} \int_{\lambda_1}^{\lambda_2} \left(\frac{(1-\alpha)^2}{q\alpha(b-a)} - \frac{a}{b-a} \right) \mathrm{d}\alpha \tag{3.71}$$

or

$$R = \frac{1}{2q(b-a)(\lambda_2 - \lambda_1)} \left[2 \ln \left(\frac{\lambda_2}{\lambda_1} \right) + \lambda_2(\lambda_2 - 4) - \lambda_1(\lambda_1 - 4) \right] - \frac{a}{(b-a)}. \tag{3.72}$$

For $\lambda_1 \in [\alpha_b, \alpha_a]$ and $\lambda_2 \in (\alpha_a, 1]$, we get

$$R = \frac{1}{\lambda_2 - \lambda_1} \left(\int_{\lambda_1}^{\alpha_a} \frac{\xi_0 - a}{b - a} \mathrm{d}\alpha + \int_{\alpha_a}^{\lambda_2} \frac{2\xi_0}{b - a} \mathrm{d}\alpha \right) \tag{3.73}$$

or

$$\begin{aligned} R = {} & \frac{1}{2(b-a)(\lambda_2 - \lambda_1)} \left[2 \ln \left(\frac{\alpha_a}{\lambda_1} \right) + \alpha_a(\alpha_a - 4) - \lambda_1(\lambda_1 - 4) - a(\alpha_a - \lambda_1) \right. \\ & \left. + 4 \ln \left(\frac{\lambda_2}{\alpha_a} \right) + 2\lambda_2(\lambda_2 - 4) - 2\alpha_a(\alpha_a - 4) \right]. \end{aligned} \tag{3.74}$$

For $\lambda_j \in (\alpha_a, 1]$, reliability equals

$$R = \frac{1}{\lambda_2 - \lambda_1} \int_{\lambda_1}^{\lambda_2} \frac{2\xi_0}{b-a} \mathrm{d}\alpha = \frac{2}{\lambda_2 - \lambda_1} \int_{\lambda_1}^{\lambda_2} \frac{(1-\alpha)^2}{q\alpha(b-a)} \mathrm{d}\alpha \tag{3.75}$$

or

$$R = \frac{1}{q(b-a)(\lambda_2 - \lambda_1)} \left[2\ln\left(\frac{\lambda_2}{\lambda_1}\right) + \lambda_2(\lambda_2 - 4) - \lambda_1(\lambda_1 - 4) \right]. \tag{3.76}$$

For $\lambda_1 \in (\alpha_b, 1]$ and $\lambda_2 > 1$, calculations leads to

$$R = \frac{1}{\lambda_2 - \lambda_1} \int_{\lambda_1}^{1} \frac{2\xi_0}{b-a} \mathrm{d}\alpha = \frac{2}{\lambda_2 - \lambda_1} \int_{\lambda_1}^{1} \frac{(1-\alpha)^2}{q\alpha(b-a)} \mathrm{d}\alpha \tag{3.77}$$

or

$$R = \frac{1}{q(b-a)(\lambda_2 - \lambda_1)} \left[2\ln\left(\frac{1}{\lambda_1}\right) - 3 - \lambda_1(\lambda_1 - 4) \right]. \tag{3.78}$$

For $\lambda_j > 1$, we finally obtain

$$R = 0. \tag{3.79}$$

8. Numerical Examples and Discussion

First, we consider the case $|a| < b$, namely $a^{(I)} = -1, b^{(I)} = 5$ and $q = 4/3$. Equations (3.51) and (3.52) provide expressions of α_a and α_b, and we get $\alpha_a^{(I)} = 0.6667$ and $\alpha_b^{(I)} = 0.2339$.

We obtain in Table 3.2 reliability values for his seven cases defined above. First, λ_j values are chosen arbitrarily in their definition area. A second study will show the effect of λ_j position in each case.

For the case $|a| > b$, we consider $a^{(II)} = -5$, $b^{(II)} = 1$ and $q = 4/3$, yielding $\alpha_a^{(II)} = 0.2339$ and $\alpha_b^{(II)} = 0.6667$. Note that in the new circumstances, we have chosen

$$|a^{(II)}| = b^{(I)}, \quad |a^{(I)}| = b^{(II)}. \tag{3.80}$$

Then by choosing the same values of λ_j as in Table 3.2, we get the same reliability values. Indeed, it is remarkable that expressions of reliability for $|a| > b$ have symmetrical expressions as $|a| < b$, if conditions in Eq. (3.80) are satisfied.

We make a special note of two observations from this study. The first one is Table 3.2 shows us reliability decreases from one to zero when λ_j increases.

Table 3.2. Reliability values as depending upon bounds of the applied load.

Case number	1	2	3	4	5	6	7
λ_1	0.1	0.1	0.4	0.4	0.7	0.7	1.5
λ_2	0.2	0.4	0.6	0.8	0.8	1.1	2
R	1	0.6457	0.2326	0.1166	0.0284	0.0097	0

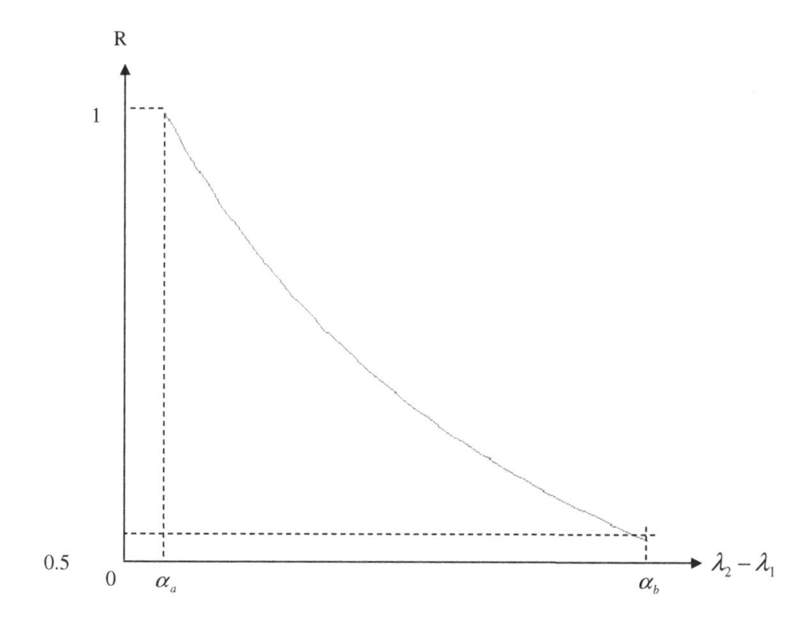

Fig. 3.9. Variation of reliability for the diameter $\lambda_2 - \lambda_1$ of the load variation.

Moreover, when the range $\lambda_2 - \lambda_1$, that is, the distance between λ_1 and λ_2 increases, the reliability decreases. To illustrate this property, consider case 2 when $|a| > b$, depicted in Fig. 3.8. To evaluate the reliability we fix λ_1 to zero, and let λ_2 vary from α_a to α_b. Then reliability follows the decreasing curve, as depicted on Fig. 3.9.

Chapter 4

Non-probabilistic Resolution

...the guaranteed approach does not require any knowledge about the disturbances... besides their bounds... Note that the probability distributions for unknown disturbances are seldom available in practical problems. Moreover, the guaranteed approach produces results which are applicable to any individual realization of the process. Thus, the guaranteed approach requires less information and is more reliable. However if there is a lot of statistical data for the uncertainties, the probabilistic approach can give better estimates.

F.L. Chernousko (1999)

This chapter deals with an alternative to probabilistic modeling. As if the probabilistic resolution of the shell stability was not sufficient, the author was blessed by yet another methodology that is not based on probabilistic modeling. The initial imperfection is that Fourier coefficients were treated as being bounded by an ellipsoidal set. Then the minimum buckling load is determined in an analytic form when the numerical code is available for buckling load evaluation. Thus, semi-analytic and seminumeric methodology was proposed to bridge the gap between theory and practice. This methodology is known as the convex modeling of uncertainty (Ben-Haim and Elishakoff, 1990), guaranteed approach (Chernousko, 1990), worst-case design (Hlavaček *et al.*, 2004), info-gap modeling (Ben-Haim, 2006), or max-min analysis (Wald, 1945, 1950; Sniedovich, 2007, 2008, 2010, 2012).

1. Resolution of Conundrum via Nonprobabilistic Convex Modeling

One of the most common and effective methods of disseminating the dominant conception of insecurity and risk is the creation of the "scenario," a postulated condition that functions both as a parable (offering a warning about a convergence of undesirable circumstances) and as a regime of activity.

S. Price (2011)

The worst does sometimes happen.

F. Dürrenmatt (1921–1990)

...it should "scenario," be borne in mind that the uncertainty set may include worst cases which never happen in reality.

X.J. Chen *et al.* (2004)

It appears instructive at this stage to quote from the author's paper (1995, p. 889):

> Here a particular personal encounter appears to be in order. The present writer's research in difficulties of probabilistic modeling was inspired in the following way. In 1980, the mastermind of modern buckling theory of structures, Professor Warner Tjardus Koiter of Delft University of Technology had organized an European Mechanics Colloquium on Buckling of Structures. There the present writer had presented a lecture entitled "Random Buckling of Shells", co-authored with Professor Johann Arbocz of Delft University of Technology. After the lecture, Professor Koiter posed a question which the present writer can recall almost verbatim: "In order for society to adopt probabilistic methods, the resulting reliability of the structure must be extremely high or, alternatively, the probability of failure must be extremely small. Yet, in order to calculate the above probabilities, one must utilize deterministic theories which are of an approximate nature. Can one be sure that such small probabilities of failure can be accurately predicted?" To reply to this question, at least partially, the present writer undertook investigation (1983) based on simple models where exact solution was available. It turned out that in some circumstances, due to the approximate deterministic theory the error in evaluating the probability of failure may be quite substantial. Professor Koiter's cardinal question, and its ramifications, was always kept in the present writer's mind; indeed, this question has had a great impact on the present writer's research since then. The present writer had an opportunity to investigate the substantial errors which can be introduced by an ergodicity assumption (Scheurkogel *et al.*, 1981; Elishakoff, 1983b; Scheurkogel and Elishakoff, 1985) or neglect of cross-correlations [both before and after Koiter posed his stimulating question] (Elishakoff, 1977; Elishakoff *et al.*, 1979; Elishakoff, 1983, 1986, 1988).

Questions associated with the validity of probabilistic methods in engineering were raised by numerous investigators. Chapter 1, titled "Probabilistic Modeling: Pros and Cons" in the monograph by Ben-Haim and Elishakoff (1990) reproduces quotes of numerous authors, along with illustrative model problems to shed more light on this issue. Whereas the probabilistic modeling was seemingly embraced in the West, it has powerful opponents in the East.

Most prominent in such an opposition was Professor Vsevolod Ivanovitch Feodosiev of Bauman Technical University in Moscow. In his subtle as well as highly informative book *Ten Lectures — Conversations on Strength of Materials*, Feodosiev (1969) states:

> Attempts at statistical processing of initial imperfections have little chance of success at present... This should be borne in mind in view of the fact that the significance of the probabilistic approach is often exaggerated, and there is a very real danger of the effort invested in urgent creation of mathematical loads for predicting loss of stability as random event — proving to be wasted, when the indispensable distribution functions of the imperfections remain unknown even in those few cases where they can be categorized as random parameters. The practicing engineer is certain, in doubtful cases, to prefer stricter quality control, or even changes in design, to costly study of "transient" distribution functions.

Whereas Feodosiev's (1969) book still awaits its availability to the English reader, it can be noted that more extensive quote from it is reproduced by Ben-Haim and Elishakoff (1990, pp. 33–37).

Not everybody agrees with Feodosiev. Lindley's (1987) conviction is that "Probability is the only satisfactory description of uncertainty." However, according to Kalman (1995):

> Whatever (intuitive, formalized, abstract) probability might be, and however the axioms might be phrased to hide the underlying assumptions, physical probability has meaning only in relation to specific systems. It is NOT an universal concept... We should take a new look at all the probabilistic and statistical analysis to see what they really amount in the R-world [real world].

Cordioli (2006) wrote:

> In a probabilistic analysis, the first step is related to the identification of the random variables and/or process associated with the structure. To perform the analysis it is thus necessary to define the statistics of those variables in the form of the joint probability density functions (jpdf). However, information concerning the statistics of the input parameters is not generally available. In fact, obtaining the joint probability density function (jpdf) of the random variable involved is almost impossible. Therefore, it is common in a probabilistic analysis that the statistical data of the random variable be assumed based on the designer experience instead of obtained experimentally. Elishakoff (1995) gives a detailed discussion on this issue. It is demonstrated that small errors in the input statistical data for a probabilistic analysis can lead to significant discrepancy in the predicted response statistic. Elishakoff (1995) then reviews some methods that he calls non-probabilistic methods, which are not based on a precise description for the random variable... The possible scenarios include the definition of intervals or envelope boundaries for random variables without considering a specific probabilistic behavior within the specified limits.

Alastair Brown (2007) notes:

> These are often difficulties in finding the statistics of the material or structural properties. Elishakoff (1995) argues that in most studies necessary information is assumed rather than obtained from extensive experimental data, which is normally not available... The full joint P.D.F. of the random variables involved is often practically impossible to obtain, and the evaluation of the response statistics extremely complicated... often the available information about the uncertainty is not of a probabilistic nature, and may just indicate that the variables lie somewhere with certain range of possible values. The objective is then to find the boundaries on the response given this set of input values. Interval analysis sets lower and upper bounds, whereas convex modeling assumes that the input variables fall within a convex region of parameter space. Elishakoff (1995) gives a review of the convex modeling type of approach.

In words of Zhiteckij (1996),

> ...the effect of disturbances on the system behavior is the important factor. The above effect is investigated from statistical considerations. This approach requires

> some knowledge of disturbance statistics. However, in serious control applications, the assumptions regarding the disturbance statistics may be invalid. In this case, the statistical approach is unsuitable. Meanwhile, in most cases the available *a priori* information about the disturbance is given not in statistical terms but as bounds on its absolute value. In the case mentioned, the bounding approaches are appropriate.

One of the architects of the probabilistic applications to mechanics, Freudenthal (1956) stressed that "...ignorance of the cause of variation does not make such variation random."

Despite the fact of this recognition that the probability must not be considered the only game in town, he didn't provide any alternative to it. Yet, this noble self-criticism of his "own" methods is not yet followed by the probabilistic analysts.

These and other considerations led Ben-Haim (1985) and Ben-Haim and Elishakoff (1990) to develop the method of *convex modeling* for applied mechanics applications. The name stems from the realization of the fact that most inequalities describing the range of variations of uncertain variables constitute convex sets. As the variables are defined by their ranges of variation only, rather than by the probability densities, the following questions can be posed:

(a) What is the maximum buckling load that the structure may experience when the initial imperfections vary in a convex set?
(b) What is the minimum buckling load may attain in these circumstances?

Once these questions are answered, it is prudent to utilize the minimum buckling load as design load.

Such analyses were performed by Ben-Haim and Elishakoff (1989, 1990), Ben-Haim (1993a,b, 1994), Lindberg (1992), Pantelides (1996a,b), Elishakoff *et al.* (1994), and Elseifi *et al.* (1999). Hereinafter, we follow the work by Ben-Haim and Elishakoff (1989, 1990). The initial imperfection vector **Q** is represented as the sum of a nominal vector $\mathbf{Q}_0$ and the deviation vector ζ. The deviation is postulated to fall within the following ellipsoidal set:

$$R(\chi, \omega) = \left\{ \zeta : \sum_{i=1}^{n} \frac{\zeta_i}{\omega_i^2} \leq \chi^2 \right\} \tag{4.1}$$

where the size parameter χ and the semiaxes $\omega_1, \omega_2, \ldots, \omega_n$ are based on experimental data, obtainable from the initial imperfection surveys or data banks. The lowest buckling load that can be obtained for any of the shells in the ensemble described by Eq. (3.1) is expressed formally as the minimum of expression (3) on the set R:

$$\mu(\chi, \omega) = \min_{\zeta \in R(\chi,\omega)} \varphi(\mathbf{Q}_0 + \zeta). \tag{4.2}$$

Hence $\mu(\chi, \omega)$ is the buckling load of the "weakest" shell in the ensemble R, constructed to represent a realistic range of shells. The limit load for an imperfection vector $\mathbf{Q}_0 + \zeta$, to the first order of ζ, is

$$\varphi(\mathbf{Q}_0 + \zeta) = \varphi(\mathbf{Q}_0) + \sum_{i=1}^{n} \frac{\partial \varphi(\mathbf{Q}_0)}{\partial Q_i} \zeta_i. \tag{4.3}$$

Thus, the problem (3.2) is replaced by the following equation:

$$\mu(\chi, \omega) = \min_{\zeta \in R(\chi, \omega)} \left[\varphi(\mathbf{Q}_0) + \varphi^T \zeta \right] \tag{4.4}$$

where

$$\varphi^{\mathrm{T}} = \left(\frac{\partial \varphi(\mathbf{Q}_0)}{\partial Q_1}, \frac{\partial \varphi(\mathbf{Q}_0)}{\partial Q_2}, \ldots, \frac{\partial \varphi(\mathbf{Q}_0)}{\partial Q_n} \right) \tag{4.5}$$

where the superscript T stands for matrix transposition. The minimum buckling load is given by the formula

$$\mu(\chi, \omega) = \varphi(\mathbf{Q}_0) - \chi \left[\sum_{i=1}^{n} \left(\omega_i \frac{\partial \varphi(\mathbf{Q}_0)}{\partial Q_i} \right)^2 \right]^{1/2} \tag{4.6}$$

From this relation, one recognizes that significant reduction of the buckling load results from high sensitivity of the nominal buckling load to those Fourier coefficients, whose semiaxes in the imperfection ellipsoid are large. We also recognize that the minimum buckling load depends linearly on the overall size χ of the imperfection ellipsoid, and nonlinearly on its shape parameters $\omega_1, \omega_2, \ldots, \omega_n$ and on the sensitivity derivatives $\partial \varphi(\mathbf{Q}_0)/\partial \mathbf{Q}_i$. Note that both the probabilistic analysis by Elishakoff *et al.* (1987) and the convex analysis by Ben-Haim and Elishakoff (1989) involve the sensitivity derivatives $\partial \varphi(\mathbf{Q}_0)/\partial \mathbf{Q}_i$ for small deviation of initial imperfection coefficients from the nominal values. The values of the sensitivity derivatives have been borrowed from our previous probabilistic study (Elishakoff *et al.*, 1987).

For large deviations ζ_i one cannot utilize the asymptotic expression (4.6); in such a case one has to resort to nonlinear programming codes, as was done by Li *et al.* (1995).

Whereas the formula (3.4) is a first-order approximation, a second-order approximation has also been derived by Ben-Haim and Elishakoff (1989) explicitly, in terms of Hessian matrix with elements $\partial^2 \varphi(\mathbf{Q}_0)/\partial \zeta_i \partial \zeta_j$. For details, an interested reader can consult Ben-Haim and Elishakoff (1990).

It is also of interest to define the variations of the imperfections in terms of a radial tolerance on the shape of the shell. One obtains the following expression of the buckling load in terms of the imperfection:

$$\varphi(\mathbf{Q}_0 + \zeta(\xi, \vartheta)) = \varphi(\mathbf{Q}_0) + \int_0^{2\pi} \int_0^{\pi} \zeta(\xi, \vartheta) S(\xi, \vartheta) d\xi d\vartheta \tag{4.7}$$

where $S(\xi, \vartheta)$ is a combination of trigonometric functions with coefficients that depend on the elements of the vector $\partial(\mathbf{Q}_0)/\partial\zeta_i$. A close examination of Eq. (3.7) reveals that the largest reduction in the buckling load is obtained from the imperfection profile, which switches between its extreme values $\hat{\zeta}$ and $-\hat{\zeta}$, where $\hat{\zeta}$ is the radial tolerance. The minimum buckling load for the ensemble of shells with radial tolerance $\hat{\zeta}$ reads:

$$\mu(\hat{\zeta}) = \min_{|\hat{\zeta}|<\hat{\zeta}}[\varphi(\mathbf{Q}_0 + \zeta(\xi, \vartheta)] = \varphi(\mathbf{Q}_0) - \hat{\zeta} \int_0^{2\pi} \int_0^{\pi} |S(\xi, \vartheta)| \mathrm{d}\xi \mathrm{d}\vartheta \tag{4.8}$$

Suppose now that one wishes to construct a radial tolerance for which the minimum buckling load takes on the value λ_0. Then one chooses $\hat{\zeta}$ as follows:

$$\hat{\zeta} = \frac{[\mu(\hat{\zeta}) - \varphi(\boldsymbol{Q}_0)]}{\int_0^{2\pi} \int_0^{\pi} |S(\xi, \vartheta)| \mathrm{d}\xi \mathrm{d}\vartheta} \tag{4.9}$$

This approach permits theoretical determination of the knockdown factor within the convex modeling (CKF). It is defined as the ratio of the minimum buckling load to the classical buckling load:

$$\mathrm{CKF} = \frac{1}{P_{\mathrm{cl}}} \left\{ \varphi(\mathbf{Q}_0) - \chi \left[\left(\sum_{i=1}^{n} \omega_i \frac{\partial \varphi(\mathbf{Q}_0)}{\partial Q_i} \right)^2 \right]^{1/2} \right\} \tag{4.10}$$

for ellipsoidally modeled initial imperfections.

For shells with radial tolerance

$$\mathrm{CKF} = \frac{1}{P_{\mathrm{cl}}} \left\{ \varphi(\mathbf{Q}_0) - \hat{\zeta} \int_0^{2\pi} \int_0^{\pi} |S(\xi, \vartheta)| \mathrm{d}\xi \mathrm{d}\vartheta \right\}. \tag{4.11}$$

As is mentioned in the review paper (Elishakoff, 2000),

> This knockdown factor is anticipated to lie *above* those provided by the NASA monograph (1969), which would imply that the existing monograph specifies too conservative estimates and thus *penalizes* carefully designed shell. As we see, convex non-probabilistic modeling of uncertainty provides theoretical means for determining the knockdown factor.

Our work was applied to both static and dynamic buckling problems by some investigators. Problems in the static setting were dealt by van der Nieuwendijk

Fig. 4.1. Professor ZhiPing Qiu (left) and Professor Xiaojun Wang of the Beihang University.

(1997), Elseifi *et al.* (1999), Qiu (2005), Wang *et al.* (2009), Baitsch and Hartmann (2006), and others.

Dynamic buckling problems were treated by Elishakoff and Ben-Haim (1990), Lindberg (1991,1992a,b), Ben-Haim (1993a,b), Qiu and Wang (2005), and others.

2. Competition between Probabilistic and Convex Analyses: Which One Wins?

The assumption of stochastic nature of uncertainty cannot be accepted at least in two cases: (a) when the volume of a priori experimental data on the nature of the uncertain factors is so small that it does not allow the conclusion of the availability of stable statistic characteristics; (b) when it is known a priori that the uncertainty basically cannot be considered to be produced by some probabilistic mechanisms.

V.M. Kunzevich and M. Lychak (1992)

The "competition" between the probabilistic and convex modeling was discussed by Elishakoff *et al.* (1994a) on the example of Fraser and Budiansky (1969) problem.

Before proceeding further, it is instructive to recall that during my presentation at the conference on "Uncertainty: Models and Measures" in 1996, I was asked by one of its organizers: "How can you compare the probabilistic and convex models of uncertainty? They have such differing premises!" My response was as follows. Assume that there is a firm that provides the same data associated with some engineering problem to two different research organizations: one organization specializes in probabilistic methods, whereas the other engages itself with nonprobabilistic, convex analysis. These organizations are not made aware that the competing organization is also solving the same problem. Once these organizations complete their analyses, they deliver the final reports to the firm that ordered the work. Naturally, the firm wants to know what the results are of the provided analyses; they compare the furnished numerical results.

Thus, comparison between competing methods can be conducted by the third party, if not by the researchers themselves. However, it is anticipated that in the future more researchers will be engaged in comparing results provided by different analyses of uncertainty. Indeed, as W.W. Soyer notes:

> To a person who is studying algebra, it is often more useful to solve the same problem with three or four different methods, than to solve three or four different problems. By solving problems by different methods, one can by the comparison clarify which of them is shorter and more effective.

This above quote pertains to the topic of algebra. How much more this quote applies to engineering disciplines, which can be approached from vastly different points of views.

Elishakoff *et al.* (1994) for their analysis chose — as a model problem — the problem of Fraser and Budiansky (1969). They chose the truncated normal density for each imperfection coefficient:

$$f(\overline{\xi_m}) = \begin{cases} c_m \exp\left(-\dfrac{\overline{\xi_m}^2}{b_m^2}\right), & \text{for } |\overline{\xi_m}| \leq A_m \\ 0, & \text{for } |\overline{\xi_m}| > A_m \end{cases} \tag{4.12}$$

where $f(\overline{\xi_m})$ is the probability density of $\overline{\xi_m}$, each A_m is a maximum possible value for the random variable ξ_m, b_m are parameters, and the normalization constants C_m are derived from

$$C_m = \left[2b_m \operatorname{erf}\left(\frac{A_m}{b_m}\right)\right]^{-1} \tag{4.13}$$

in which the error function $\operatorname{erf}(x)$ is defined as

$$\operatorname{erf}(x) = \frac{2}{\sqrt{\pi}} \int_0^x e^{-t^2} dt. \tag{4.14}$$

The following conclusions were arrived at in the study by Elishakoff et al. (1994):

> For the case of large deviation of initial imperfections, the minimum buckling load derived by non-probabilistic, convex modeling and the admissible loads corresponding to different values of the required reliability do not exhibit much difference. Therefore, design can be made based on non-stochastic approach since it is much simpler than the stochastic one. However, if the deviation of the initial imperfections is small and the boundary value of the initial imperfection is large, the admissible value for the axial

> load obtained from the stochastic approach may be well above the minimum buckling load obtainable by the non-stochastic approach. This implies, that the use of the non-probabilistic method may lead to a conservative design.

It is remarkable that in some circumstances, both approaches, although being of cardinally different nature, may yield close values for the design axial loads. If probabilistic information is unavailable, one should not propose a probabilistic model that is based on arbitrary assumption about the distribution of Fourier coefficients. Rather, in many circumstances one should use the nonstochastic approach to uncertainty. When the full probabilistic information is available and the initial imperfection's Fourier coefficients have a relatively small deviation, use of the nonstochastic approach will be inadvisable and purely stochastic analysis should be performed. Even when probabilistic information is available to substantiate the probabilistic analysis, if the distribution of the initial imperfections is close to uniform, one may prefer a simpler, nonstochastic, convex analysis, since it yields admissible loads comparable with the results of stochastic approach.

Convex modeling was utilized by this author not only to model uncertain initial imperfections. The studies by Elishakoff *et al.* (1994) and Li *et al.* (1997) deal with *nontraditional* uncertain variables, specifically with uncertain material properties in the composite shells. This work was adopted at the Delft University of Technology in the master's thesis of van der Nieuwendijk (1997). He built the methodology of Elishakoff *et al.* (1994) into the computer program DISDECO (Delft Interactive Shell Design Code), "for variations in the elastic moduli and layer angle θ of laminate shells and for variations in the initial amplitude of axisymmetric imperfections." Special modules, designated as PERFCT and AXBIF, were incorporated into DISDECO to introduce our analysis into it. Van der Nieuwendijk (1997, p. 27) writes:

> To use the method of Convex Modelling DISDECO needs extra input. At the beginning of a normal run of PERFCT or AXBIF the program first asks if the user wants to use variations in the material properties. If the question is confirmed the user must specify the variations of the material properties. If one or more material properties are constant, the user just specifies the variation as zero. The input can be given in three different ways. First one can give an upper and lower bound. The program uses the average of these two values as the nominal material property (E_i^0). Second method is that one specifies a percentage, which is the maximum variation of the average value. The material properties already defined in the model are used as the nominal values. Last, one can specify an absolute variation... After specifying the variations in the material properties DISDECO asks if the layer angle θ has a variation.

Van der Nieuwendijk (1997, pp. 31–32) concludes:

> The method of Convex Modeling is successfully built in DISDECO. The results obtained with DISDECO are the same as described in [Elishakoff *et al.*, 1994]. The method of Convex Modeling is able to calculate the upper and lower bounds of the buckling load

> of a shell due to uncertainties in the geometry of the shell. The method is able to predict these bounds even when there is very little experimental data available. The validity of the method is tested for a variation in the axisymmetric imperfection amplitude. The test showed that a variation of 30% of the initial imperfection amplitude still gave accurate results.

As a curious anecdote, I would like to record the question, asked at the AIAA SDM meeting, by the prominent buckling specialist: "Was the probabilistic analysis that you developed, for the shell buckling, not enough?! Why did you also need the convex modeling?!" The answer to this slightly provocative question is better given by a student of convex modeling; according to van der Nieuwendijk (1997, p. 54), "... method of Convex Modeling is a suitable method to do a sensitivity analysis on shell parameters. The method is simple and quick, and it gives good insight in the role of the different shell parameters on the buckling load."

Note that for large variations of parameters Li *et al.* (1997) suggested to use the nonlinear programming methodology. They concluded (pp. 151–152):

> ... when the lamination angle θ is less than 45°, the results from convex modeling almost overlap with those from nonlinear programming. When θ is greater than 45°, convex modeling predicts a slightly bigger variability than nonlinear programming. As far as the natural frequency and the critical external pressure are concerned, a design based on convex modeling results seems more conservative... the proposed non-stochastic, convex modeling, method may serve as a variable alternative to both deterministic and probabilistic methods.

Other papers on convex modeling in the initial imperfections context include works by Adali *et al.* (1994), Adali *et al.* (1995, 1997), Baitsch and Hartmann (2006), Ben-Haim (1993, 1995, 1996), Elishakoff and Ben- Haim (1990), Lindberg (1991, 1992a,b), Elseifi *et al.* (1999), Pantelides (1996), Qiu *et al.* (1996), Qiu *et al.* (2006), Qiu and Wang (2005), Wang *et al.* (2009), and others. Papers by Faria (2002, 2004, 2007), Faria and Hansen (2001a,b), and Faria and Almeida (2003a,b, 2004, 2006) also are definitive contributions to the convex modeling in buckling context.

3. Brief History of Nonprobabilistic Uncertainty Modeling

> *If a guy tells me the probability of failure is 1 in 10^5, I know he's full of crap.*
>
> R. Feynman (1988)

I met Dr. Yakov Ben-Haim, then Senior Lecturer and currently Yitzhak Moda'i Chair of Technology and Economics at the Technion-Israel Institute of Technology for the first time in 1982. My mother Margarita Leah had passed away, and I attended the house of prayer daily for the mourning ritual. After some weeks, during the intermission between the Minha and Ma'ariv services on a Sabbath evening, we struck up a conversation. He told me that he was associated with the then existing Department of Nuclear Engineering. I asked what he thought of my gut feeling that nuclear power plants, however promising, were too hazardous, especially in small countries. He responded that the probability of accidents in a nuclear power plant was negligibly small. I commented that as a layman, though far away from apocalyptic visions, I would have preferred a zero probability: a terrorist attack, or a combination of unforeseeable events during an environmental upheaval, could have disastrous effects. He reassured me that all contingencies were allowed for by the designers and authorities. Almost 30 years later, on March 11, 2011, the world witnessed the triple disaster at Fukushima — an earthquake, tsunami, and crippling of the nuclear reactors. I wrote to Professor Ben-Haim, asking whether he still thought in the same way. His reply was: "I have been very far from nuclear engineering for many years. Clearly, the unfolding Japanese nuclear disaster needs to be carefully studied and conclusions need to be drawn." At the writing of this book, out of 54 reactors covering 30% of Japan's requirements, only two are operational, the matter being under review and the country split on this issue. According to Yukiya Amano, head of the International Atomic Energy Agency, the Fukushima accident had proved "an important wake-up call" that itself triggered "a nuclear safety renaissance" (Pfeifer, 2012).

To digress, some two years before our first meeting I attended a seminar at Delft University of Technology, given by a prominent Austrian probabilistic mechanician. He informed the audience that the failure probability of a nuclear reactor was not more than 10^{-42}. Stunned by the amazing figure, I voiced a doubt as to its plausibility. He assured me that it was valid. Only years later I had a chance to read Richard Feynman's (1918–1988) undiplomatic take on such small probabilities, as given in the motto above (*Quod licet* a Nobel Prize Winner *non licet* ordinary mortals . . .).

Some years later I again met the same lecturer, who told me that Austria and Germany were abandoning their plans for constructing nuclear power plants, for fear of the extreme hazards. I opined that this was the right thing to do, to which he responded quite violently. Eventually he broke off all contact with me. He even removed me from the international scientific committee of the conference that he organized every now and then. While Sander Bais (2010) claims that "science can liberate us from our cultural straitjacket of prejudice and intolerance," scientists, just like "regular" folk, can be quite intolerant of others' views.

Next time I met Ben-Haim was in November 1987, after returning to the Technion from a sabbatical at the Naval Postgraduate School in Monterey, California. We both attended a seminar at the Aerospace Department, and asked closely related questions. After the seminar, Ben-Haim invited me to discuss certain scientific ideas that he developed in the nuclear engineering context. I immediately recalled the so-called *Landau principle*, attributed to Nobel laureate Lev Landau (1908–1968). According to it, if one approaches you with the intention of discussing his scientific ideas, you should immediately counter with one of your own; otherwise, you risk ending up engrossed in someone else's problems. Ben-Haim informed me that he had contacted three colleagues from my department for the same purpose, but had been turned down for "lack of time." I replied that even though I was not less busy than those colleagues, I would make time for him, to "atone" as it were for my colleagues, and suggested that he prepare a seminar lasting about an hour.

At that seminar Ben-Haim presented the ellipsoidal model developed by Schweppe (1973) at M.I.T. and his extension of it as described in his book (Ben-Haim, 1985) on issues of nuclear engineering. *Inter alia*, he noted that Schweppe must have been an egg farmer, because of his ellipsoidal approach. Both these methodologies were nonprobabilistic in nature.

My criticism of certain probabilistic assumptions, like disregard of cross-correlations (Elishakoff, 1977; Elishakoff *et al.*, 1979) or the ergodicity assumptions (Scheurkogel *et al.*, 1981; Elishakoff, 1983; Scheurkogel and Elishakoff, 1985), made me receptive to Ben-Haim's approach. (By the way, note that our conversations of 1982 seem rather paradoxical: me, a probabilistic analyst, resorting to worst-case scenarios, and Ben-Haim, the "nonprobabilicist" using probabilistic arguments in defence of the nuclear power plants!)

I raised numerous questions and suggested further meetings. While Yakov was eager to identify the problem of joint interest and start the joint work, I preferred to proceed more slowly. Then he told me that in the course of a search in the Aerospace Department library, he found my two papers on the role of initial imperfections in the impact buckling problem of bars (Elishakoff 1978a,b). Thus I was introduced

Fig. 4.2. Attending the IUTAM Congress in Copenhagen, 1984 (Professor Y. Ben-Haim is second from left).

Fig. 4.3. Professor R.E. Drenick (in the middle) at Florida Atlantic University (Professor Y.K. Lin on the right).

to convex modeling, and Ben-Haim to imperfection sensitivity. Later I gave him some books in applied mechanics and suggested that we extend our first convex modeling paper (Ben-Haim and Elishakoff, 1989) dealing with the linear problem, to the nonlinear one. Actually, our first paper generalized my previous probabilistic studies (Elishakoff 1978a, b) to convex modeling, and my idea was to extend my previous "nonlinear" study (Elishakoff *et al.*, 1987) in the same direction.

Our joint studies were first made public at the conference in honor of Professor Yu Kweng (Mike) Lin's 65th birth anniversary, held in 1988 at the University of Illinois at Urbana, where the participants were all "stochasticists." After my presentation Professor Stephen Crandall of M.I.T. told me (a "censored" rendering): "You are certainly right but they may hate the message." Crandall also informed me that he has heard the arguments for nonprobabilistic treatment of uncertainty from Chernousko (1981) of the Institute of Problems in Mechanics, of the Russian Academy of Sciences, in Moscow. I was able to locate Chernousko's articles published in the journal *Technical Cybernetics.*

Later I recalled that both Crandall and I had attended a lecture in the same vein by Drenick (1977) of the Polytechnic Institute of New York at the IUTAM Symposium on Stochastic Problems in Dynamics in Southampton, England, a year earlier. I was even able to locate my comments from that occasion, to the effect that this methodology ought to be employed for highly responsible systems, with attendant determination of the worst-case scenarios. In the course of writing our book (Ben-Haim and Elishakoff, 1990) I asked Professor Drenick to contribute the Foreword. He gladly agreed. Here is an excerpt

> Professors Ben-Haim and Elishakoff... point out in this book that the designer should make clear to himself what he knows on a sufficiently high confidence level, and then

> use this knowledge to place bounds on the design data and parameters that he can properly utilize. In effect, this will isolate for him a set of possible designs. The book deals with the case in which this set is convex and demonstrates that a very broad range of problems in applied mechanics actually satisfies the convexity requirement. Their approach is novel and highly welcome. In my opinion, it is inevitable that it, and its extensions, will dominate the future practice in engineering.

Later on, Ben-Haim obtained from me Professor Drenick's address and invited him to the Technion as a visiting professor. At Florida Atlantic University, we were able to invite Drenick as a Visiting Distinguished Professor at the then-existing Center for Applied Stochastics Research.

Ben-Haim and I published another joint study on imperfection sensitivity (Elishakoff and Ben-Haim, 1990) and had a joint paper presented at the International Conference on ModalAnalysis (Elishakoff and Ben-Haim, 1990). Prior to that, at the IUTAM Congress in Grenoble (France), I presented our joint paper on imperfection-sensitivity in nonlinear shells. Professors Koiter, Budiansky, Singer, and Arbocz, who also attended the conference, reacted to it very warmly. These and other researches culminated in a monograph (Ben-Haim and Elishakoff, 1990). In the Preface to it we wrote:

> Except for the first two chapters, the material is exclusively based on our joint research, which continuously stimulates us to further the development of the subject, in nearly perfect cooperation and harmony (not free from an occasional "fight" on this or that idea, sentence or word).

Fig. 4.4. Professor B.V. Bulgakov.

It should be noted that in our research we had been preceded not only by Schweppe (1973), Drenick (1970, 1977), Shinozuka (1970), and Chernousko (1981), but also by Bulgakov (1940, 1942) as I found later. The latter author dealt with *de facto* convex models long before either I or Yakov had been born.

This acknowledgment is important also in light of the following legend. A certain sage toiled over a difficult conundrum for many months before he succeeded in cracking it. Afterward, while visiting an elementary school, he heard his own question posed by a pupil to a young teacher, who promptly provided exactly the same answer that the sage had arrived at after many months of exertion. The sage was both amazed and depressed by this, until he figured out that once he had resolved the problem, its solution became "available" to all humankind. Edgar Allan Poe's (1809–1849) statement, made in 1845, "no thought can perish", is in keeping with the above legend.

In his book, Nahin (1998) refers to the above phenomenon as the Bannister effect, although in a different context:

> This is an interesting example of how, once a problem is *known* to have a solution, others quickly find it too, a phenomenon related, I think, to sports records. E.g., within a month of Roger Bonnister breaking the four-minute mile it seemed as though every good runner in the world started doing it.

Another reason for these comments on priority is the questionable tendency on the part of some researchers to characterize convex models of uncertainty in its numerous forms as "radically different from all current theories of decision under uncertainty." The research that Ben-Haim and I (1990) presented in our monograph was generously supported by the Technion-Niedersachsen (Federal Republic of Germany) Research Cooperation Fund. This happened as follows. At a reception organized by the Technion's Research and Development Foundation, I met Dr. Christian Hodler who served as "Ministerialdirigent" in the Niedersachsen (Lower Saxony, Federal Republic of Germany), Ministry of Science and Art. We had a long conversation on various scientific and nonscientific topics. He suggested that I submit a grant application to the Fund, mentioning that many decades earlier, Einstein and his colleagues sought financial help for the fledgling Technikum (as the projected Technical College was then called in German) in Haifa, thereby supporting resettlement of the old–new Land of Israel with agricultural know-how, training electricians and telephone and telegraph workers. This is how the present-day Technion-I.I.T. came into being. Dr. Hodler advised me to enlist a German counterpart whom I soon found in Professor Günther Natke (1933–2002) of the University of Hanover, and contacted him. He kindly agreed to cooperate. The grant was thus approved. Yakov and I worked frantically both on our researches and on our joint book, seeking to communicate to the world at large our excitement over the nonprobabilistic analysis of uncertainty. We voiced "our gratitude to the Technion — Israel Institute of Technology for providing an atmosphere which stimulates fruitful cooperation among its staff." Incidentally, the Google citation of our book far exceeds its counterparts for our respective "solo" books.

I lucked out to realize that the convex models of uncertainty may cause excessive overestimation of the response characteristics of the system. In fact, it looks for the worst-case scenarios. This is why Drenick's (1970, 1977) researches have not been implemented. In practice, engineers look for the optimal solutions, whereas the nonstochastic uncertainty directs toward the worst ones. Accordingly, I understand that what I called as antioptimization (looking for the worst) (Elishakoff, 1990) ought to be balanced by optimization (looking for the best). As the popular saying goes, "make the best out of the worst." This philosophy was expounded in much detail in the book by Elishakoff and Ohsaki (2010) in various contexts.

Ben-Haim also continued his research in imperfection sensitivity, publishing several papers (Ben-Haim, 1993, 1995), which are also summarized in his book (Ben-Haim, 1996).

This writer's own adventures in convex modeling resulted in books on acoustically excited structures in space shuttle applications (Elishakoff *et al.*, 1993) based on our research for the NASA Kennedy Space Center; on stochastic and convex finite elements (Elishakoff and Ren, 1994) based on research for the National Center for Earthquake Engineering; on stochastic and convex safety factors (Elishakoff, 2004) supported by the NASA Glenn Research Center; on stochastic and convex instability problems (Elishakoff *et al.*, 2001) sponsored by the NASA Langley Research Center. Specifically, in the space shuttle domain, combination of the probabilistic and convex-modeling approaches was accomplished (Elishakoff and Colombi, 1993). We also succeeded in realizing an extensive comparison of stochastic and convex modeling in the context of imperfection sensitivity (Elishakoff *et al.*, 1994a,b).

In 1989, I and my family found ourselves in the United States, having accepted the offer of Florida Atlantic University. Yaakov and I were engrossed with our own researches; our collaboration ended due to the additional need for me to write many research proposals. The grant from the Technion-Niedersachen Fund sponsored a one-month visit to the Technion; Professor Natke was also visiting the Technion at that time. I was gratified that Yakov and he were collaborating on a variety of topics, and that convex modeling of uncertainty in applied mechanics that was developed by Yakov and me was finding new adherents. This led to separate visits by Yakov and myself to the University of Hanover, where Professor Natke not only provided warm hospitality to me and my research, but also demonstrated his skills as pilot of a small airplane.

At the International Workshop on Uncertainty Models and Measures, organized by Professors Ben-Haim and Natke (1997) in Lambrecht, I was asked by one of the organizers: "How can such dissimilar modelings as stochasticity and convexity be compared?" In reply, I offered a situation where a wealthy firm wants to conduct an extensive uncertainty analysis for its problems. Being in doubt as to the preferable approach, it outsources the job to competing analysts and is then able to compare their input and make the choice.

Later on we showed (Elishakoff, 1999; Wang *et al.*, 2011) that there is no contradiction between the probabilistic and convex modelings, the latter paper being titled "Probability and Convexity are Not Antagonistic." I recalled that already in my first probabilistic book (1983, p. 15) I compared the probabilistic and worst-case analyses.

Still later, Ben-Haim (2006, p. 249) too advocated combination of the two approaches, writing: "Often . . . the probabilistic information is insufficient to cover all the facets of the problem, in which case one can combine it with an info-gap model to create a hybrid decision algorithm. This has been recognized for a long time (Elishakoff and Colombi, 1993)."

In contrast to my hybrid optimization/antioptimization extension of convex modeling, Ben-Haim adopted a different philosophy, geographical distance necessarily leading to different thinking. According to Norbert Wiener (1894–1964) "there is not only room, but a definitive need in different books" and thinking (quoted by Gleick, 2011).

Ben-Haim proposed the notion of information gap, apparently introduced by Galbraith (1973, p. 5) who defines it as "the difference between the amount of information required to perform the task and the amount of information already possessed by the organization."

According to Mack (1971, p. 1), "Uncertainty is the complement of knowledge. It is the gap between what is known and what needs to be known to make decisions."

Ben-Haim (2006, p. 2) states, "Info-gap models concentrate on the disparity between what *is known* and what *could be* known while making very little commitment what the structure of the uncertainty." He adds (2006, p. 12): "The ignorance which is important to the decision maker is the disparity between [what] *is known* and what *needs to be known* in order to make a responsible decision; ignorance is an info-gap," and, finally (2006, p. 18), he asks:

"What is the gap between what is known and what could be known?" These definitions raise some legitimate questions: How to measure "the amount of information"? Why does one not know "what could be known?" It is because of not enough effort? Does this happen due to negligence? The latter contingency is likely to be prominent in the juridical domain. Additional questions may persist — what is that body of knowledge that "could be known"? Why is there no connection made with Shannon's information theory (Gleick, 2011), Shannon being credited by "his development of *information theory* as a full-blown subject

(Davis, 2013)"? Or, perhaps, there is no connection between information theory and the term information gap?

And in the case of imperfection sensitivity, does the information gap comprise the probabilistic characteristics of the imperfections, or detailed values of the imperfections of each shell? How can potentially (and hopefully) knowable and known quantities be differentiated in the beginning of the research? These immediate questions appear to be in need of answer, in the humble opinion of the present writer.

The "info-gap" concept and terminology found both supporters (Hemez and Ben-Haim, 2004; Takewaki and Ben-Haim, 2005) and opponent (Sniedovich, 2007–2012).

In the general context, it is pleasing to report that the nonprobabilistic approach to uncertainty is attracting an increasing member of researchers around the world, in various branches of mechanics, including elastic stability. Indeed, the following quotation from Arthur Schopenhauer (1788–1860) appears most appropriate: "All truth passes through three stages. First, it is ridiculed. Second, it is violently opposed. Third, it is accepted and self-evident."

This methodology became known around the world under different names. It is referred to as the ellipsoidal approach (Schweppe, 1973), the guaranteed approach (Chernousko, 1994), convex modeling (Ben-Haim and Elishakoff, 1990; Elishakoff *et al.*, 1993; Elishakoff and Ren, 2001), robust modeling (Ben-Haim, 1995), optimization and antioptimization (Elishakoff, 1990, 2004; Elishakoff and Ohsaki, 2010), and min-max approach (Wald, 1949, 1950; Sniedovich, 2007, 2012).

In 1989, Professor Walter Wedig of the University of Karlsruhe (Federal Republic of Germany) visited us at Florida Atlantic University. After I described to him the convex modeling of uncertainty and the contents of the forthcoming book by Ben-Haim and myself, he informed me that his colleagues at the Mathematics Department in Karlshruhe are utilizing the interval analysis methods for guaranteed mathematical evaluations. I was excited by this information and got acquainted with the books by Moore (1979), Alefeld and Herzberger (1983), and Adams and Kulisch (1993). This led me to recognize that interval analysis was nothing else but the simplest form of convex modeling, and I prepared a paper that propagated this idea (1995). In that year I also visited for the first time the Peoples's Republic of China, where I met a young researcher from Zhilin University, Dr. Zhi Ping Qiu. I suggested to him the possibility of collaboration which he eagerly accepted. Later, Dr. Qiu moved to Beijing University of Aeronautics and Astronautics along with our teamwork, in which Professor Xiaojun Wang now also took part. We put our heads together also on imperfection sensitivity topics, proceeding from purely theoretical studies to incorporation of experimental scatter data in studying the effect of nontraditional imperfections, namely the uncertainty in material properties.

Later, during a sabbatical stay when I divided time between Japan and China, I had an opportunity to collaborate also with Professor Makoto Ohsaki of Kyoto University, in addition to earlier research conducted with Professors Shigeru Nakagiri and Nobuhiro Yoshikawa of Tokyo University. We summarized our efforts in a joint monograph (Elishakoff and Ohsaki, 2010). Lately, Ben-Haim (2010) pursues, among other applications, the info gap concept in economics. I cannot offer any sensible comment in this field, which remains for me, to borrow from Eli Sternberg (1985) "the richest store of ignorance conceivable," except wishing Professor Yakov Ben-Haim well in his endeavor at implementing uncertainty in economics, for the benefit of us all. It was pleasing to see that the *Engineering Design Reliability Handbook* (Nikolaidis *et al.*, 2005) contains a chapter on convex modeling, so that engineers who want to engage in structural reliability will be able to get acquainted with alternative probabilistic methodologies, specifically with fuzzy sets on the one hand, and convex modeling on the other.

Chapter 5

Nontraditional Imperfections in Shells

It may safety be said that all real structural systems are imperfect in form, imperfect in material properties, imperfect in the sense of residual stresses and imperfect in the way loads are applied.

J. Roorda (1980)

The theory of structural stability made further progress in order to embrace physical effects, called "imperfections," not considered in the classic theory. The first was suggested by the experimental evidence that the buckling loads measured in laboratories were systematically lower than those predicted by the theory. The reason was that unavoidable geometric deviations from the ideal shape of a structure drastically reduce its buckling load. A second extension was the setting of the classic equations in terms of probabilistic variables in order to account for the uncertainties about the loads.... All these topics are now subject to fervent studies, but the results are still partial.

P. Villaggio (2011)

This chapter investigates the combined effect of three quantities, namely, randomness in initial geometric imperfections, thickness variations, and the applied loading on the reliability of axially compressed cylindrical shells. Some cooperative work with Warner T. Koiter and James H. Starnes, Jr. are first described. In order to gain insight we consider the simplest possible case when the initial imperfections, thickness variations, and the applied loads are uniformly distributed. It is shown that hybrid randomness may increase or decrease the reliability of the shell in contrast to the case when the latter is treated as experiencing the sole randomness in initial imperfections.

1. Spatial Parametric Resonance and Other Novel Buckling Problems Inspired by James H. Starnes, Jr.

For me Jim Starnes was amazingly effective in the role that I believe is central to a researcher in a government laboratory. He provided a vital link between industry needs and long range research by combining resources of industry, government and academe.

R.T. Hafka (2006)

This section is devoted to the apparently new phenomenon in buckling of shells, that is, a spatial parametric resonance due to thickness variation in isotropic and composite cylindrical shells. This problem was posed and inspired by late Dr. James H. Starnes, Jr. whose tremendous impact on the research on thin-walled structures in the United States is yet to be properly ascertained.

Dr. James H. Starnes, Jr., as the head of Aircraft Structures Branch, had a great influence on the research that was conducted at the universities and within National Aeronautical and Space Administration (NASA).

I recall that after I approached him with a proposal to obtain a research grant from NASA, he suggested to me that I give him a "private" lecture at the SDM conference about my research in stochastic analysis of initial imperfections, which culminated in papers (Elishakoff 1983, 1998), paving the way for introducing the initial-imperfection concept, developed by Koiter (1945, 1963) in the deterministic context, into design. Indeed, we skipped one of the lunches and had a working session instead, where during one hour I expanded on the importance of stochastic analysis of shell imperfections. Discussions with Dr. Starnes were always fruitful and motivating as he had the grand vision for shell buckling and structural mechanics in his mind. Following our discussion, he mentioned that he would think about some problem that would be closely related to the research that I was proposing. After a week he contacted me and informed that he was interested in the influence of thickness imperfections in isotropic and composite shells. I suggested that I would contemplate about this problem, and if I could come with an appropriate idea I would communicate it to him.

After some thoughts, the following germ of an idea occurred, during the discussion of this topic with Professor W.T. Koiter. Since cylindrical shells are extremely sensitive to initial geometric surface imperfections that are co-configurational to the mode shape for classical buckling, one could anticipate that if the thickness distribution of the shell has a component that is proportional to such a buckling mode shape, then the thickness variation (or thickness imperfection) ought to influence the buckling behavior of the shells.

After getting a research grant on this topic, we embarked on this research intensively. Fortunately, I was invited as a visiting professor at the Delft University of Technology during that summer of 1992 and had a chance to discuss this matter extensively with Professor Warner T. Koiter. He expressed his interest in participating in the research. Since then we had a very intense exchange of ideas between Dr. Starnes, Professor Koiter, as well as my then-student, now Dr. Yiwei Li, and myself. Derivations and calculations have been conducted on the both sides of Atlantic. I must note that Professor Koiter performed all derivations and calculations by hand, whereas we extensively used the symbolic algebraic packages.

It turned out that our anticipation that the unfavorable thickness imperfections were proportional to the classical buckling mode were not fully incorrect; still, later on we characterized our conjecture as naïve. The effect of thickness imperfections on shell buckling turned out to be more subtle than anticipated. It was found that when the thickness variation took the form similar to the classical buckling mode shape, it may have a remarkable effect on the classical buckling load, that is, the classical buckling load is decreased by over 6% when the imperfection amplitude is 15%.

In other words, if the axisymmetric thickness imperfection pattern is

$$h(x) = h_0\left(1 - \varepsilon\cos\frac{2px}{R}\right) \tag{5.1}$$

where h_0 = nominal thickness, ε = thickness variation amplitude, p_0 corresponds to the top of Koiter's semicircle, then the expression for the buckling load reduction factor α is

$$\alpha = 1 - \frac{1}{2}\nu\varepsilon - \frac{832 + 464\nu - 23\nu^2}{512}\varepsilon^2 \tag{5.2}$$

where υ = Poisson's ratio. The first two terms correspond to the formula obtained by Koiter by hand evaluation.

However, if

$$h(x) = h_0\left(1 - \varepsilon\frac{2p_0x}{R}\right) \tag{5.3}$$

then the buckling load reduction formula takes a different form:

$$\alpha = 1 - \varepsilon - \frac{25}{32}\varepsilon^2. \tag{5.4}$$

The symbolic algebra derivation provided an additional term whose influence is very small for small thickness variation amplitudes. Again, expression (5.4) was obtained by using symbolic calculation package, while Professor Koiter obtained the first two terms by hand calculation.

The formula (5.4) signifies that the most detrimental thickness variation is the one with wave number twice that corresponding to the classical buckling mode. In this situation even if the amplitude of the thickness variation is as small as 0.1, the thickness variation reduces the buckling load by 10% from its counterpart of the shell with uniform thickness. Thus, it turned out in the absence of initial geometric surface imperfection, this particular kind of thickness variation may constitute the most important factor in the buckling load reduction for axially compressed cylindrical shells.

A key finding was that the thickness imperfection whose wave number is twice the wave number of the classical buckling mode is reminiscent of a parametric resonance. It has a spatial character, which is in contrast to classical parametric resonance problems. The details can be found in the paper by Koiter, Elishakoff, Li and Starnes (1994a).

At this stage, Dr. Starnes suggested further work on the presence of both geometric and thickness imperfections. His style was using the word "and" instead of the word "but." He mentioned to the effect that "and now we should move to the geometrically imperfect shell case." This type of approach in leadership is usually referred as neurolinguistic programming. Thus, I found that Dr. Starnes had some natural, magnificent leadership skills that allowed to "squeeze," as it were, more and more useful results from ourselves.

Let us return to the effect of initial imperfections. The fundamental study on the effect of a geometric imperfection in cylindrical shells was conducted by Professor Koiter in 1963. Hence I suggested that we take the Koiter solution as the basic one and superimpose on it the terms stemming due to deviation due to the thickness imperfections. The resulting work (Koiter, Elishakoff, Li and Starnes, 1994b) had two components: asymptotic analysis via the energy criterion and numerical analysis.

The following two asymptotic formulas were derived:

$$(1-\lambda)\left(1-\frac{1}{2}\nu\varepsilon-\lambda\right)-\frac{3c}{2}\lambda\mu=0 \tag{5.5}$$

and

$$(1-\lambda)(1-\varepsilon-\lambda)-\frac{3c}{2}\lambda\mu=0 \tag{5.6}$$

where $\mu =$ amplitude of the thickness coefficient, $c=[3(1-\nu^2)]^{1/2}$ (ν is the Poisson's ratio). Equation (5.5) corresponds to the case when thickness imperfections are co-configurational to the buckling mode, whereas Eq. (5.6) is associated with the case when thickness imperfection wave number is twice the wave number of the axisymmetric buckling mode.

It turned out that when initial geometric surface imperfections are present, the combination of initial geometric surface imperfections and thickness imperfections reduce the buckling load even more drastically. When the thickness variation amplitude is 0.2 times the nominal thickness while initial geometric surface imperfection amplitude is just 0.02, the thickness variation causes a further 11% decrease in the critical buckling load. This reduction is in addition to the reduction from the initial geometric imperfection, which is 20%. Thus, the decrease in load-carrying capacity of the shell due to both initial geometric surface imperfections and thickness imperfections amounts to 31%.

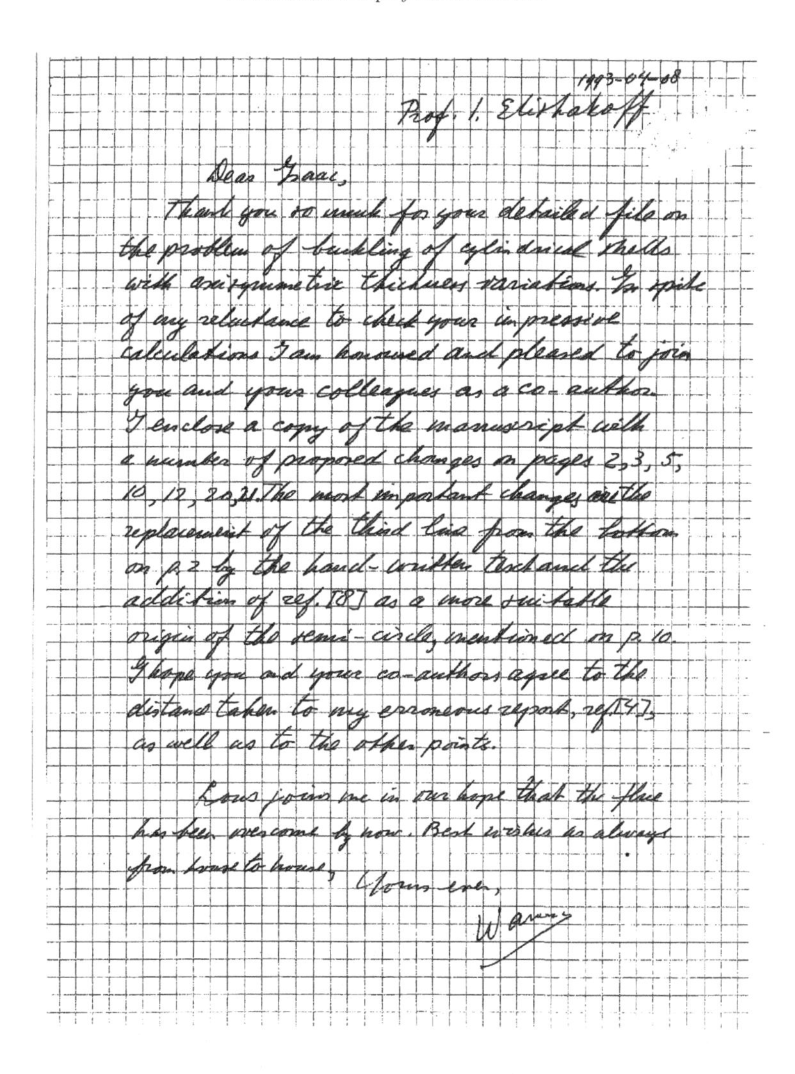

1993-04-08

Prof. I. Elishakoff

Dear Isaac,

Thank you so much for your detailed file on the problem of buckling of cylindrical shells with axisymmetric thickness variations. In spite of my reluctance to check your impressive calculations I am honoured and pleased to join you and your colleagues as a co-author. I enclose a copy of the manuscript with a number of proposed changes on pages 2, 3, 5, 10, 12, 20, 21. The most important changes are the replacement of the third line from the bottom on p. 2 by the hand-written text and the addition of ref. [8] as a more suitable origin of the semi-circle, mentioned on p. 10. I hope you and your co-authors agree to the distance taken to my erroneous report, ref. [4], as well as to the other points.

Lous joins me in our hope that the flue has been overcome by now. Best wishes as always from house to house,

Yours ever,

Warner

This study illustrated that despite the fact that the initial geometric surface imperfections stand out as the main factor for the reduction of the critical buckling load and the effect of thickness variation of certain patterns is less significant in many cases, the thickness variations of certain kinds may cause further notable decrease in the critical buckling load. Neglect of thickness variation, therefore, is not on the safe side, for design purposes. This one example illustrates the tremendous engineering foresight of the late Dr. Starnes.

At this stage Dr. Starnes posed the problem of comparison of the results with commercially available software. This task was accomplished by our cooperation with the computer genius Dr. David Bushnell — another pillar of the shell buckling research (Li *et al.*, 1997). Further inspiration was again provided by Starnes, suggesting on generalizing the results to composite shells. This was performed in the paper by Li *et al.* (1995). In his private communication to the author, Professor W.T. Koiter wrote me about his "happiness" with the results derived in the paper by Elishakoff *et al.* (1995).

From the subsequent studies, the paper by Gusik *et al.* (2000) stands out. They used FE analysis to study the "influence of harmonic thickness variations in the circumferential direction on the bifurcation pressure of thin cylindrical shell." They also derived analytical formulas, extending our Eq. (5.4) to the external pressure case:

$$\alpha = 1 - b\varepsilon \tag{5.7}$$

where

$$\begin{aligned} b &= 0.071(\log Z)^3 - 0.616(\log Z)^2 + 1.551(\log Z) + 0.784 \\ b' &= 0.126\ln(Z^{-1/2}) + 2.25 \end{aligned} \tag{5.8}$$

for long wavelength thickness imperfections, and

$$\begin{aligned} b &= -0.086(\log Z)^2 - 0.716(\log Z) - 0.181 \\ b' &= 21.37(Z^{-1/2}) - 7.20(Z^{-1/2}) + 1.39 \end{aligned} \tag{5.9}$$

for short wavelength thickness imperfections. Note that parameter b is evaluated as a function of $\log(Z)$ and of $Z^{-1/2}$ in order to cover two limiting values of Z ($Z \to 0$ and $Z \to \infty$).

The authors found that the most detrimental imperfection mode depends on the Batdorf parameter,

$$Z = \sqrt{1 - \nu^2}\left(\frac{l}{R}\right)^2 \frac{R}{h}. \tag{5.10}$$

They also showed that there is a transitional value of the thickness imperfection amplitude at which the worst decrease of the buckling load takes place. The critical imperfection amplitude, corresponding to the transition reads:

$$\varepsilon_{cr} = 0.0032 + \frac{2.45}{(n_w - 1)^{2.5}} \tag{5.11}$$

where n_w is the Windenburg parameter

$$n_w = \sqrt{6\pi^2\sqrt{1 - \nu^2}\frac{R}{L}\sqrt{\frac{R}{h}}}. \tag{5.12}$$

The interested reader can also consult papers by Gusic *et al.* (1998), Gusic (1999), and Michel *et al.* (2000). Subsequent works on thickness imperfections include papers by Combescure and Gusik (2001), Nguyen and Dang (2006), Nguyen and Thach (2006a, 2006b) and Nguyen *et al.* (2009).

Specifically, the latter work develops coupled linearized governing stability equations for cylindrical shell with variable thickness under external pressure. The formulas for the buckling load are derived by using the hybrid perturbation, the Bubnov–Galerkin method. It was concluded that the variable thickness can cause the reduction of the load-carrying capacity of cylindrical shells. Therefore, this effect ought to be taken into account in the design of cylindrical shell structures.

Other important papers include those by Papadopoulos and Papadrakakis (2005) and, Bathe *et al.* (2003). Stam and Arbocz (1997) utilized formulas (4.5) and (4.6) to investigate their effect on reliability estimates.

On many other occasions, our joint research and my personal views were tremendously affected by his experience and great questions that were posed by Dr. Starnes; he had an incomparable ability to pose most pertinent questions. Our cooperation resulted in the monograph titled *Non-Classical Problems in the Theory of Elastic Stability* published by the Cambridge University Press (2001), co-authored with Dr. Li and Dr. Starnes. We were both gratified and humbled by the extremely positive reviews it has received in the open literature; to salute Dr. Starnes's memory and in curious mixture of modesty and lack of it, some quotes will be reproduced here. The journal *Current Engineering Practice* wrote: "This substantial and attractive volume is well-organized and superbly written one that should be warmly welcomed by both theorists and practitioners... [Authors] have given us a jewel of a book." *AIAA Journal* wrote:

> ... excellent presentation ... It is well written, with the material presented in an informational fashion as well as to raise questions related to unresolved or questionable challenges ... In the vernacular of film critics, "thumbs up"."

Journal *Ocean Engineering* characterized this monograph as follows: "...outstanding book...elegant...unique book...will be of enormous use..."

Without the timely questions, constant encouragement, critical attitude, and the divine gifts of Dr. Starnes, of feeling what is important and what is not, the monograph would not have had the same impact.

Many other research efforts were inspired by Dr. Starnes. Some of these efforts include: studying the effect of limited data on the prediction of the variability of the buckling loads and natural frequencies; efforts related to the enhancement of safety factors approach based upon reliability concept; and others.

The author can humbly testify about many research programs that Dr. Starnes's branch supported resulted in extremely beautiful works in mechanics, with attendant feeling of a need to attend a lecture whose author or co-author was Dr. Starnes — and these were many. Dr. Starnes was a true lighting lamp for many research activities conducted at many universities via the association with NASA Langley Research Center. May this light continue to inspire more and more collaborative research activities between the academia and NASA. NASA, NSF, ONR, DOE, DARDA, and other agencies are in urgent need of people who could enter into Dr. Starnes's shoes.

2. Scatter in Load: Problem Description

Loads on structural members that carry primarily compressive loads, such as columns and arches, are not necessarily centered. Furthermore they may unmodeled effects such as small lateral loads, fluctuations, nonconservativeness, and so on.

L.A. Godoy (1996)

While it is generally recognized that initial geometric shell-wall imperfections are a major contributor to the difference between the predicted shell buckling loads and the experimentally measured loads, the traditional sources of design knockdown factors do not include data or information related to the sensitivity of the response of a shell to other forms of imperfections.

B. Kriegesmann *et al.* (2012)

It appears imperative to present a hybrid deterministic/probabilistic study of the imperfection sensitivity that takes into account uncertainty in both thickness variations and the initial geometric imperfections. Moreover, in some circumstances, the applied load also has a scatter, and it ought to be modeled as an uncertain quantity. This study represents such a hybrid investigation of the imperfection sensitivity. Whereas at some laboratories experimental data banks have been compiled for the initial imperfections, such an experimental data is not generally pursued. To illustrate the possible effects of such uncertainties we deal with the case when all uncertain random variables have a uniform distribution.

Consider the governing equation (Koiter *et al.*, 1994b) that connects the buckling load with the magnitudes of the critical imperfection μ and thickness variation amplitude ε

$$(1-\lambda)(1-\lambda=\varepsilon)-\frac{3c}{2}\lambda\mu=0 \tag{5.13}$$

where $c=\sqrt{3(1-\nu^2)}$, $\nu=$ Poisson ratio, $\lambda=$ load factor. The reliability is given by the following expression:

$$R(p)=\text{Prob}(\lambda>p) \tag{5.14}$$

where p is the arbitrary, preselected load coefficient. To evaluate this probability in Eq. (5.14), we express from Eq. (5.13) the value of λ as a function of μ and ε:

$$\lambda_{1,2}=\frac{1}{2}\left[-\varepsilon+2+\frac{3c}{2}\mu\pm\sqrt{\left(\varepsilon-\frac{3c}{2}\mu\right)^2+6c\mu}\right]. \tag{5.15}$$

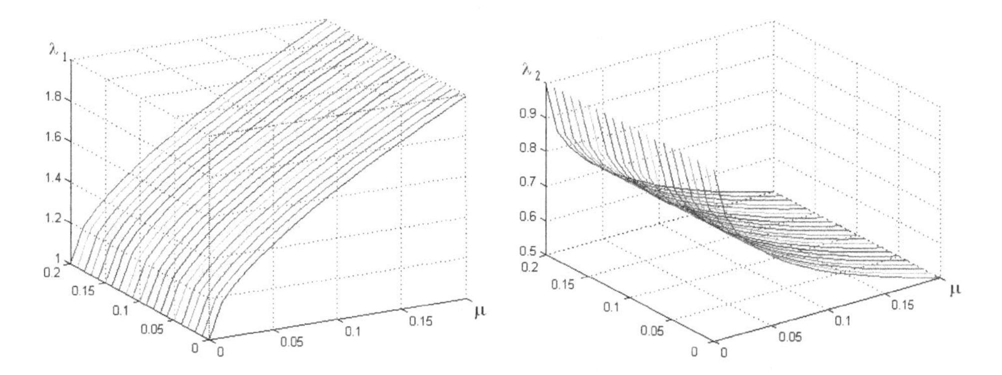

Fig. 5.1. Variation of λ_1 and λ_2 versus initial imperfection μ and thickness ε.

According to Fig. 5.1, λ_1 corresponds to values $\lambda > 1$, which does not describe the physical phenomenon at hand. Only the curve λ_2 with values $\lambda < 1$ corresponds to the imperfection sensitivity. Hence the second root, λ_2 in Eq. (5.15), is of interest for us hereinafter. The reliability becomes

$$R(p) = \text{Prob}[2(p-1)(p-1+\varepsilon) - 3cp\mu) > 0]. \tag{5.16}$$

At this juncture we should specify which quantities are treated as random variables and which ones are considered as deterministic ones for attendant derivation of the reliability.

3. Combined Randomness in Imperfection and Thickness Variation for Deterministic Load

It is instructive first to consider the case when ε and μ are random, but p is deterministic. The probability density of thickness variation magnitude ε is taken to be uniform:

$$f_\varepsilon(\varepsilon) = \begin{cases} 0, & \text{for } \varepsilon \le \varepsilon_1 \quad \text{or} \quad \varepsilon \ge \varepsilon_2 \\ (\varepsilon_2 - \varepsilon_1)^{-1}, & \text{for } \varepsilon_1 < \varepsilon < \varepsilon_2 \end{cases}. \tag{5.17}$$

The probability density of initial imperfections μ is also taken as uniform:

$$f_\mu(\mu) = \begin{cases} 0, & \text{for } \mu \le \mu_1 \quad \text{or} \quad \mu \ge \mu_2 \\ (\mu_2 - \mu_1)^{-1}, & \text{for } \mu_1 < \mu < \mu_2 \end{cases}. \tag{5.18}$$

Reliability can be also expressed as

$$R(p) = \text{Prob}\left[\mu < \frac{2(p-1)(p-1+\varepsilon)}{3cp}\right]. \tag{5.19}$$

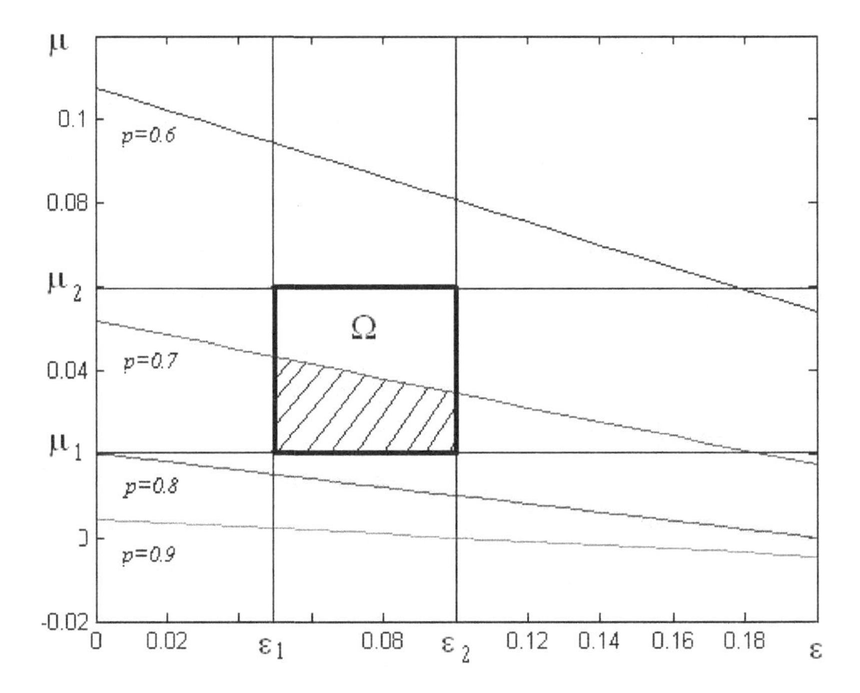

Fig. 5.2. Imperfection magnitude μ as function of thickness variation amplitude ε for several values of the load p.

Figure 5.2 portrays μ as a function of the load parameter p and the thickness parameter ε

$$\mu = \frac{2(p-1)(p-1+\varepsilon)}{3cp} \tag{5.20}$$

with $\nu = 0.3$. Thus, $c = 1.65$, and μ and ε are both taking values not in excess of 0.2. The box of variation of ε and μ is denoted by Ω. It is shown in boldface.

The reliability for $p = 0.7$ is represented by the hatched area in Fig. 5.2, since in this region the inequality in Eq. (5.19) is satisfied. In order to calculate the expression for the reliability as the function of p, we need to consider six different cases depending on the governing parameters that are needed for the evaluation, as drawn in the Fig. 5.3:

The point of intersection of the curve $\mu(\varepsilon)$ associated with fixed p with the line $\mu = \mu_2$ is denoted as (ε_y, μ_2); the point of intersection of that curve with line $\mu = \mu_1$ is designated by (ε_x, μ_1). Six different cases are illustrated in Fig. 5.4. Note that for fixed p, the curve (8), is represented by a line.

The case (I) corresponds to the situation when the straight line does not cross the rectangular domain Ω in the plane (ε, μ) and, moreover, is above that domain. The case (II) corresponds to the case when the straight line crosses the variation domain Ω and is passing through the segments AB and BC. The case (III) is associated with

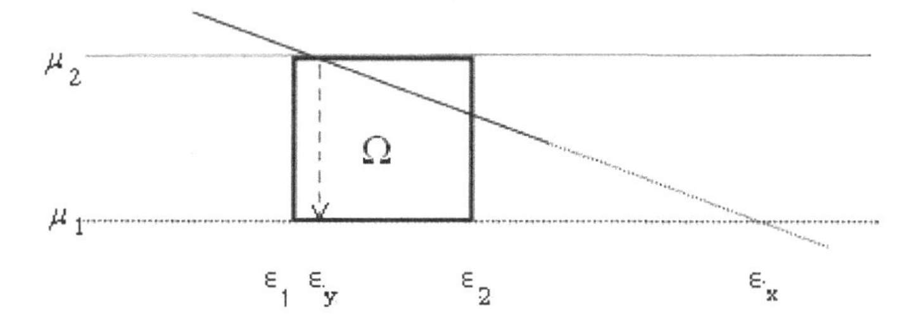

Fig. 5.3. Parameters for the calculation of reliability.

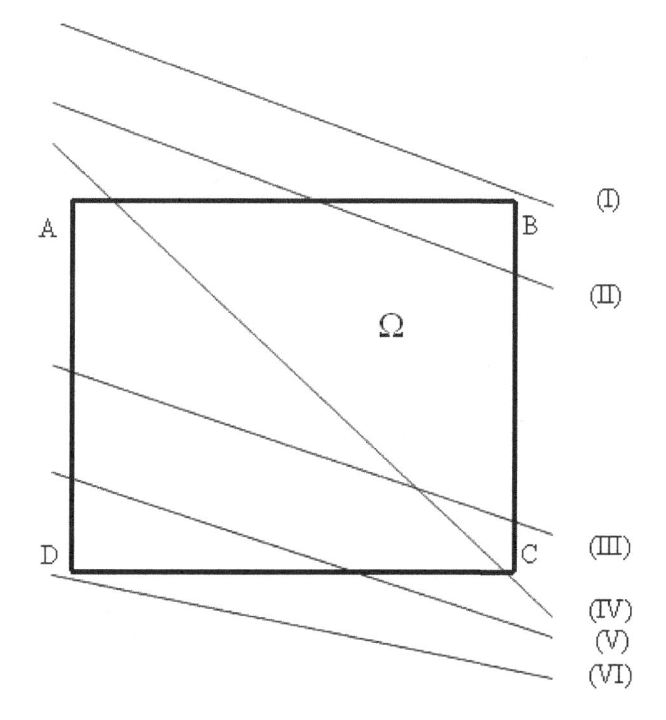

Fig. 5.4. Different cases for the evaluation of reliability.

the case when straight line crossing the variation domain Ω and the segments AD and BC. The case (IV) corresponds to the situation when the straight line crosses the variation domain Ω as well as the segments AB and CD. The case (V) deals with the case when the straight line crosses the variation domain Ω and the segments AD and CD. Finally, the case (VI) corresponds to the situation when the straight line does not cross the variation domain, and moreover, stays below it.

We introduce the notations

$$\varepsilon_x = 1 - p + \frac{3cp\mu_1}{2(p-1)}, \quad \varepsilon_y = 1 - p + \frac{3cp\mu_2}{2(p-1)} \tag{5.21}$$

as shown in Fig. 5.3. Note that, since $p < 1$, $\varepsilon_x > \varepsilon_y$. In case (I) i.e. $\varepsilon_x > \varepsilon_2$ and $\varepsilon_y > \varepsilon_2$, the system is absolutely reliable, i.e.,

$$\text{Case (I)} \quad R(p) = 1. \tag{5.22}$$

For case (II) i.e. $\varepsilon_x > \varepsilon_2$ and $\varepsilon_1 \leq \varepsilon_y \leq \varepsilon_2$, the reliability reads

$$R(p) = \frac{\varepsilon_2 - \varepsilon_y}{(\varepsilon_2 - \varepsilon_1)(\mu_2 - \mu_1)} \left[\frac{2}{3cp}(p-1)^2 + \frac{(\varepsilon_2 + \varepsilon_y)}{3cp}(p-1) - \mu_1 \right] + \frac{\varepsilon_y - \varepsilon_1}{\varepsilon_2 - \varepsilon_1} \tag{5.23}$$

or

$$R(p) = \frac{(\mu_2 - \mu_1)^{-1}}{(\varepsilon_2 - \varepsilon_1)} \left[\frac{p^2}{3c} + \left(\frac{2\varepsilon_2}{3c} - \frac{1}{c} - \mu_2 \right) p + \frac{1}{c} - \frac{4\varepsilon_2}{3c} + \frac{\varepsilon_2^2}{3c} + \mu_2 - \mu_1\varepsilon_2 - \varepsilon_1(\mu_2 - \mu_1) + \frac{3cp\mu_2^2}{4(p-1)} - \frac{(\varepsilon_2 - 1)^2}{3cp} \right]. \tag{5.24}$$

Case(II)

In case (III) i.e. $\varepsilon_1 \leq \varepsilon_x \leq \varepsilon_2$ and $\varepsilon_1 \leq \varepsilon_y \leq \varepsilon_2$, we obtain

$$R(p) = \frac{\varepsilon_x - \varepsilon_y}{(\varepsilon_2 - \varepsilon_1)(\mu_2 - \mu_1)} \left[\frac{2}{3cp}(p-1)^2 + \frac{(\varepsilon_x + \varepsilon_y)}{3cp}(p-1) - \mu_1 \right] + \frac{\varepsilon_y - \varepsilon_1}{\varepsilon_2 - \varepsilon_1} \tag{5.25}$$

or

$$R(p) = \frac{1}{(\varepsilon_2 - \varepsilon_1)} \left[1 - p - \varepsilon_1 + \frac{3c(\mu_2 + \mu_1)}{4} \frac{p}{1-p} \right]. \tag{5.26}$$

Case (III)

In case (IV), i.e. $\varepsilon_x > \varepsilon_2$ and $\varepsilon_y < \varepsilon_1$, we derive

$$R(p) = \frac{1}{(\mu_2 - \mu_1)} \left[\frac{2}{3cp}(p-1)^2 + \frac{(\varepsilon_2 + \varepsilon_1)}{3cp}(p-1) - \mu_1 \right] \tag{5.27}$$

or

$$R(p) = \frac{1}{(\mu_2 - \mu_1)} \left[\frac{2}{3c}p - \frac{4}{3c} + \frac{\varepsilon_2 + \varepsilon_1}{3c} - \mu_1 + \left(\frac{2}{3c} - \frac{\varepsilon_2 + \varepsilon_1}{3c} \right) \frac{1}{p} \right]. \tag{5.28}$$

Case (IV)

For case (V), i.e., $\varepsilon_1 \le \varepsilon_x \le \varepsilon_2$ and $\varepsilon_y < \varepsilon_1$, we are left with

$$R(p) = \frac{\varepsilon_x - \varepsilon_1}{(\varepsilon_2 - \varepsilon_1)(\mu_2 - \mu_1)}\left[\frac{2}{3cp}(p-1)^2 + \frac{(\varepsilon_x + \varepsilon_1)}{3cp}(p-1) - \mu_1\right] \tag{5.29}$$

or

$$R(p) = \frac{(\mu_2 - \mu_1)^{-1}}{(\varepsilon_2 - \varepsilon_1)}\left[-\frac{p^2}{3c} + \left(-\frac{2\varepsilon_1}{3c} + \frac{1}{c} + \mu_1\right)p - \frac{1}{c} + \frac{4\varepsilon_1}{3c} - \frac{\varepsilon_1^2}{3c}\right.$$
$$\left. - \mu_1 + \mu_1\varepsilon_1 - \frac{3c\mu_1^2 p}{4(p-1)} + \frac{(\varepsilon_1 - 1)^2}{3cp}\right]. \tag{5.30}$$

Case (V)

Finally, in case (VI), we get

$$\text{Case(VI)} \quad R(p) = 0. \tag{5.31}$$

We observe that the expressions ε_x and ε_y are functions of p. It appears instructive to find the conditions on p that decide which values of p correspond to above six cases. First we consider the case (I) for which $\varepsilon_x > \varepsilon_2$ and $\varepsilon_y > \varepsilon_2$. Following Eq. (5.21), we get

$$\frac{3cp\mu_1}{2(p-1)} - (p-1) > \varepsilon_2, \quad \frac{3cp\mu_2}{2(p-1)} - (p-1) > \varepsilon_2 \tag{5.32}$$

which leads to conditions

$$\frac{p^2 + \left(\varepsilon_2 - 2 - \frac{3}{2}c\mu_2\right)p + (1 - \varepsilon_2)}{p-1} < 0 \tag{5.33}$$

$$\frac{p^2 + \left(\varepsilon_2 - 2 - \frac{3}{2}c\mu_1\right)p + (1 - \varepsilon_2)}{p-1} < 0 \tag{5.34}$$

Since p belongs to the interval [0, 1], the denominators are negative. To be able to judge about the signs of the numerator in Eq. (5.33), we consider the quadratic equation

$$p^2 + \left(\varepsilon_2 - 2 - \frac{3}{2}c\mu_2\right)p + (1 - \varepsilon_2) = 0. \tag{5.35}$$

It has two roots $p_{\alpha1} < p_{\alpha2}$, which have the analytical expressions coincident with those given in Eq. (5.15). Thus there is one root, namely, $p_{\alpha1}$, which is smaller than unity and another one, $p_{\alpha2}$, which is greater than unity. According to the curves of Fig. 5.1, we can state that the numerator in Eq. (5.33)

$$p^2 + \left(\varepsilon_2 - 2 - \frac{3}{2}c\mu_2\right)p + (1 - \varepsilon_2) = (p - p_{\alpha1})(p - p_{\alpha2}) \tag{5.36}$$

is positive when p is less than $p_{\alpha 1}$ or more than $p_{\alpha 2}$. As p belongs to the interval [0, 1], the inequality in Eq. (5.33) is valid if the numerator is positive. This implies that Eq. (5.33) holds when p belongs to the interval $[0; p_{\alpha 1}]$.

In perfect analogy, we arrive at a conclusion that the condition in Eq. (5.34) is fulfilled when p belongs to the interval $[0; p_{\beta 1}]$, where

$$p_{\beta 1} = \frac{1}{2}\left[-\varepsilon_2 + 2 + \frac{3c}{2}\mu_1 - \sqrt{\left(\varepsilon_2 - \frac{3c}{2}\mu_1\right)^2 + 6c\mu_1}\right] \tag{5.37}$$

obtainable by equating the numerator in Eq. (5.34) to zero. We introduce the following convenient notation,

$$p_{ij} = \frac{1}{2}\left[-\varepsilon_i + 2 + \frac{3c}{2}\mu_j - \sqrt{\left(\varepsilon_i - \frac{3c}{2}\mu_j\right)^2 + 6c\mu_j}\right]. \tag{5.38}$$

Thus, for the case (I), $p_{\alpha 1} = p_{22}$ and $p_{\beta 1} = p_{21}$. To recapitulate, the conditions in Eq. (5.32) are satisfied for $p < p_{22}$ and $p < p_{21}$, respectively. We conclude that both conditions in Eqs. (5.33) and (5.34) are met in the interval $[0; \min(p_{22}, p_{21})]$.

Consider now the case (II): p satisfies the following three inequalities:

$$\frac{3cp\mu_1}{2(p-1)} - (p-1) > \varepsilon_2, \quad \frac{3cp\mu_2}{2(p-1)} - (p-1) \geq \varepsilon_1,$$
$$\frac{3cp\mu_2}{2(p-1)} - (p-1) \leq \varepsilon_2. \tag{5.39}$$

In the perfect analogy with the above reasoning we conclude that these inequalities are met if p belongs to the interval $[p_{22}, \min(p_{21}, p_{12})]$. In the case (III) p ought to belong to the interval $[\max(p_{21}, p_{22}), \min(p_{11}, p_{12})]$, whereas in case (IV) it should lie in the interval $[p_{12}, p_{21}]$. Note that the cases (III) and (IV) are incompatible with each other: $p_{12} > p_{21}$ when $\varepsilon_2 + \varepsilon_1 > 3/2c(\mu_2 + \mu_1)$; therefore, depending on the values of p_{21} and p_{12}, we have either case (III) or (IV). In the case (V) p should belong to the interval $[\max(p_{12}, p_{21}), p_{11}]$. In the final, sixth case the inequality $p > \max(p_{11}, p_{12})$ should be satisfied.

4. Numerical Examples

We consider two possible sets of values for the parameters ε_1, ε_2, μ_1 and μ_2. The first set involves the parameter set in which p_{21} is greater than p_{12}; the second set deals with the parameter set in which p_{12} is greater than p_{21}; note that the equality $p_{12} = p_{21}$ does not take place because it would correspond to $\varepsilon_1 = \varepsilon_2$ and $\mu_1 = \mu_2$, contrary to Eqs. (5.17) and (5.18).

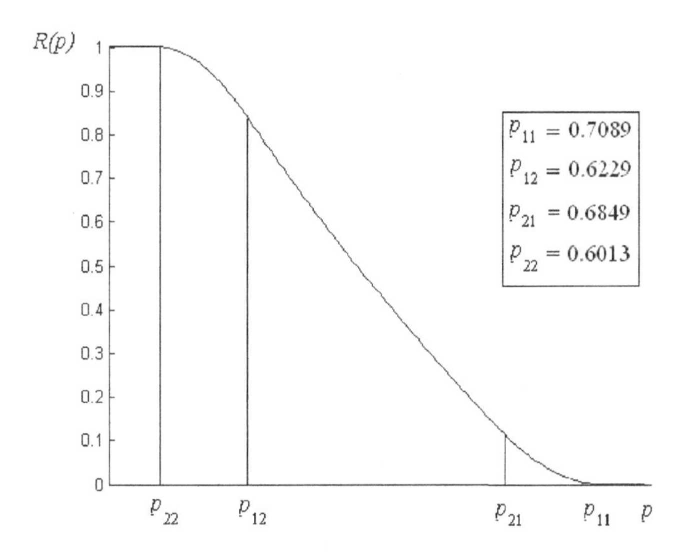

Fig. 5.5. Variation of reliability as a function of p.

The variation of reliability versus p for this first choice of parameters is shown in Fig. 5.5. Here parameters are fixed at $\varepsilon_1 = 0.05$, $\varepsilon_2 = 0.1$, $\mu_1 = 0.04$, $\mu_2 = 0.08$, $\nu = 0.3$, $c = 1.65$. For these values, $p_{11} = 0.7089$, $p_{12} = 0.6229$, $p_{21} = 0.6849$ and $p_{22} = 0.6013$. Thus p_{12} is less than p_{21}. Therefore, the analytical expressions for $R(p)$ are as follows.

For $p < 0.6013$, we get from Eq. (5.22)

$$R(p) = 1. \tag{5.40}$$

For $0.6013 \leq p < 0.6229$, Eq. (5.24) yields

$$R(p) \approx 101.01\, p^2 - 322.83p + 300.64 + 3.96\frac{p}{p-1} - \frac{81.82}{p}. \tag{5.41}$$

For $0.6229 \leq p < 0.6849$, Eq. (5.28) leads to

$$R(p) \approx 10.10p - 20.44 + \frac{9.34}{p}. \tag{5.42}$$

For $0.6849 \leq p < 0.7089$, Eq. (5.30) results in

$$R(p) = -101.01p^2 + 312.93p - 302.08 - 0.99\frac{p}{(p-1)} + \frac{91.16}{p}. \tag{5.43}$$

Finally, for $0.7089 \leq p < 1$, we get from Eq. (5.31)

$$R(p) = 0. \tag{5.44}$$

The variation of $R(p)$ versus p is depicted in Fig. 5.5.

Table 5.1. Comparison between the theoretical and the simulated reliabilities, for $\varepsilon_1 = 0.05$, $\varepsilon_2 = 0.1$, $\mu_1 = 0.04$ and $\mu_2 = 0.08$.

Case	I	II	IV	V	VI
p	0.6	0.61	0.64	0.69	0.71
R_{TH}	1	0.9750	0.6193	0.0713	0
R_{MC}	1	0.9746	0.6194	0.0705	0
Cov	na	1.97%	0.40%	0.09%	0%

The crude Monte Carlo simulations have also been conducted to check the compatibility between the reliabilities obtained by the theoretical and the simulation evaluations; theoretical reliability is denoted R_{TH}, whereas the experimental reliability is indicated by R_{MC}. The sample of 10^5 shells was considered. Table 5.1 lists nearly coincident results between the theoretical and simulated values.

Let us now consider the second choice of parameter set in which $p_{21} < p_{12}$: $\varepsilon_1 = 0.01$, $\varepsilon_2 = 0.2$, $\mu_1 = 0.01$ and $\mu_2 = 0.02$. The case (IV) is not actualized in these circumstances. The p_{ij} values are $p_{11} = 0.8499$, $p_{12} = 0.7964$, $p_{21} = 0.7323$, and $p_{22} = 0.6899$. The analytical expressions for $R(p)$ are listed below.

For $p < 0.6899$, Eq. (5.22) becomes

$$R(p) = 1. \tag{5.45}$$

For $0.6899 \leq p < 0.7323$, Eq. (5.24) results in

$$R(p) \approx 106.32p^2 - 286.97p + 247.59 + 2.6 \cdot 10^{-1}\frac{p}{p-1} - \frac{68.05}{p}. \tag{5.46}$$

For $0.7323 \leq p < 0.7964$, Eq. (5.26) leads to

$$R(p) \approx -5.26p - 5.21 + 1.95 \cdot 10^{-1}\frac{p}{p-1}. \tag{5.47}$$

For $0.7964 \leq p < 0.8499$, Eq. (5.30) yields

$$R(p) \approx 106.33p^2 + 322.12p - 319.95 + 6.5 \cdot 10^{-2}\frac{p}{p-1} - \frac{104.21}{p}. \tag{5.48}$$

Finally for $0.8499 \leq p < 1$, one obtains

$$R(p) = 0 \tag{5.49}$$

from the Eq. (5.31). The theoretical dependence of $R(p)$ is represented in Fig. 5.6.

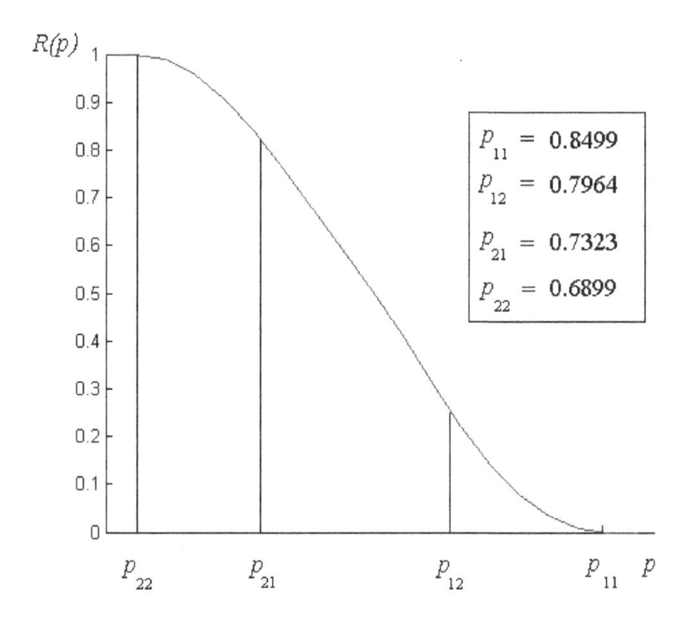

Fig. 5.6. Variation of reliability as a function of p, for $\varepsilon_1 = 0.01$, $\varepsilon_2 = 0.2$, $\mu_1 = 0.01$ and $\mu_2 = 0.02$.

Table 5.2. Comparison between the theoretical and the simulated reliabilities, for $\varepsilon_1 = 0.01$, $\varepsilon_2 = 0.2$, $\mu_1 = 0.01$ and $\mu_2 = 0.02$.

Case	I	II	III	V	VI
p	0.68	0.71	0.76	0.83	0.85
R_{TH}	1	0.9578	0.5918	0.0334	0
R_{MC}	1	0.9587	0.5929	0.0322	0
Cov	na	1.51%	0.38%	0.06%	0%

The results of the crude Monte Carlo simulations are summarized in Table 5.2, which lists nearly coincident results between theoretical and simulation evaluations.

5. Combined Randomness in Imperfection, Thickness Variation and Load

In the following section we consider the case where p is also a random variable, in addition to the initial imperfection μ and the thickness variation ε.

We take a probability density $f_p(p)$ for the load p as follows:

$$f_P(p) = \begin{cases} 0, & \text{for } p \leq p_1 \quad \text{or} \quad p \geq p_2 \\ (p_2 - p_1)^{-1}, & \text{for } p_1 < p < p_2 \end{cases}. \tag{5.50}$$

Reliability is evaluated by the formula

$$R = \int_{-\infty}^{+\infty} [1 - F_\Lambda(p)] f_P(p) \mathrm{d}p \tag{5.51}$$

or, since P varies in the interval $[p_1, p_2]$,

$$R = \int_{p_1}^{p_2} [1 - F_\Lambda(p)] f_P(p) \mathrm{d}p = \frac{1}{p_2 - p_1} \int_{p_1}^{p_2} R(p) \mathrm{d}p. \tag{5.52}$$

To evaluate the reliability, consider first the case $p_{12} < p_2$. There are five intervals in which the load P can take its possible values: $[0; \min(p_{22}, p_{21})]$, $[p_{22}; \min(p_{21}, p_{12})]$, $[p_{12}; p_{21}]$, $[\max(p_{12}, p_{21}); p_{11}]$ and $[\max(p_{11}, p_{12}); 1]$. The value p_1 belongs to one of the five above intervals; therefore, p_2 belongs to either the same interval as p_1 or to an interval with greater values of P. Thus, for each value of p_1, there are one to five possible intervals for the choice of the p_2 value. The number of possible cases for different cases for the positions of p_1 and p_2 is thus the sum of an arithmetic sequence with common difference of unity, i.e., $5(5+1)/2 = 15$. The positions of p_1 and p_2 are represented by circle (●) and star (★), respectively, in Fig. 5.7.

For $p_j \in [0, \min(p_{21}, p_{22})]$, we have

$$R = \frac{1}{p_2 - p_1} \int_{p_1}^{p_2} 1 \mathrm{d}p = 1. \tag{5.53}$$

For $p_1 \in [0, \min(p_{21}, p_{22})]$ and $p_2 \in [p_{22}, \min(p_{21}, p_{12})]$, we find

$$\begin{aligned} R = \frac{p_{22} - p_1}{p_2 - p_1} + \frac{1}{\Delta(p_2 - p_1)} \Bigg[& \frac{p_2^3 - p_{22}^3}{9c} + A(p_2^2 - p_{22}^2) + B(p_2 - p_{22}) \\ & - \frac{(\varepsilon_2 - 1)^2}{3c} \ln \frac{p_2}{p_{22}} + \frac{3c\mu_2^2}{4} \ln \frac{p_2 - 1}{p_{22} - 1} \Bigg] \end{aligned} \tag{5.54}$$

where

$$A = \left(\frac{2\varepsilon_2}{3c} - \frac{1}{c} - \mu_2 \right) \Big/ 2 \quad \text{and}$$

$$B = \frac{1}{c} + \varepsilon_2 \frac{\varepsilon_2 - 4}{3c} + \mu_2 - \mu_1 \varepsilon_2 - \varepsilon_1(\mu_2 - \mu_1) + \frac{3c\mu_2^2}{4}. \tag{5.55}$$

For $p_1 \in [0, \min(p_{21}, p_{22})]$ and $p_2 \in [p_{12}, p_{21}]$, we get

$$\begin{aligned} R = & \frac{p_{22} - p_1}{p_2 - p_1} \\ & + \frac{1}{\Delta(p_2 - p_1)} \left[\frac{p_{12}^3 - p_{22}^3}{9c} + A(p_{12}^2 - p_{22}^2) + B(p_{12} - p_{22}) \right. \end{aligned}$$

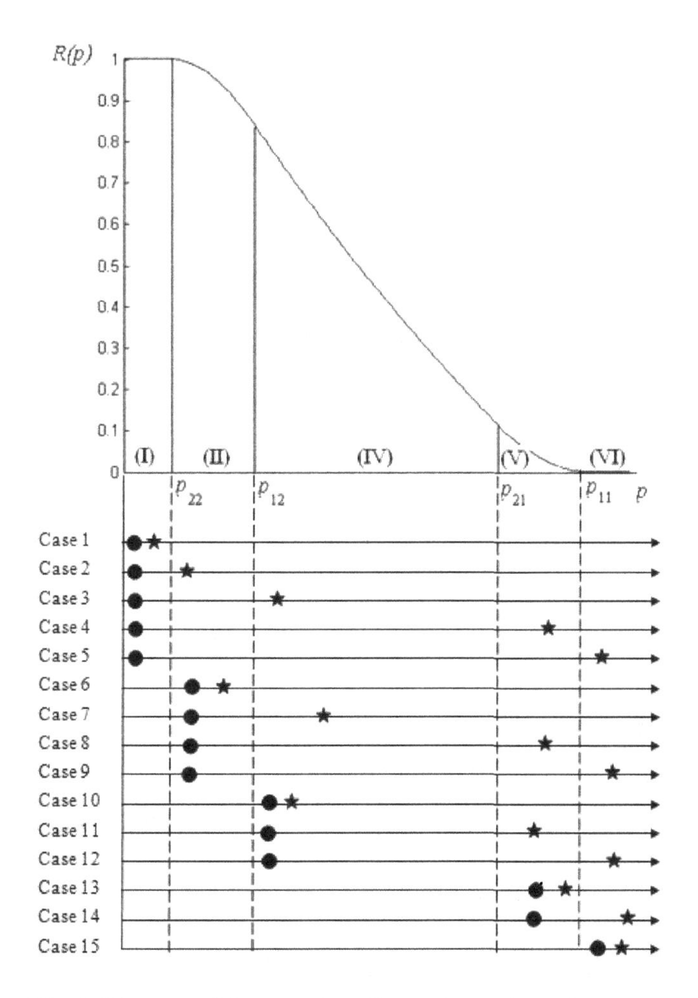

Fig. 5.7. Different possible cases of positions of p_1 and p_2 when $p_{12} < p_{21}$.

$$
\left. - \frac{(\varepsilon_2 - 1)^2}{3c} \ln \frac{p_{12}}{p_{22}} + \frac{3c\mu_2^2}{4} \ln \frac{p_{12} - 1}{p_{22} - 1} \right]
$$

$$
+ \frac{1}{(p_2 - p_1)(\mu_2 - \mu_1)} \left[\frac{p_2^2 - p_{12}^2}{3c} + D(p_2 - p_{12}) + \frac{2 - (\varepsilon_2 + \varepsilon_1)}{3c} \ln \frac{p_2}{p_{12}} \right] \tag{5.56}
$$

where

$$
D = \frac{\varepsilon_2 + \varepsilon_1 - 4}{3c} - \mu_1. \tag{5.57}
$$

For $p_1 \in [0, \min(p_{21}, p_{22})]$ and $p_2 \in [\max(p_{12}, p_{21}), p_{11}]$, we obtain

$$
R = \frac{p_{22} - p_1}{p_2 - p_1}
$$

$$+\frac{1}{\Delta(p_2-p_1)}\left[\frac{p_{12}^3-p_{22}^3}{9c}+A(p_{12}^2-p_{22}^2)+B(p_{12}-p_{22})\right.$$

$$\left.-\frac{(\varepsilon_2-1)^2}{3c}\ln\frac{p_{12}}{p_{22}}+\frac{3c\mu_2^2}{4}\ln\frac{p_{12}-1}{p_{22}-1}\right]$$

$$+\frac{1}{(p_2-p_1)(\mu_2-\mu_1)}\left[\frac{p_{21}^2-p_{12}^2}{3c}+D(p_{21}-p_{12})+\frac{2-(\varepsilon_2+\varepsilon_1)}{3c}\ln\frac{p_{21}}{p_{12}}\right]$$

$$+\frac{1}{\Delta(p_2-p_1)}\left[-\frac{p_2^3-p_{21}^3}{9c}+E(p_2^2-p_{21}^2)+F(p_2-p_{21})\right.$$

$$\left.+\frac{(\varepsilon_1-1)^2}{3c}\ln\frac{p_2}{p_{21}}-\frac{3c\mu_1^2}{4}\ln\frac{p_2-1}{p_{21}-1}\right] \tag{5.58}$$

where

$$E=\left(-\frac{2\varepsilon_1}{3c}+\frac{1}{c}+\mu_1\right)\Big/2 \quad \text{and}$$

$$F=-\frac{1}{c}-\varepsilon_1\frac{\varepsilon_1-4}{3c}-\mu_1+\mu_1\varepsilon_1-\frac{3c\mu_1^2}{4}. \tag{5.59}$$

For $p_1 \in [0, \min(p_{21}, p_{22})]$ and $p_2 \in [\max(p_{11}, p_{12}), 1]$, we arrive at

$$R=\frac{p_{22}-p_1}{p_2-p_1}$$

$$+\frac{1}{\Delta(p_2-p_1)}\left[\frac{p_{12}^3-p_{22}^3}{9c}+A(p_{12}^2-p_{22}^2)+B(p_{12}-p_{22})\right.$$

$$\left.-\frac{(\varepsilon_2-1)^2}{3c}\ln\frac{p_{12}}{p_{22}}+\frac{3c\mu_2^2}{4}\ln\frac{p_{12}-1}{p_{22}-1}\right]$$

$$+\frac{1}{(p_2-p_1)(\mu_2-\mu_1)}\left[\frac{p_{21}^2-p_{12}^2}{3c}+D(p_{21}-p_{12})+\frac{2-(\varepsilon_2+\varepsilon_1)}{3c}\ln\frac{p_{21}}{p_{12}}\right]$$

$$+\frac{1}{\Delta(p_2-p_1)}\left[-\frac{p_{11}^3-p_{21}^3}{9c}+E(p_{11}^2-p_{21}^2)+F(p_{11}-p_{21})\right.$$

$$\left.+\frac{(\varepsilon_1-1)^2}{3c}\ln\frac{p_{11}}{p_{21}}-\frac{3c\mu_1^2}{4}\ln\frac{p_{11}-1}{p_{21}-1}\right]. \tag{5.60}$$

For $p_j \in [p_{22}, \min(p_{21}, p_{12})]$, the reliability is

$$R=\frac{1}{\Delta(p_2-p_1)}\left[\frac{p_2^3-p_1^3}{9c}+A(p_2^2-p_1^2)+B(p_2-p_1)\right.$$

$$\left.-\frac{(\varepsilon_2-1)^2}{3c}\ln\frac{p_2}{p_1}+\frac{3c\mu_2^2}{4}\ln\frac{p_2-1}{p_1-1}\right]. \tag{5.61}$$

For $p_1 \in [p_{22}, \min(p_{21}, p_{12})]$ and $p_2 \in [p_{12}, p_{21}]$, we have

$$R = \frac{1}{\Delta(p_2 - p_1)} \left[\frac{p_{22}^3 - p_1^3}{9c} + A(p_{22}^2 - p_1^2) + B(p_{22} - p_1) \right.$$
$$\left. - \frac{(\varepsilon_2 - 1)^2}{3c} \ln \frac{p_{22}}{p_1} + \frac{3c\mu_2^2}{4} \ln \frac{p_{22} - 1}{p_1 - 1} \right]$$
$$+ \frac{1}{(p_2 - p_1)(\mu_2 - \mu_1)} \left[\frac{p_2^2 - p_{22}^2}{3c} + D(p_2 - p_{22}) + \frac{2 - (\varepsilon_2 + \varepsilon_1)}{3c} \ln \frac{p_2}{p_{22}} \right]. \tag{5.62}$$

For $p_1 \in [p_{22}, \min(p_{21}, p_{12})]$ and $p_2 \in [\max(p_{12}, p_{21}), p_{11}]$, the reliability is

$$R = \frac{1}{\Delta(p_2 - p_1)} \left[\frac{p_{12}^3 - p_1^3}{9c} + A(p_{12}^2 - p_1^2) + B(p_{12} - p_1) \right.$$
$$\left. - \frac{(\varepsilon_2 - 1)^2}{3c} \ln \frac{p_{12}}{p_1} + \frac{3c\mu_2^2}{4} \ln \frac{p_{12} - 1}{p_1 - 1} \right]$$
$$+ \frac{1}{(p_2 - p_1)(\mu_2 - \mu_1)} \left[\frac{p_{21}^2 - p_{12}^2}{3c} + D(p_{21} - p_{12}) + \frac{2 - (\varepsilon_2 + \varepsilon_1)}{3c} \ln \frac{p_{21}}{p_{12}} \right]$$
$$+ \frac{1}{\Delta(p_2 - p_1)} \left[-\frac{p_2^3 - p_{21}^3}{9c} + E(p_2^2 - p_{21}^2) + F(p_2 - p_{21}) \right.$$
$$\left. + \frac{(\varepsilon_1 - 1)^2}{3c} \ln \frac{p_2}{p_{21}} - \frac{3c\mu_1^2}{4} \ln \frac{p_2 - 1}{p_{21} - 1} \right]. \tag{5.63}$$

For $p_1 \in [p_{22}, \min(p_{21}, p_{12})]$ and $p_2 \in [\max(p_{11}, p_{12}), 1]$, we obtain

$$R = \frac{1}{\Delta(p_2 - p_1)} \left[\frac{p_{12}^3 - p_1^3}{9c} + A(p_{12}^2 - p_1^2) + B(p_{12} - p_1) \right.$$
$$\left. - \frac{(\varepsilon_2 - 1)^2}{3c} \ln \frac{p_{12}}{p_1} + \frac{3c\mu_2^2}{4} \ln \frac{p_{12} - 1}{p_1 - 1} \right]$$
$$+ \frac{1}{(p_2 - p_1)(\mu_2 - \mu_1)} \left[\frac{p_{21}^2 - p_{12}^2}{3c} + D(p_{21} - p_{12}) + \frac{2 - (\varepsilon_2 + \varepsilon_1)}{3c} \ln \frac{p_{21}}{p_{12}} \right]$$
$$+ \frac{1}{\Delta(p_2 - p_1)} \left[-\frac{p_{11}^3 - p_{21}^3}{9c} + E(p_{11}^2 - p_{21}^2) + F(p_{11} - p_{21}) \right.$$
$$\left. + \frac{(\varepsilon_1 - 1)^2}{3c} \ln \frac{p_{11}}{p_{21}} - \frac{3c\mu_1^2}{4} \ln \frac{p_{11} - 1}{p_{21} - 1} \right]. \tag{5.64}$$

For $p_j \in [p_{12}, p_{21}]$, we have

$$R = \frac{1}{(p_2 - p_1)(\mu_2 - \mu_1)} \left[\frac{p_2^2 - p_1^2}{3c} + D(p_2 - p_1) + \frac{2 - (\varepsilon_2 + \varepsilon_1)}{3c} \ln \frac{p_2}{p_1} \right]. \tag{5.65}$$

For $p_1 \in [p_{12}, p_{21}]$ and $p_2 \in [\max(p_{12}, p_{21}), p_{11}]$, we obtain

$$\begin{aligned} R = &\frac{1}{(p_2 - p_1)(\mu_2 - \mu_1)} \left[\frac{p_{21}^2 - p_1^2}{3c} + D(p_{21} - p_1) + \frac{2 - (\varepsilon_2 + \varepsilon_1)}{3c} \ln \frac{p_{21}}{p_1} \right] \\ &+ \frac{1}{\Delta(p_2 - p_1)} \left[-\frac{p_2^3 - p_{21}^3}{9c} + E(p_2^2 - p_{21}^2) + F(p_2 - p_{21}) \right. \\ &\left. + \frac{(\varepsilon_1 - 1)^2}{3c} \ln \frac{p_2}{p_{21}} - \frac{3c\mu_1^2}{4} \ln \frac{p_2 - 1}{p_{21} - 1} \right]. \end{aligned} \tag{5.66}$$

For $p_1 \in [p_{12}, p_{21}]$ and $p_2 \in [\max(p_{11}, p_{12}), 1]$, we have

$$\begin{aligned} R = &\frac{1}{(p_2 - p_1)(\mu_2 - \mu_1)} \left[\frac{p_{21}^2 - p_1^2}{3c} + D(p_{21} - p_1) + \frac{2 - (\varepsilon_2 + \varepsilon_1)}{3c} \ln \frac{p_{21}}{p_1} \right] \\ &+ \frac{1}{\Delta(p_2 - p_1)} \left[-\frac{p_{11}^3 - p_{21}^3}{9c} + E(p_{11}^2 - p_{21}^2) + F(p_{11} - p_{21}) \right. \\ &\left. + \frac{(\varepsilon_1 - 1)^2}{3c} \ln \frac{p_{11}}{p_{21}} - \frac{3c\mu_1^2}{4} \ln \frac{p_{11} - 1}{p_{21} - 1} \right]. \end{aligned} \tag{5.67}$$

For $p_j \in [\max(p_{12}, p_{21}), p_{11}]$, we get

$$\begin{aligned} R = &\frac{1}{\Delta(p_2 - p_1)} \left[-\frac{p_2^3 - p_1^3}{9c} + E(p_2^2 - p_1^2) + F(p_2 - p_1) \right. \\ &\left. + \frac{(\varepsilon_1 - 1)^2}{3c} \ln \frac{p_2}{p_1} - \frac{3c\mu_1^2}{4} \ln \frac{p_2 - 1}{p_1 - 1} \right]. \end{aligned} \tag{5.68}$$

For $p_1 \in [\max(p_{12}, p_{21}), p_{11}]$ and $p_2 \in [\max(p_{11}, p_{12}), 1]$, we obtain

$$\begin{aligned} R = &\frac{1}{\Delta(p_2 - p_1)} \left[-\frac{p_{11}^3 - p_1^3}{9c} + E(p_{11}^2 - p_1^2) + F(p_{11} - p_1) \right. \\ &\left. + \frac{(\varepsilon_1 - 1)^2}{3c} \ln \frac{p_{11}}{p_1} - \frac{3c\mu_1^2}{4} \ln \frac{p_{11} - 1}{p_1 - 1} \right]. \end{aligned} \tag{5.69}$$

For $p_j \in [\max(p_{11}, p_{12}), 1]$, finally we obtain $R = 0$.

The above expressions for reliability pertain to the case $p_{12} < p_{21}$. We consider now the case $p_{21} \leq p_{12}$. The positions of p_1 and p_2 are represented by circle ● and star ★, respectively, in Fig. 5.8.

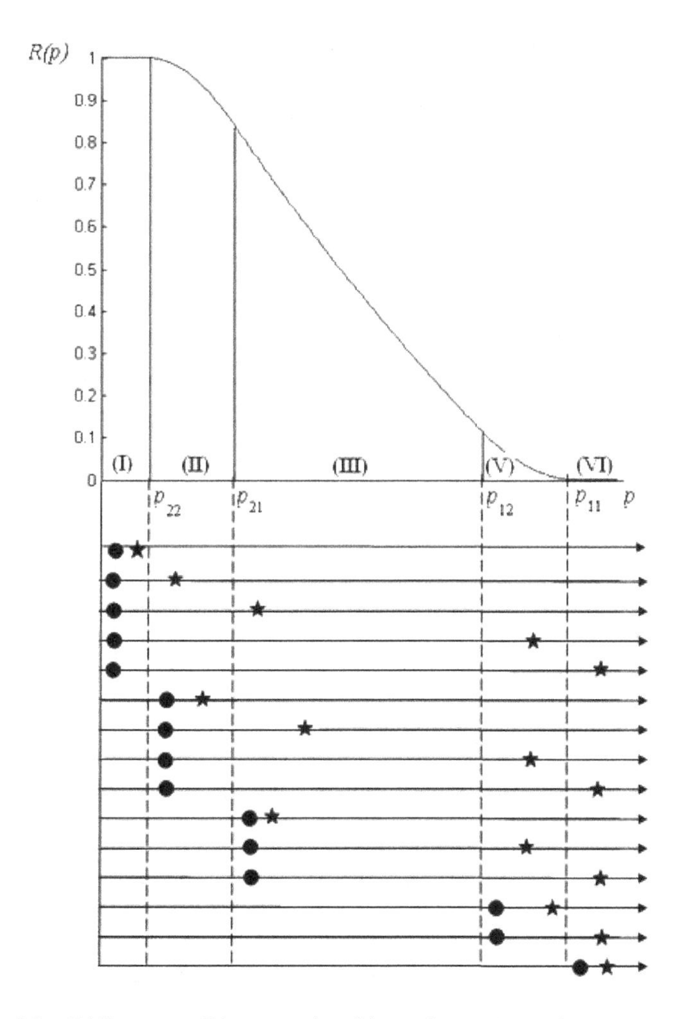

Fig. 5.8. Different possible cases of positions of p_1 and p_2 when $p_{12} > p_{21}$.

For $p_j \in [0, \min(p_{21}, p_{22})]$,

$$R = \frac{1}{p_2 - p_1} \int_{p_1}^{p_2} 1 \mathrm{d}p = 1. \tag{5.70}$$

For $p_1 \in [0, \min(p_{21}, p_{22})]$ and $p_2 \in [p_{22}, \min(p_{21}, p_{12})]$, we find

$$\begin{aligned} R = {} & \frac{p_{22} - p_1}{p_2 - p_1} \\ & + \frac{1}{\Delta(p_2 - p_1)} \left[\frac{p_2^3 - p_{22}^3}{9c} + A(p_2^2 - p_{22}^2) + B(p_2 - p_{22}) \right. \\ & \left. + \frac{(\varepsilon_2 - 1)^2}{3c} \ln \frac{p_2}{p_{22}} + \frac{3c\mu_2^2}{4} \ln \frac{p_2 - 1}{p_{22} - 1} \right]. \end{aligned} \tag{5.71}$$

For $p_1 \in [0, \min(p_{21}, p_{22})]$ and $p_2 \in [p_{21}, p_{12}]$, we get

$$\begin{aligned}
R = {} & \frac{p_{22} - p_1}{p_2 - p_1} \\
& + \frac{1}{\Delta(p_2 - p_1)} \left[\frac{p_{21}^3 - p_{22}^3}{9c} + A(p_{21}^2 - p_{22}^2) + B(p_{21} - p_{22}) \right. \\
& \left. - \frac{(\varepsilon_2 - 1)^2}{3c} \ln \frac{p_{21}}{p_{22}} + \frac{3c\mu_2^2}{4} \ln \frac{p_{21} - 1}{p_{22} - 1} \right] \\
& + \frac{1}{(p_2 - p_1)(\varepsilon_2 - \varepsilon_1)} \left[-\frac{p_2^2 - p_{21}^2}{3c} + C(p_2 - p_{21}) \right. \\
& \left. - \frac{3c(\mu_2 + \mu_1)}{4} \ln \frac{p_2 - 1}{p_{21} - 1} \right]
\end{aligned} \tag{5.72}$$

where

$$C = 1 - \varepsilon_1 - \frac{3c(\mu_2 + \mu_1)}{4}. \tag{5.73}$$

For $p_1 \in [0, \min(p_{21}, p_{22})]$ and $p_2 \in [\max(p_{12}, p_{21}), p_{11}]$, we find

$$\begin{aligned}
R = {} & \frac{p_{22} - p_1}{p_2 - p_1} \\
& + \frac{1}{\Delta(p_2 - p_1)} \left[\frac{p_{21}^3 - p_{22}^3}{9c} + A(p_{21}^2 - p_{22}^2) + B(p_{21} - p_{22}) \right. \\
& \left. - \frac{(\varepsilon_2 - 1)^2}{3c} \ln \frac{p_{21}}{p_{22}} + \frac{3c\mu_2^2}{4} \ln \frac{p_{21} - 1}{p_{22} - 1} \right] \\
& + \frac{1}{(p_2 - p_1)(\varepsilon_2 - \varepsilon_1)} \left[-\frac{p_{12}^2 - p_{21}^2}{3c} + C(p_{12} - p_{21}) \right. \\
& \left. - \frac{3c(\mu_2 + \mu_1)}{4} \ln \frac{p_{12} - 1}{p_{21} - 1} \right] \\
& + \frac{1}{\Delta(p_2 - p_1)} \left[-\frac{p_2^3 - p_{12}^3}{9c} + E(p_2^2 - p_{12}^2) + F(p_2 - p_{12}) \right. \\
& \left. + \frac{(\varepsilon_1 - 1)^2}{3c} \ln \frac{p_2}{p_{12}} - \frac{3c\mu_1^2}{4} \ln \frac{p_2 - 1}{p_{12} - 1} \right].
\end{aligned} \tag{5.74}$$

For $p_1 \in [0, \min(p_{21}, p_{22})]$ and $p_2 \in [\max(p_{11}, p_{12}), 1]$, we find

$$\begin{aligned}
R = {} & \frac{p_{22} - p_1}{p_2 - p_1} \\
& + \frac{1}{\Delta(p_2 - p_1)} \left[\frac{p_{21}^3 - p_{22}^3}{9c} + A(p_{21}^2 - p_{22}^2) + B(p_{21} - p_{22}) \right.
\end{aligned}$$

$$- \frac{(\varepsilon_2 - 1)^2}{3c} \ln \frac{p_{21}}{p_{22}} + \frac{3c\mu_2^2}{4} \ln \frac{p_{21} - 1}{p_{22} - 1}\Bigg]$$

$$+ \frac{1}{(p_2 - p_1)(\varepsilon_2 - \varepsilon_1)} \Bigg[-\frac{p_{12}^2 - p_{21}^2}{3c} + C(p_{12} - p_{21})$$

$$- \frac{3c(\mu_2 + \mu_1)}{4} \ln \frac{p_{12} - 1}{p_{21} - 1}\Bigg]$$

$$+ \frac{1}{\Delta(p_2 - p_1)} \Bigg[-\frac{p_{11}^3 - p_{12}^3}{9c} + E(p_{11}^2 - p_{12}^2) + F(p_{11} - p_{12})$$

$$+ \frac{(\varepsilon_1 - 1)^2}{3c} \ln \frac{p_{11}}{p_{12}} - \frac{3c\mu_1^2}{4} \ln \frac{p_{11} - 1}{p_{12} - 1}\Bigg]. \tag{5.75}$$

For $p_j \in [p_{22}, \min(p_{21}, p_{12})]$, the reliability is

$$R = \frac{1}{\Delta(p_2 - p_1)} \Bigg[\frac{p_2^3 - p_1^3}{9c} + A(p_2^2 - p_1^2) + B(p_2 - p_1)$$

$$- \frac{(\varepsilon_2 - 1)^2}{3c} \ln \frac{p_2}{p_1} + \frac{3c\mu_2^2}{4} \ln \frac{p_2 - 1}{p_1 - 1}\Bigg]. \tag{5.76}$$

For $p_1 \in [p_{22}, \min(p_{21}, p_{12})]$ and $p_2 \in [p_{21}, p_{12}]$, we have

$$R = \frac{1}{\Delta(p_2 - p_1)} \Bigg[\frac{p_{21}^3 - p_1^3}{9c} + A(p_{21}^2 - p_1^2) + B(p_{21} - p_1)$$

$$- \frac{(\varepsilon_2 - 1)^2}{3c} \ln \frac{p_{21}}{p_1} + \frac{3c\mu_2^2}{4} \ln \frac{p_{21} - 1}{p_1 - 1}\Bigg]$$

$$+ \frac{1}{(p_2 - p_1)(\varepsilon_2 - \varepsilon_1)} \Bigg[-\frac{p_2^2 - p_{21}^2}{3c} + C(p_2 - p_{21})$$

$$- \frac{3c(\mu_2 + \mu_1)}{4} \ln \frac{p_2 - 1}{p_{21} - 1}\Bigg]. \tag{5.77}$$

For $p_1 \in [p_{22}, \min(p_{21}, p_{12})]$ and $p_2 \in [\max(p_{12}, p_{21}), p_{11}]$, the reliability is

$$R = \frac{1}{\Delta(p_2 - p_1)} \Bigg[\frac{p_{21}^3 - p_1^3}{9c} + A(p_{21}^2 - p_1^2) + B(p_{21} - p_1)$$

$$- \frac{(\varepsilon_2 - 1)^2}{3c} \ln \frac{p_{21}}{p_1} + \frac{3c\mu_2^2}{4} \ln \frac{p_{21} - 1}{p_1 - 1}\Bigg]$$

$$+\frac{1}{(p_2-p_1)(\varepsilon_2-\varepsilon_1)}\left[-\frac{p_{12}^2-p_{21}^2}{3c}+C(p_{12}-p_{21})\right.$$
$$\left.-\frac{3c(\mu_2+\mu_1)}{4}\ln\frac{p_{12}-1}{p_{21}-1}\right]$$
$$+\frac{1}{\Delta(p_2-p_1)}\left[-\frac{p_2^3-p_{12}^3}{9c}+E(p_2^2-p_{12}^2)+F(p_2-p_{12})\right.$$
$$\left.+\frac{(\varepsilon_1-1)^2}{3c}\ln\frac{p_2}{p_{12}}-\frac{3c\mu_1^2}{4}\ln\frac{p_2-1}{p_{12}-1}\right]. \tag{5.78}$$

For $p_1 \in [p_{22}, \min(p_{21}, p_{12})]$ and $p_2 \in [\max(p_{11}, p_{12}), 1]$, we obtain

$$R=\frac{1}{\Delta(p_2-p_1)}\left[\frac{p_{21}^3-p_1^3}{9c}+A(p_{21}^2-p_1^2)+B(p_{21}-p_1)\right.$$
$$\left.-\frac{(\varepsilon_2-1)^2}{3c}\ln\frac{p_{21}}{p_1}+\frac{3c\mu_2^2}{4}\ln\frac{p_{21}-1}{p_1-1}\right]$$
$$+\frac{1}{(p_2-p_1)(\varepsilon_2-\varepsilon_1)}\left[-\frac{p_{12}^2-p_{21}^2}{3c}+C(p_{12}-p_{21})\right.$$
$$\left.-\frac{3c(\mu_2+\mu_1)}{4}\ln\frac{p_{12}-1}{p_{21}-1}\right]$$
$$+\frac{1}{\Delta(p_2-p_1)}\left[-\frac{p_{11}^3-p_{12}^3}{9c}+E(p_{11}^2-p_{12}^2)+F(p_{11}-p_{12})\right.$$
$$\left.+\frac{(\varepsilon_1-1)^2}{3c}\ln\frac{p_{11}}{p_{12}}-\frac{3c\mu_1^2}{4}\ln\frac{p_{11}-1}{p_{12}-1}\right]. \tag{5.79}$$

For $p_j \in [p_{21}, p_{12}]$, we have

$$R=\frac{1}{(p_2-p_1)(\varepsilon_2-\varepsilon_1)}\left[-\frac{p_2^2-p_1^2}{3c}+C(p_2-p_1)-\frac{3c(\mu_2+\mu_1)}{4}\ln\frac{p_2-1}{p_1-1}\right]. \tag{5.80}$$

For $p_1 \in [p_{21}, p_{12}]$ and $p_2 \in [\max(p_{12}, p_{21}), p_{11}]$, we obtain

$$R=\frac{1}{(p_2-p_1)(\varepsilon_2-\varepsilon_1)}\left[-\frac{p_{12}^2-p_1^2}{3c}+C(p_{12}-p_1)\right.$$
$$\left.-\frac{3c(\mu_2+\mu_1)}{4}\ln\frac{p_{12}-1}{p_1-1}\right]$$

$$+\frac{1}{\Delta(p_2-p_1)}\left[-\frac{p_2^3-p_{12}^3}{9c}+E(p_2^2-p_{12}^2)+F(p_2-p_{12})\right.$$

$$\left.+\frac{(\varepsilon_1-1)^2}{3c}\ln\frac{p_2}{p_{12}}-\frac{3c\mu_1^2}{4}\ln\frac{p_2-1}{p_{12}-1}\right]. \tag{5.81}$$

For $p_1 \in [p_{21}, p_{12}]$ and $p_2 \in [\max(p_{11}, p_{12}), 1]$, we have

$$R=\frac{1}{(p_2-p_1)(\mu_2-\mu_1)}\left[\frac{p_{21}^2-p_1^2}{3c}+D(p_{21}-p_1)+\frac{2-(\varepsilon_2+\varepsilon_1)}{3c}\ln\frac{p_{21}}{p_1}\right]$$

$$+\frac{1}{\Delta(p_2-p_1)}\left[-\frac{p_{11}^3-p_{21}^3}{9c}+E(p_{11}^2-p_{21}^2)+F(p_{11}-p_{21})\right.$$

$$\left.+\frac{(\varepsilon_1-1)^2}{3c}\ln\frac{p_{11}}{p_{21}}-\frac{3c\mu_1^2}{4}\ln\frac{p_{11}-1}{p_{21}-1}\right]. \tag{5.82}$$

For $p_j \in [\max(p_{12}, p_{21}), p_{11}]$, we get

$$R=\frac{1}{\Delta(p_2-p_1)}\left[-\frac{p_2^3-p_1^3}{9c}+E(p_2^2-p_1^2)+F(p_2-p_1)\right.$$

$$\left.+\frac{(\varepsilon_1-1)^2}{3c}\ln\frac{p_2}{p_1}-\frac{3c\mu_1^2}{4}\ln\frac{p_2-1}{p_1-1}\right]. \tag{5.83}$$

For $p_1 \in [\max(p_{12}, p_{21}), p_{11}]$ and $p_2 \in [\max(p_{11}, p_{12}), 1]$, we obtain

$$R=\frac{1}{\Delta(p_2-p_1)}\left[-\frac{p_{11}^3-p_1^3}{9c}+E(p_{11}^2-p_1^2)+F(p_{11}-p_1)\right.$$

$$\left.+\frac{(\varepsilon_1-1)^2}{3c}\ln\frac{p_{11}}{p_1}-\frac{3c\mu_1^2}{4}\ln\frac{p_{11}-1}{p_1-1}\right]. \tag{5.84}$$

For $p_j \in [\max(p_{11}, p_{12}), 1]$, finally we obtain $R = 0$.

6. Numerical Examples

First we consider the case $p_{12} < p_{21}$; we take the following set of parameters $\varepsilon_1 = 0.05$, $\varepsilon_2 = 0.1$, $\mu_1 = 0.04$ and $\mu_2 = 0.08$. In this case $p_{11} = 0.7089$, $p_{12} = 0.6229$, $p_{21} = 0.6849$ and $p_{22} = 0.6013$. Table 5.3 lists the reliability values for the above 15 different cases. The theoretical results R_{TH} and those obtained by the crude Monte Carlo R_{MC} with 10^5 simulations are reported. The coefficient of variation $\mathrm{cov}(R)$ of the reliability is evaluated as

$$\mathrm{cov}(R)=\sqrt{\frac{R}{N(1-R)}} \tag{5.85}$$

is also reported for the crude Monte Carlo-method.

Table 5.3. Reliability values as depending upon bounds of the applied load $p_{12} < p_{21}$.

Case number	p_1	p_2	RTH	RMC	cov(R)
1	0.1	0.2	1	1	na
2	0.1	0.61	0.9999	0.9999	na
3	0.1	0.64	0.9895	0.9894	3.12%
4	0.1	0.69	0.9336	0.9330	1.19%
5	0.1	0.71	0.9037	0.9051	0.98%
6	0.61	0.62	0.9354	0.9351	1.20%
7	0.61	0.64	0.8126	0.8128	0.66%
8	0.61	0.69	0.5108	0.5100	0.32%
9	0.61	0.71	0.4131	0.4141	0.27%
10	0.64	0.65	0.5569	0.5551	0.35%
11	0.64	0.69	0.3298	0.3387	0.22%
12	0.64	0.71	0.2420	0.2425	0.18%
13	0.69	0.70	0.0401	0.0399	0.07%
14	0.69	0.71	0.0223	0.0224	0.05%
15	0.71	0.75	0	0	0%

Table 5.4. Reliability values as depending upon bounds of the applied load in the case $p_{12} > p_{21}$.

Case number	p_1	p_2	RTH	RMC	cov($\Lambda - p$)
1	0.6	0.68	1	1	na
2	0.6	0.71	0.9974	0.9974	6.17%
3	0.6	0.76	0.9329	0.9344	1.18%
4	0.6	0.83	0.7367	0.7360	0.53%
5	0.6	0.85	0.6755	0.6779	0.46%
6	0.71	0.72	0.9342	0.9343	1.20%
7	0.71	0.76	0.7912	0.7928	0.62%
8	0.71	0.83	0.4957	0.4953	0.31%
9	0.71	0.85	0.4270	0.4259	0.27%
10	0.76	0.78	0.5038	0.5037	0.32%
11	0.76	0.83	0.2825	0.2825	0.20%
12	0.76	0.85	0.2207	0.2220	0.17%
13	0.83	0.84	0.0195	0.0194	0.04%
14	0.83	0.85	0.0105	0.0110	0.03%
15	0.85	0.87	0	0	0%

The results of the crude Monte Carlo are nearly coincident to the theoretical results.

As for the second case $p_{12} > p_{21}$, we fix the parameters at $\varepsilon_1 = 0.01$, $\varepsilon_2 = 0.2$, $\mu_1 = 0.01$ and $\mu_2 = 0.02$. In this case $p_{11} = 0.8499$, $p_{12} = 0.7964$, $p_{21} = 0.7323$, $p_{22} = 0.6899$. Table 5.4 presents reliability values R_{TH} and R_{MC} for 15 different cases.

The results of the crude Monte Carlo are nearly coincident to the theoretical results.

7. Implications on Design Criteria

We deal with the design of the shell, demanding the structural reliability level to be at least r.

Consider the case where the applied load is a deterministic quantity (Sec. 3 in this chapter). The value of p corresponding to required reliability r is denoted hereinafter as p_r. The latter value p_r can be referred to as *design load*, implying that if $p \leq p_r$, then the shell reliability is at least r. Naturally p_r can belong to either of several intervals. For example, for the set of parameters $\varepsilon_1 = 0.095$, $\varepsilon_2 = 0.1$, $\mu_1 = 0.001$, $\mu_2 = 0.5$ the value p_r could belong to one of the following possible intervals: [0; 0.3194] (hereinafter referred to as the first interval), [0.3194; 0.3207] (second interval), [0.3207; 0.8816] (third interval), [0.8816; 0.8858] (fourth interval), or [0.8858, 1] (fifth interval) depending on the required reliability r. Indeed, if required reliability equals unity, p_r belongs to the first interval. For $r = 0.9999$, $p_r = 0.3196$ and thus the latter belongs to the second interval; for $r = 0.9975$, $p_r = 0.3204$ and is contained in the third interval; for $r = 0.99$, $p_r = 0.321625$ and p_r belongs to the third interval. It should be emphasized that for the deterministic applied load p less than p_r the reliability is greater than or equal to r whereas for p greater than p_r the reliability is less than or equal to r. Now we consider a more realistic case when the applied load also has a scatter and can be modeled as a random variable. Following Sec. 4 in this chapter, we consider the case when it is representable by a uniform random variable between values p_1 and p_2. The question arises if the randomness of P causes the reliability to be reduced, as it was maintained in the literature. The results of calculation are listed in Table 5.5 for various values of p_1 and p_2. The actual reliability, calculated from Eqs. (5.53), (5.54), and from (5.61) to (5.69) is denoted as R_{act}. It is instructive to compare it with the reliability that was demanded in the idealized circumstances, namely, when the load was treated as a deterministic quantity. For the random applied load with $p_1 = 0$ and $p_2 = 0.31$, the actual reliability R_{act} equals unity; this value exceeds the required reliability $r = 0.9999$ associated with the deterministically applied load. This implies that as a result of the load's randomness the reliability of the structure was increased. Now, if $p_1 = 0.3195$ and $p_2 = p_r = 0.3196$, the actual reliability equals the required reliability p_r. In this case both the deterministic and random applied leads yield the same structural reliability. If, however, $p_1 = 0.32$ $(p_1 > p_r)$ and $p_2 = 0.3206$ $(p_2 > p_r)$ then the actual reliability is $R_{\text{act}} = 0.9979$, which is less than the required reliability r. These and other cases for p_1 and p_2 are listed in Table 5.5.

Table 5.5. Variation of the reliability for $p_r = 0.3196$ belonging to the second interval, $r = 0.9999$.

p_1	0	0.25	0.3195	0.32	0.32	0.32	0.32	0.34	0.34	0.34	0.882	0.882	0.886
p_2	0.31	0.32	0.3196	0.3206	0.3232	0.882	0.886	0.36	0.882	0.886	0.883	0.886	0.888
R_{act}	1	1	0.9999	0.9979	0.9901	0.2625	0.2606	0.8294	0.2375	0.2357	0.0002	0.0001	0
	$R_{act} > r$	$R_{act} > r$	$R_{act} = r$	$R_{act} < r$	$R_{act} < r$	$R_{act} < r$	$R_{act} < r$	$R_{act} < r$	$R_{act} < r$	$R_{act} < r$	$R_{act} < r$	$R_{act} < r$	$R_{act} < r$

Table 5.5 shows that when the random load takes values in the interval contained in [0; p_r], the resulting reliability is greater than or equal to the demanded value r; on the other hand, when the random load takes values outside the interval [0; p_r], then the reliability is less than the codified value r.

8. Conclusion

This chapter deals with the effect of the combined thickness and initial imperfections in axially loaded cylindrical shell under axial compression. Two cases are considered: the applied load is treated either as a deterministic quantity or as a random variable. Special emphasis is placed on the change that occurs in the structural reliability when the applied load is also treated as a random variable. It turns out that there is no single mechanism of behavior that exists for the structural reliability; specifically, three scenarios may take place as a result of load's randomness: the reliability may increase, decrease, or stay unaffected.

Chapter 6

Lower Bound for Buckling Load in Presence of Uncertainty

Am I a circus rider on 2 horses? (Bin Ich ein circusreiter auf 2 Pferden?)

F. Kafka (1967)

To tackle such bounded-but-unknown uncertainties, a technique based on anti-optimization (a term dubbed by Elishakoff (1990) is proposed [by Elishakoff et al., 1995]. In this technique, uncertainty-based optimization is basically split in two parts, namely, main- and anti-optimization. The main optimization is treated as a standard minimization problem which searches for the best design in the design domain. The design domain is typically specified by upper- and lower bounds on design variables. The anti-optimization consists of performing numerical searches for the combination of uncertainties which yields the worst response for a given design and a particular response function. In the worst case scenario, an antioptimization for every constraint is required. Within these anti-optimizations, the uncertainties are set as design variables, whereas the design domain is specified by the bounds on the uncertainties. Thus, anti-optimizations are nested within the main optimization, making it a two-level optimization problem..."

S.P. Gurav *et al.* (2004)

This chapter deals with hybrid optimization and antioptimization of the buckling load of composite cylindrical shells. The methodology, which has been developed in the previous works, is applied to a set of cylindrical composite shells, tested at German Aerospace Center (DLR). Furthermore, the existing approach is enhanced to fit within the design optimization scheme. The shells possess traditional imperfections in the form of Fourier series coefficients of their initial imperfection profile. Additionally, two nontraditional imperfections are included in the analysis. The available experimental data is enclosed by either 11-dimensional hyper-rectangle or hyper-ellipsoid. The minimum buckling load of the ensemble of such shells is determined by the antioptimization procedure. Then, this minimum load is maximized by varying the laminate angle. It is shown that the proposed method is a viable and relatively simple alternative to probabilistic approaches and successfully supplements them. It is shown that the proposed method is a successful supplement to probabilistic methods and the deterministic single buckle approach (SBA), since it is deterministic in nature and thus could appeal the engineers and investigators alike, and it takes into account the actual scatter of input data.

1. Introduction

The notion of imperfection-sensitivity — that is, of drastic reduction in the load-carrying capacity of cylindrical shells due to presence of small imperfections like deviations from the ideal, nominal values of the parameters — was introduced in the pioneering Ph.D. dissertation by Koiter (1945). It was soon recognized rigorously by Bolotin (1962) that in order to be practical, the imperfection sensitivity concept should be combined with recognition of the uncertain nature of imperfections. Bolotin (1967) argued that

> a real structure differs from the idealized structure designated by an engineer. This difference is connected with the great amount of small imperfections and defects. An engineer must be sure that in spite these differences the real structure will behave approximately in the same manner as the corresponding idealized scheme. In the absence of such confidence the engineering design would lose all its sense.

In his breakthrough paper Bolotin (1962, 1969) argued that engineers ought to investigate reliability of the structure. "However," as Arbocz (1981) noted, "it was not until 1979, when Elishakoff (1979) published his reliability study of the buckling of stochastically imperfect finite column on a nonlinear elastic foundation, that a method has been proposed, which made it possible to introduce the results of the initial imperfection surveys into the analysis." It is notable that analogous statement is repeated in several works by Arbocz, see for example Singer *et al.* (1997) and recently published ECSS *Buckling Handbook* (2010).

The probabilistic analysis of the imperfection sensitivity has been extensively reviewed by Chryssanthopoulos (1998), Elishakoff (1983), Arbocz and Stam (2004), and other investigators. It is outside the scope of the current discussion to review again the probabilistic approach to the imperfection sensitivity of structures. In his review, Elishakoff (2000) poses the following question: "Do probabilistic methods have disadvantages, or do they constitute a panacea for fully closing the chasm between theory and practice?" Moreover, to quote from Ref. (2000),

> Initial imperfection data banks, even when compiled, may still contain insufficient information for rigorous probabilistic processing of all variables. In such circumstances, researchers "randomize" the problem by assigning the probability distributions. By doing so they "make something out of nothing" and create the illusion of availability of information, while in actuality it is lacking. Is such a procedure a necessary evil, and should one just live with it? At least, many investigators felt uncomfortable with this situation.

This quote correlates well with the statement made by Arbocz *et al.* (1998):

> . . . the vast majority of practicing engineers agree that the true reliability must be demonstrated and not simply estimated from analysis. It is the authors' opinion that before the

> engineering community will begin in large numbers to accept the current generation of probabilistic analysis tools two conditions must be satisfied. First, there must be test-conducted data bases which can help in mapping the input parameter uncertainties into probabilistic density functions. In addition, there must be failure and failure rate data bases, which can be used for test verification of the probabilistic failure based predictions.

In the paper by Ben-Haim and Elishakoff (1989) and in another by Elishakoff and Ben-Haim (1990) an alternative to the probabilistic method (see also its application in work by van der Nieuwendijk (1997)) was proposed, namely the convex model of uncertainty. Specifically, the initial imperfection vector was represented as a sum of the nominal vector and the deviation vector. The deviation was postulated to fall within the ellipsoidal set, whose size parameter and the semiaxes were based on experimental data, obtainable from initial imperfection data bank compiles due to indefatigable efforts of Arbocz (1979). The design buckling load was identified with the lowest buckling load which can be obtained for any of the shells in the ensemble ellipsoid of imperfections. Further details can be found in the monographs by Ben-Haim and Elishakoff (1990) and Elishakoff *et al.* (2001). This approach of searching for the minimum buckling load of structures under interval or ellipsoidal uncertainties was dubbed by Elishakoff (1990) as an antioptimization. It was recognized that antioptimization process may yield conservative results. Therefore, it was later suggested to optimize the system designed by the antioptimization process, providing the hybrid optimization and antioptimization. As Elishakoff (2000) notes, "one is interested in maximizing the minimum buckling load the structure can carry due to uncertainty in the system." For the current state of the art of optimization and antioptimization the reader can consult the recent monograph by Elishakoff and Ohsaki (2010).

In engineering practice it is still a common way to reduce the buckling load of the perfect shell by knockdown factors (KDF) as given by guidelines like NASA-SP 8007 (1968) in order to obtain a design load. This principle has the advantage of simplicity and it does not require measurements. However, it is a very coarse way to account for imperfection sensitivity and it turned out to be overly conservative for modern shells. Furthermore, all guideline have been derived for metallic shells and are not applicable to composite shells in general.

Beside the knockdown factor concept, probabilistic methods and convex antioptimization another approach has been followed in the recent years. Motivated by similar reasons as quoted above in (2000) and (1998), Hühne *et al.* (2008) and Rolfes *et al.* (2004) looked for an alternative to probabilistic design, which is less conservative than NASA-SP 8007. They preformed thousands of buckling tests on several elastically buckling shells, reusing the shells to repeat tests and to test with modified configurations. Within tests and simulations a lateral concentrated load has been applied in the radial direction. This perturbation load was varied but kept constant within each test while the axial load or displacement was increased. Hühne *et al.* (2008) and Rolfes *et al.* (2004) discovered that the buckling load of a shell with a single prebuckle imperfection created by a radial perturbation load decreases with increasing perturbation load, but reaches a minimum for a certain magnitude of the perturbation load and the single prebuckle respectively. The reason was found to be that beyond the so detected minimum buckling load, the shell with single prebuckle tends to behave like a shell with cutout. Consequently, an even deeper prebuckle would not decrease the buckling load any more, since the cutout is the worst case. Furthermore, the authors show that a single buckle is a stimulating as well as a worst imperfection. They defined the obtained lower bound of the buckling load as design load and optimized the design of the same shells as used by Zimmermann (1992) by maximizing the lower bound. Interestingly, the design load of this so-called single buckle approach (SBA), although at the first glance appearing to be quite rigorous, turned out to be significantly higher than the one provided by NASA-SP 8007. Based on an excellent data base of tests on 10 nominally identical shells provided by Degenhardt *et al.* (2010), Kriegesmann *et al.* (2011) compared the SBA with a probabilistic design covering traditional as well as nontraditional imperfections. They revealed that the SBA corresponds to a reliability level of 99.9%. However, the generalization of this result to arbitrary shells is pending.

Assuming that a reliable deterministic design method can be found the question remains which value of the knockdown factor has to be applied. Even though this question has not to be answered by the design engineer, it must be decided by the relevant standardization committees, via the reliability or probability of failure. If well-founded probabilistic investigations exist, it will be a very good decision support for these committees. In view of high computational cost, which is usually associated with probabilistic methods, it would be desirable to develop a deterministic method for the design engineer. In cases where appropriate data about scatter of input parameters are available and it seems to be economic, probabilistic design can be performed also in engineering practice.

This gives rise to the conclusion that both probabilistic as well as deterministic methods are valid, though in different circumstances: The former ought to be used when sufficient data is available; the latter may prove preferable when scarce data is present.

In the context of design of structures it is not only required to find a robust design, but it is also desired to find the optimal design. Optimizing the design load given by the KDF concept is equivalent to optimizing the perfect structure. Optimization of the buckling load of axially compressed composite cylinders without invoking uncertain parameters by varying the laminate setup has been performed for example, by Hirano (1983) and Zimmermann (1992), who optimized buckling load of perfect shell.

Hühne *et al.* (2008) showed that such an approach might not lead to the optimum design for real application, since the optimization of the perfect shell does not take into account the sensitivity to imperfections. He optimized the design load given by the SBA and found a significantly different design to be optimal. Optimization of the design load of axially cylindrical shells that is given by probabilistic methods or convex antioptimization is still lacking.

In the present section the convex modeling method proposed by Ben-Haim and Elishakoff (1989) (later dubbed as antioptimization by Elishakoff) is used to evaluate the lower bound of buckling load for a set of cylindrical shells. The procedure is applied to the ensemble of shells given in (2010) taking into account traditional as well as nontraditional imperfections. The derived lower bound of buckling load is then compared to the results of probabilistic methods and the SBA. Furthermore, the lower bound of buckling load given by convex optimization is maximized by optimizing the laminate setup.

2. Convex Antioptimization

In order to find a lower bound of buckling load it is desired to find the minimum buckling load, which constitutes a classical optimization problem. If the buckling load is interpreted as a function of only one scattering input parameter, the domain could, for example, be given by a tolerance interval, and the minimum buckling load can be found using standard optimization algorithms (see Fig. 6.1).

By taking into account more scattering input parameters the definition of the domain space may become difficult. Furthermore, since the buckling load function is not given analytically, but can only be evaluated for a certain set of input parameters numerically, the use of standard gradientbased optimization algorithms is computationally costly. Therefore, this approach is not useable as design procedure. The goal of convex modeling (antioptimization) is the same: finding the minimum buckling load in the domain of scattering input parameters. However, by making use of some

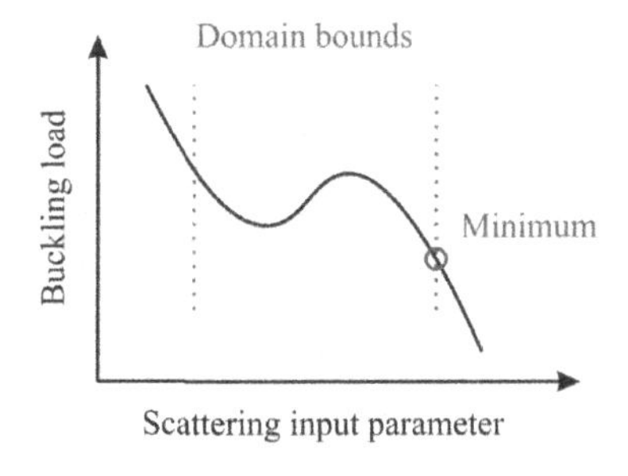

Fig. 6.1. Sketch of a one-dimensional minimization problem.

simplifying assumptions, a domain of input parameters can be found on the basis of measurements, and lower bound of buckling load can be found by performing only a small number of buckling analyses. For the antioptimization the following 11 parameters were regarded as uncertain variables, specifically nine geometric parameters, as well as wall-thickness and bending angle (for more details, see Appendix C.1 and Kriegesmann *et al.* (2011)):

$$\mathbf{x} = (\underbrace{z_1, \ldots, z_9}_{\text{geometry}}, \underbrace{t}_{\text{wall thickness}}, \underbrace{\theta}_{\substack{\text{bending}\\ \text{angle}}})^{\mathrm{T}}. \tag{6.1}$$

In order to find the worst-case combination of the input parameters, the domain of possible combinations must be defined. For this, each measurement is regarded as a point in an appropriate dimensional space and the set of points is enclosed by a specified convex figure.

According to Zhu L.P. *et al.* (1996) it is important to normalize the parameters, since parameters may have different dimensions. Otherwise, problems of invariance may occur when determining the enclosing geometric figure. The parameters in original units $\hat{x}_i^{(j)}$ are normalized as follows:

$$x_i^{(j)} = \frac{\hat{x}_i^{(j)} - \frac{1}{2}[\max(\hat{x}_i) + \min(\hat{x}_i)]}{\frac{1}{2}[\max(\hat{x}_i) - \min(\hat{x}_i)]}. \tag{6.2}$$

Figures 6.2 and 6.3 show the minimum area enclosing rectangle and the minimum area enclosing ellipse for the two-dimensional case that only the bending angle θ and the wall thickness t are considered, as an example. In this case x_1 is the normalized bending angle and x_2 equals the normalized wall-thickness. In higher dimensions, the

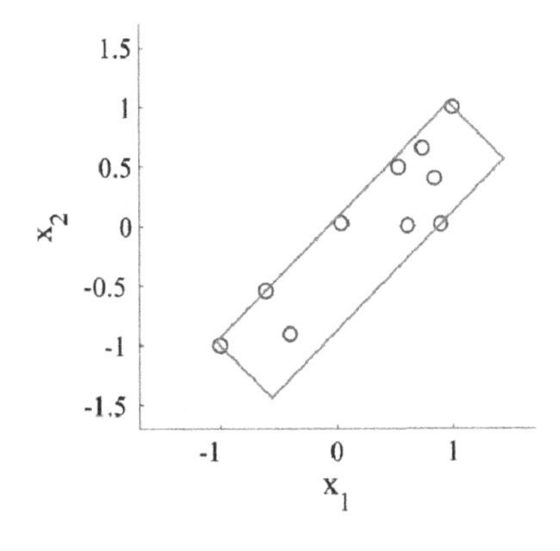

Fig. 6.2. Minimum area enclosing rectangle.

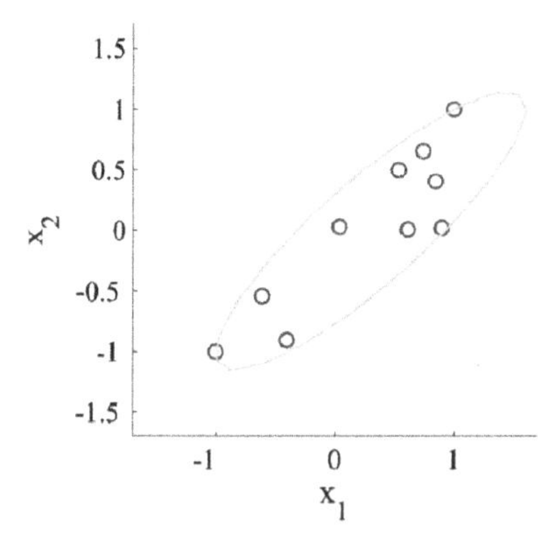

Fig. 6.3. Minimum area enclosing ellipse.

measurement points are enclosed by the minimum volume enclosing hyper-rectangle (MVER) or the minimum volume enclosing hyper-ellipsoid (MVEE), respectively.

Theoretically, any geometric figure that includes an enclosing space can be used to define the domain of uncertainty parameters and there is no physical justification to prefer one or the other except the fact that the areas that yield higher maximum buckling load are less preferable, because of the attendant overdesign. However, assuming a hyper-rectangle or a hyper-ellipsoid makes it easier to find the minimum buckling load.

The procedures for determining the MVER and the MVEE of a set of points are described in detail in Appendix C.2 and 3.3.

2.1. *Antioptimization procedure*

If the function of the buckling load is convex and the domain of input parameters is given by a hyper-ellipsoid, the minimum buckling load can be determined directly by using Lagrange multiplier method (for detailed information, see Ben-Haim and Elishakoff (1990)).

For the derivation of the minimum buckling load the buckling load function is approximated by a Taylor expansion in ζ at the center $\mathbf{x}_c$:

$$x_i^{(j)} = \frac{\hat{x}_i^{(j)} - \frac{1}{2}[\max(\hat{x}_i) + \min(\hat{x}_i)]}{\frac{1}{2}[\max(\hat{x}_i) - \min(\hat{x}_i)]} \tag{6.3}$$

with the gradient $\varphi = \nabla\lambda$ and the Hessian $\Xi = \nabla\nabla\lambda$. If the following inequality

$$\varphi^{\mathrm{T}}\,\Xi^{-1}\boldsymbol{\Omega}\,\Xi^{-1}\varphi < 1 \tag{6.4}$$

holds, the minimum buckling load can be found inside the minimum volume enclosing hyper-ellipsoid and is given by the expression

$$\lambda_{\min} = \lambda(\mathbf{x}_c) - \frac{1}{2}\varphi^{\mathrm{T}}\Xi\varphi. \tag{6.5}$$

If inequality (6.4) does not hold, the input parameters that lead to the minimum buckling load are on the boundary of the minimum volume enclosing hyper-ellipsoid. Then the second-order approximation of the minimum buckling load is given by

$$\lambda_{\min} = \lambda(\mathbf{x}_c) - \varphi^{\mathrm{T}}\mathbf{M}^{-1}\varphi + \frac{1}{2}\varphi^{\mathrm{T}}\mathbf{M}^{-1}\Xi\mathbf{M}^{-1}\varphi. \tag{6.6}$$

with

$$\mathbf{M} = \Xi + 2\gamma\mathbf{\Omega}, \quad \gamma^2 = \frac{1}{4}\varphi^{\mathrm{T}}\mathbf{G}\varphi \quad \text{and} \quad \mathbf{G} = \mathbf{\Omega}^{-1}$$
$$= \mathrm{diag}(g_1^2, \ldots, g_{\mathrm{d}}^2). \tag{6.7}$$

Using the first-order Taylor expansion the minimum buckling load always can be determined by

$$\lambda_{\min} = \lambda(\mathbf{x}_{\mathrm{c}}) - \sqrt{\varphi^{\mathrm{T}}\mathbf{G}\varphi}. \tag{6.8}$$

Since function of buckling load is not given analytically, the derivatives of the buckling load have to be evaluated numerically, as is discussed in Appendix C.5. The computational cost of the methodology results from the estimation of the derivatives. As derived in Appendix C.5, the first-order approximation requires at least $d + 1$ buckling analyses, where for the second- order approach $1+2d+{}^1\!/_2(d^2–d)$ buckling load calculations have to be performed.

If a hyper-rectangle is used to define the domain of input parameters, the minimum buckling load cannot be determined directly. In order determine it still in reasonable time, it will be assumed that the worst-case scenario is given by one of the vertices of the minimum volume enclosing hyper-rectangle. The number of vertices of a minimum volume enclosing hyper-rectangle with the dimension d equals 2^d. Hence, this method becomes prohibitive for high-dimensional problem. Assuming that the buckling load function is monotonically increasing or decreasing along one axes of the minimum volume enclosing hyper-rectangle, the values of the derivatives give a sense in which vertex the minimum buckling load is to be expected.

The method of searching the worst-case scenario in a minimum volume enclosing hyper-rectangle seems to be inefficient, if every vertex is checked, or connected to significant assumptions, if the decisive vertex is determined by the derivatives at the center point. Note that in higher dimensions finding the minimum volume enclosing hyper-rectangle is easier than finding the minimum volume enclosing hyper-ellipsoid, and the practicality is tested for the current example.

2.2. *Results of antioptimization*

If the number of points that shall be enclosed is smaller than or equal to the dimension of the points, the MVER and MVEE have a volume of zero, because the length of one edge or semiaxis will tend to zero. For example, if two points should be enclosed by a two-dimensional rectangle, one edge of the rectangle would have zero length, the length of the other one would equal the distance of the two points and the area of the rectangle would be zero. In the current example, 10 measurement points with the dimension 11 are given and hence the MVER and MVEE should have a volume equal to zero. However, in case the volume is bigger than zero, the domain is bigger than the real minimum volume enclosing convex figure and more combinations are considered. Thus, the approximation can be regarded as conservative.

The obtained approximation of the MVER has a volume of 151.9 (dimensionless) and the MVEE approximation has a volume of 1454.8. The approximation of the MVEE seems to be dramatically worse than the MVER approximation, since the volume is almost 10 times higher. Actually, both results are of the same quality, which is demonstrated by considering the volume of a hyper-sphere in $\square^{11}$. The volume of a hyper-sphere in $\square^{d}$ is given by $V = C_d r^d$, with the volume of the unit hyper-sphere C_d and the radius r. Consider the volume of two 11-dimensional hyper spheres with the volumes V_1 and V_2, with $V_1 = 10V_2$. This leads to

$$C_{11} r_1^{11} = 10 C_{11} r_2^{11} \Rightarrow r_1 = \sqrt[11]{10} r_2 = 1.23 r_2. \tag{6.9}$$

An increase of the radius of about 23% leads to 10 times greater volume. The semiaxis of the minimum volume enclosing hyper-ellipsoid have the same order of magnitude as the edges of the minimum volume enclosing a hyper-rectangle. Which of the two geometric figures has the smaller volume depends on the shape of the point set. However, the obtained approaches are used to define the worst-case scenario and the minimum buckling load.

First, the derivatives of the buckling load at the center of the MVER are determined in order to conclude which vertex gives the worst case. The associated buckling load equals 21.7 kN. Checking all vertices leads to a minimum buckling load of 18.4 kN. Hence, the proposed procedure to determine the decisive vertex does not work in the current example.

Assuming a MVEE as domain of input parameters and determining the minimum buckling load according to (6.8) leads to a minimum buckling load of 17.1 kN. The second-order approximation (6.6) of the minimum buckling load equals 21.0 kN. Here, the step size Δx_i for estimating the derivatives was chosen equal to 0.75 times the length of the associated semiaxis g_i. This relatively large step size does not lead to an accurate estimation of the derivatives at the center point, but to a good approximation of the buckling load function in the hyper-ellipsoid.

Using the derivatives of buckling load at the center of the hyper-rectangle in order to find the decisive vertex, the obtained lower bound exceeds the lowest experimentally determined buckling load (Table 6.1). Checking all vertices delivers a lower bound below all test results, but this procedure requires an unacceptably large amount of buckling load calculations. The first-order approximation in the minimum volume enclosing hyper-ellipsoid leads to the most conservative approach of the minimum buckling load. The second-order approximation is close to the minimum test result. It was expected that the second order approximation would lead to a more accurate estimation of the minimum than the first-order approximation, but depending on the actual shape of the buckling load function this is not necessarily the case. For example, for the nonconvex objective function plotted in Fig. 6.4 the second-order appoximation leads to an estimation of the minimum (red circle) that exceeds the actual minimum (blue circle) and the first-order approximation (green circle) of the minimum in the enclosing ellipsoid.

Table 6.1. Comparison of minimum buckling load approaches.

Approach	Minimum buckling load	Number of buckling load calculations
MVER, decisive vertex given by derivatives	21.7 kN	$1 + 2d = 23$
MVER, all vertices checked	18.4 kN	$2^d = 2048$
MVEE, first order	17.1 kN	$1 + 2d = 23$
MVEE, second order	21.0 kN	$1 + 2d + {}^1/_2(d^2 - d) = 78$
Experimental tests	21.3 kN	

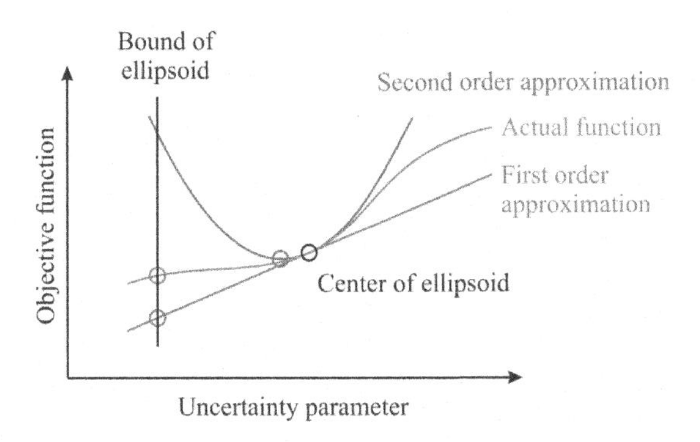

Fig. 6.4. Example of a (nonconvex) objective function that leads to a nonconservative second-order approximation of the minimum.

Of course, for a different objective function the second-order approach might lead to a more conservative approximation of the minimum than the first-order approach. Hence, the smaller value of both approaches should be used.

2.3. *Comparison with other approaches*

Presently, the most frequently used guideline to handle the effect of imperfections on the buckling load of cylindrical shells is NASA SP-8007 (1968). It proposes the reduction of the buckling load of a perfect shell by multiplying it with the knockdown factor γ given by

$$\gamma = 1 - 0.901(1 - e^{-\Phi}), \quad \Phi = \frac{1}{16}\sqrt{\frac{R}{t}} \text{ for } \frac{R}{t} < 1500. \tag{6.10}$$

The knockdown factor is defined as a function of the ratio of radius and wall thickness. It is based on a multitude of test results, obtained in the 1960s and earlier. It has been defined in the way that the obtained design load is lower than all experimental test results that were available at that time. For the given set of shells, γ equals 0.322 and the linear buckling load of the perfect shell equals 38.6 kN.

Another deterministic design procedure has been proposed by Hühne *et al.* (2008). As it already has been shown by Esslinger (1974), the onset of buckling of imperfect cylindrical shells always starts with a single buckle, and hence, a single buckle can be regarded as stimulating or triggering imperfection. Furthermore, Demel (1997) showed that a single buckle also can be regarded as worst-case geometric imperfection. A single buckle can easily be applied by applying a lateral perturbation load. As the amplitude of the buckle or the size of the perturbation load, respectively, increases, the buckling load decreases. But if the amplitude of the buckle exceeds a certain value, the buckle has the same effect as a cutout. Then, the buckling load is not decreased further as the perturbation load is increased. Hence, there is a physically reasonable lower bound of buckling load for cylindrical shells with an initial single buckle. Hühne *et al.* (2008) defined this lower bound as design load and showed that it leads to a less conservative design than NASA SP-8007 (1968).

Probabilistic analysis of the given set of shells has been performed by Kriegesmann *et al.* (2011). They used a semianalytical method and validated it with Monte Carlo simulation. The design loads given by all mentioned approaches are given in Table 6.2.

2.4. *Sensitivity analysis*

When dealing with uncertainty it should always be checked whether it is worthwhile to use probabilistic methods, fuzzy sets based methodology, or antioptimization.

Table 6.2. Comparison of design loads given by different approaches.

	Design approach		Design load
Deterministic	NASA SP-8007 (knockdown factor $\gamma = 0.322$)		12.4 kN
	Single buckle approach[a], (see Hühne *et al.* (2008))		17.4 kN
	Convex antioptimization, MVEE, first order		17.1 kN
	Convex antioptimization, MVEE, second order		21.0 kN
Probabilistic (see Kriegesmann *et al.* (2011))	Semianalytical probabilistic approach	for reliability of 99 %	18.1 kN
		for reliability of 99.9 %	16.3 kN
	Monte Carlo simulation	for reliability of 99 %	18.6 kN
		for reliability of 99.9 %	17.2 kN
	Experimental tests		21.3 kN

[a] Determined with slightly different material parameters.

Fuzzy sets theory is beyond the scope of the current discussion. In these circumstances the basic question is: Does the choice of the methodology between probabilistic and antioptimization methods affect the result? Furthermore, it is of interest to determine which uncertainty parameters have a significant influence on the output. Once the answer to the latter question is obtained, one may conclude that only a few parameters have to be considered in the uncertainty analysis.

In order to answer the above question in the context of convex antioptimization, the sensitivity vector $\mathbf{s}$ with the entries s_i shall be defined:

$$s_i = \left| g_i \frac{\partial \lambda}{\partial \xi_i} \right| . \tag{6.11}$$

The minimum buckling load given by the first-order approach can be written as an algebraic sum:

$$\lambda_{\min} = \lambda(\mathbf{x}_c) - \sqrt{\sum_{i=1}^{d} g_i^2 \left(\frac{\partial \lambda(\mathbf{x}_c)}{\partial x_i} \right)^2} = \lambda(\mathbf{x}_c) - \|\mathbf{s}\|. \tag{6.12}$$

Hence, the contribution of each parameter to the reduction of the minimum buckling load can be determined. For the given example the dimensionless sensitivity vector is $\mathbf{s} = (1.323, 0.147, 0.101, 0.373, 0.664, 0.924, 0.398, 0.251, 0.135, 6.962, 1.043)^{\mathrm{T}}$.

Obviously, there is one parameter that has a dominant influence on the result. It must be checked, whether the influence of this parameter is large enough to neglect all other parameters or if there are an essential number of parameters that have to be taken into account. If, for example, all parameters with a sensitivity of less than 0.5 would be neglected, only five parameters would be considered. In these circumstances the minimum buckling load equals 17.4 kN instead of

Table 6.3. Influence of reducing the number of uncertainty parameters according to sensitivity analysis.

Considered parameters	Number of parameters	Minimum buckling load approach
All	11	17.34 kN
Only $s_i > 0.5$	5	17.37 kN
Only $s_i > 1$	3	17.46 kN
Only $s_i > 1.5$	1	17.66 kN

17.1 kN. It should be noted that $\mathbf{s}$ describes the sensitivity to parameters in the coordinate system of the MVEE. Hence, these considerations do not allow a conclusion about the sensitivity to the original physical parameters as given in Eq. (6.1).

The approach of the minimum buckling load determination obtained with only one parameter is sufficiently accurate for optimization purposes. Unfortunately, if the structure is to be changed within an optimization procedure the derivatives of the buckling load and hence the sensitivity vector changes.

3. Optimization of the Laminate Setup

If the imperfections are assumed to be independent of the laminate setup, the MVEE and MVER minimum volume enclosing hyper-ellipsoid and minimum volume enclosing hyper-rectangle are independent from the fiber orientations as well. Hence, once the domain of the uncertain parameters is determined, the laminate setup can be optimized regarding the fiber orientations $[\beta_1, \beta_2, \beta_3, \beta_4]$ as design parameters. The minimum buckling load can be regarded as a function of $\boldsymbol{\beta} = (\beta_1, \beta_2, \beta_3, \beta_4)$. The fiber angles are related to the shell axis and β_1 is the angle of the innermost ply:

$$\lambda_{\min}(\boldsymbol{\beta}) = \lambda\,(\mathbf{x}_c, \boldsymbol{\beta}) - \sqrt{\varphi(\boldsymbol{\beta})^{\mathrm{T}}\mathbf{G}\varphi(\boldsymbol{\beta})} \tag{6.13}$$

with

$$\varphi = \left(\frac{\partial\lambda}{\partial x_1}, \ldots, \frac{\partial\lambda}{\partial x_{\mathrm{d}}}\right)^{\mathrm{T}}, \quad \mathbf{G} = \boldsymbol{\Omega}^{-1} = \operatorname{diag}(g_1^2, \ldots, g_{\mathrm{d}}^2). \tag{6.14}$$

For simplification a two-dimensional optimization problem shall be treated first. For this, the laminate setup shall be defined by $[\beta_1, -\beta_1, \beta_2, -\beta_2]$ and $\beta_1, \beta_2 \in [0°, 90°]$.

3.1. *Naive optimization*

Restricting the laminate setup to $[\beta_1, -\beta_1, \beta_2, -\beta_2]$, the optimization problem with two design parameters can be solved numerically by determining the minimum

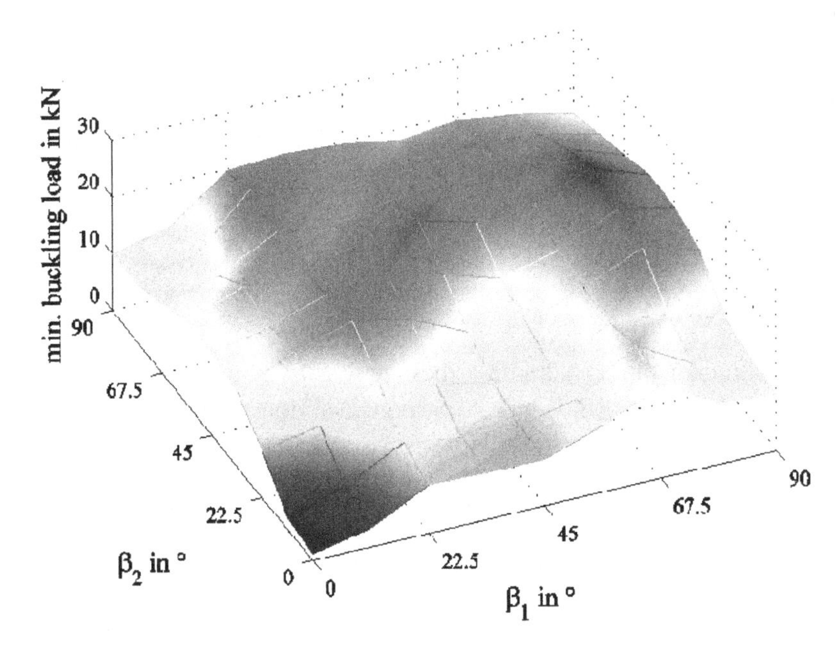

Fig. 6.5. Minimum buckling load from convex anti-optimization for different laminate setups $[\beta_1, -\beta_1, \beta_2, -\beta_2]$.

buckling load for each combination, as plotted in Fig. 6.5. The step size for varying β_1 and β_2 was chosen to 11.25°. The maximum of minimum buckling loads equals 23.25 kN and the associated laminate setup is [±78.75°, ±56.25°].

3.2. *Derivatives of the minimum buckling load*

The current objective function is the minimum buckling load given by Eq. (6.13). Since gradient-based optimization algorithms require the gradient and some the Hessian of the objective function, the first and second derivatives of the minimum buckling load with respect to the design parameters β_j have to be calculated:

$$\varphi = \left(\frac{\partial\lambda}{\partial x_1}, \ldots, \frac{\partial\lambda}{\partial x_d}\right)^{\mathrm{T}}, \quad \frac{\partial\lambda_{\min}}{\partial\beta_j} = \frac{\partial\lambda}{\partial\beta_j} - \frac{1}{\sqrt{\varphi^{\mathrm{T}}\mathbf{G}\varphi}}\varphi^{\mathrm{T}}\mathbf{G}\frac{\partial\varphi}{\partial\beta_j} \tag{6.15}$$

$$\begin{aligned} &\varphi = \left(\frac{\partial\lambda}{\partial x_1}, \ldots, \frac{\partial\lambda}{\partial x_d}\right)^{\mathrm{T}}, \\ &\frac{\partial\lambda_{\min}}{\partial\beta_j\,\partial\beta_k} \approx \frac{\partial^2\lambda}{\partial\beta_j\,\partial\beta_k} + \frac{1}{\sqrt{\varphi^{\mathrm{T}}\,\mathbf{G}\,\varphi}}\left[\frac{2}{\varphi^{\mathrm{T}}\,\mathbf{G}\,\varphi}\varphi^{\mathrm{T}} - \left(\frac{\partial\varphi}{\partial\beta_k}\right)^{\mathrm{T}}\right]\mathbf{G}\,\frac{\partial\varphi}{\partial\beta_j} \end{aligned} \tag{6.16}$$

with

$$\frac{\partial \varphi}{\partial \beta_j} = \left(\frac{\partial^2 \lambda}{\partial x_1 \, \partial \beta_j}, \ldots, \frac{\partial^2 \lambda}{\partial x_d \, \partial \beta_j} \right)^T . \tag{6.17}$$

The estimations of the required derivatives of the buckling load, which drive the number of buckling analyses per iteration step and hence, the computational cost, are given in Appendix C.4.

3.3. *Results of design optimization*

The naive optimization (Sec. 3.1) delivers an optimum of the lower bound of buckling load of 23.25 kN for the laminate setup [±78.75°, ±56.25°]. Using this laminate setup as start vector for gradient-based optimization leads to a design load of 23.74 kN and an optimum design given by the layup [±79.4°, ±56.8°]. The start vectors [±45°, ±56.3°] and [±45°, ±0°] did not lead to a higher design load.

Due to the shape of the objective function the success of gradient-based optimization heavily depends on the choice of the start vector. It is to be expected that this problem also appears in the four-dimensional design space. A naive optimization as executed in Sec 3.1 does not come into consideration due to the computational costs. Hence, randomly chosen start vectors will be used for the gradient-based optimization. One reasonable choice for the start vector is the optimal design of the two-dimensional optimization. This leads to an (possibly local) optimum of 23.78 kN and [84.1°, –75.2°, 57.5°, –55.5°]. Other start vectors did not deliver a higher design load. An overview of the obtained optimum designs is given in Table 6.4

Hühne *et al.* (2008) optimized the design load given by the single buckle approach (see Fig. 6.6). Hühne *et al.* (2008) found the maximum imperfect buckling load of 17.7 kN for a laminate setup of [±25°, 90°_2]. It is worth mentioning that Hühne used

Table 6.4. Number of buckling load calculations in each iteration step.

Procedure		Number of design variables	Design load	Optimal layup
Convex antioptimization	Naïve/discrete optimization	2	23.25 kN	[±78.75°, ±56.25°]
	Simple gradient optimization	2	23.74 kN	[±79.4°, ±56.8°]
		4	23.78 kN	[84.1°, −75.2°, 57.5°, −55.5°]
Hühne *et al.* (2008)		2	17.7 kN	[±25°, 90°_2]

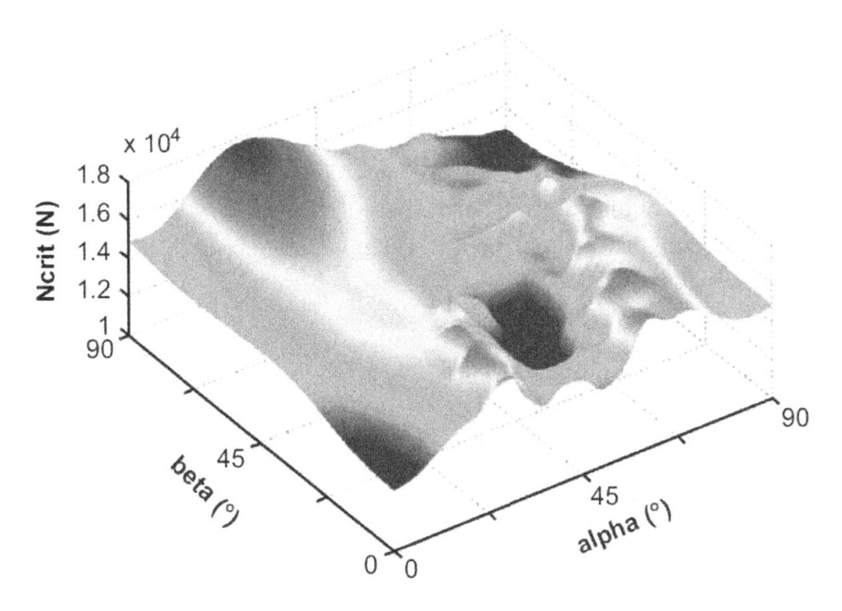

Fig. 6.6. Buckling load of a shell with laminat setup ($\pm\alpha$, $\pm\beta$) and single perturbation load of 15 N, from (2008).

different material properties. However, the differences are not that significant that they cause the difference in the optimal design.

At first glance, the response surface determined by Hühne *et al.* (2008) optimization differs significantly from the response surface given by convex antioptimization (compare Fig. 6.5. and Fig. 6.6, which is not surprising, since the approaches are completely different and the SBA does not consider measurements of input parameters. However, both approaches find local minima in the vertices of the design domain ([0°, 0°], [90°, 0°], [0°, 90°], [90°, 90°] and local maxima around [22.5°, 90°] and [67.5°, 45°]. Obviously, a robust design can be reached either by combining axially orientated inner plies with circumferentially orientated outer plies or by combining circumferentially orientated inner plies with about ±45° outer plies. It appears that optimizing the laminate layup under consideration of uncertainties by convex antioptimization leads to a higher design load. The reliability level however cannot be checked by the present approach.

4. Conclusion and Outlook

In their definitive article Arbocz and Starnes (1998) pose the following pertinent question: "Why, after so many years of concentrated research effort on shell stability, has an improved design philosophy not emerged that is better than the empirical lower-bound design approach?" It appears that the probabilistic methodology of introducing the imperfection sensitivity concept in the design (as pioneered by Elishakoff (1983), and further developed by Kriegesmann *et al.* (2011)), the single buckle approach (see Hühne *et al.* (2008)) and the current hybrid optimization and antioptimization provide the needed deterministic and probabilistic design tools that will rigorously fit the needs for design and standardization.

The current methodology uses three major elements. First, it utilizes the existing initial imperfection data surveys, along with information on scatter of other, nontraditional imperfections. These values are then enclosed into convex sets, namely either in multi-dimensional hyper-rectangle or multi-dimensional ellipsoid.

The second element consists in the deterministic analysis of buckling loads. In terminology of Arbocz and Starnes (1998), "modern computational facilities provided by today's high-powered computers, offer the opportunity to address complex response phenomena with robust nonlinear procedures." Indeed, the suggested philosophy utilizes the FEM capabilities in the form of ABAQUS.

The third crucial step is *combined* use of antioptimization and optimization. Antioptimization provides with determination of minimum buckling load of shells whose combined traditional and nontraditional imperfections are enclosed by multi-dimensional hyper-rectangle, or hyper-ellipsoids. However, this minimum buckling load evaluation differs drastically from the ideas of Weingarten *et al.* (1965) and NASA SP-8007 (1968). The latter approach finds the minimum buckling loads in the entire set of then-available shells, produced by different manufacturing processes, mixing as it were, apples and oranges. The proposed methodology delivers the minimum buckling load numerically in the ensemble of shells produced by the same manufacturing procedure. Therefore, the enclosing geometric figures are intimately characterizing the manufacturing process at hand.

The current work goes beyond determining of minimum buckling loads, realizing that these may still may turn out to be conservative though not as conservative as those reported by Seide or NASA SP-8007. Within the current methodology, the structure is optimized so as to maximize the minimum buckling load. Thus the proposed design philosophy combines experimental, numerical and theoretical analyses, and appears to be attractive, since in contrast to the SBA it takes into account the real scatter of input parameters, and is as time efficient as probabilistic analyses. It is hoped that the suggested *Weltanschauung* will be embraced and further developed by engineers and researchers alike. Application of this novel methodology to imperfection-sensitive stiffened shells which were already subject of intensive research in large coordinated projects (see Zimmermann *et al.* (2006) and Degenhardt *et al.* (2006)) ought to be investigated.

Chapter 7

Miscellaneous Topics

Science is the belief in the ignorance of experts.

R. Feynman (1969)

Elishakoff drew attention ... to the need for a probabilistic approach to the buckling loads of imperfection — sensitive structures, a subject that he has pioneered over many years. He continues to argue with conviction (Elishakoff, 2012) that from engineering standpoint the final target for the analyst must be to study imperfect shells ... and to express the results in probabilistic terms.

J.M.T. Thompson and G.H.M. van der Heijden (2014)

This chapter is devoted to various topics that appear to be outside of the main thrust of this book, namely of resolution of the conundrum created by the "curse of imperfection sensitivity." Some cooperative work with Warner T. Koiter and James H. Starnes, Jr., are first described. Then we follow with a brief story of establishment of the ASME Medal in honor of Professor Koiter. Finally, the latest works are briefly reviewed along with the discussion of the ever-present question of priority.

1. Comments on Some Later Deterministic Works

Elastic stability is indeed still a somewhat controversial topic, a property which it has in common with other fields of engineering science.

W.T. Koiter (1985)

1.1. *Deterministic resolutions of the knockdown factors*

It was most natural to seek for the deterministic resolution of the 20th-century conundrum in elastic stability. Various avenues were pursued by the researchers. Apparently the first investigator, who put a goal of determining the lower bounds of buckling loads, was Pogorelov (1960, 1966, 1967, 1971) via his geometric theory of shells. Pogorelov and Babenko (1991) wrote that

> as early as 1908, Mallock, investigating postcrictical deformations of long cylindrical tubes under axial compression, which may be accompanied by significant changes in the shape of the shell and small tension-compression deformations in the middle surface, first advanced the hypothesis of the proximity of the shape of the deformed shell to

> the isometric transformation of its undeformed median surface (bends were considered as a class of polyhedrons). This idea in the problem of stability of cylindrical shells was subsequently employed by Kirste (1954), Isohimura (1955) [see Koryo (1969)], Pugsley (1960) and others.

Pogorelov further developed this idea in his articles and books. In addition to the above references, for a description of Pogorelov's geometric theory of shells one can consult papers by Goldenveizer (1983) and Palmer (2002). As Palmer (2002) mentions, that Pogorelov's "research is not well known in the English-speaking world. It seems also to be disconnected from mainsteam shell theory research in the former Soviet Union." Goldenveizer (1983) addressed the geometric theory in his paper. For the detailed review the interested reader can consult the paper by Pogorelov and Babenko (1992). El Naschie (1977) characterizes Pogorelov's approach as "an ingenious solution of shell buckling based on geometric arguments." Pogorelov (1966) derived the following formula (see Pogorelov (1966), p. 191) for the lower bound:

$$\sigma_{cr} = fE\frac{h}{R} \tag{7.1}$$

where the numerical factor $f = 0.18$. To digress let us note that in another paper, some years later, Pogorelov (1971) gives another value, namely 0.16 instead of 0.18 for factor f in Eq. (7.1), without relating to Eq. (7.1), and comments that the modified formula is in good agreement with experimental data for the cylindrical shells with radius-to-thickness ratio R/h between 500 to 1300. El Naschie (1975) suggests a value of $f = 0.093$, but in another paper El Naschie (1977) advocates for a value of $f = 0.187$. Volmir (1963) utilizes the value 0.186. Note also that Volmir (1963) provides for $R/h = 1000$ with probability between 0.9 and 0.99 the experimental buckling load as given by Eq. (7.1) with factor $f = 0.13$, this value being in agreement with that of Kirste (1954).

After listing Eq. (7.1), Pogorelov (1966, p. 193), comments: "Our all preceding examinations were related to the perfectly elastic, geometrically perfect shells. Real shell possesses a limited elasticity and its form is far from being perfect. Both of these facts may influence the results, in particular on the value of the critical load." He then investigates the influence of initial imperfections on the upper critical load, and that of the inelasticity on the lower buckling load, given in Eq. (7.1). As Pogorelov and Babenko (1991) write (see also Pogorelov, 1966),

> a number of control experiments with spherical and cylindrical shells were set up to confirm the theoretical results obtained by Pogorelov. The shells were produced by the method of vacuum deposition of copper on a steel backing, the effective surface of which was maintained with a high accuracy. The fact that experimental values [were] close to their classical for freely supported cylindrical shells under axial compression . . . [was] established.

This finding is in agreement with those of Almroth and Tennyson on closeness of the buckling loads of carefully manufactured cylinders with the classical buckling load.

Figures in Pogorelov (1966) and Pogorelov and Babenko (1991) indicate closeness of experimental results for the postbuckling behavior of shells in their experiments with the lower bound formula (7.1). Various applications of Pogorelov's theory are given by Ivanova and Pastrone (1988, 2002). Ivanova and Trendafilova (1992) utilized it in conjunction with random imperfection sensitivity. Pogorelov and Babenko (1991) do not comment on the statement made by Koiter and Hutchinson (1970): "It is quite possible that a paper by Hoff *et al.* (1966) has put an end to the quest for the minimum load which the buckled shell can support." On this topic, Hunt (2011) mentions: "This has turned out not to be the case." We will return to this point when discussing papers by Hunt later on in this section.

The paper by Calladine (2001) about the deterministic resolution of the conundrum appears to be a must of mentioning. He writes Eq. (1.2) in the following form:

$$\frac{\sigma_{\text{cl}}}{E} = \frac{0.6\,h}{R} \tag{7.2}$$

where σ_{cl} is the compressive buckling stress. Empirical observation, however, can be approximated as

$$\frac{\sigma_{\text{mean}}}{E} \approx 5\left(\frac{h}{R}\right)^{1.5}. \tag{7.3}$$

Two standard deviations of the measurement above or below the mean require the constant to be changed from 5 to 10 or 25, respectively.

Calladine (2001) writes:

> Koiter (1945) made a seminal early study in this field. He took an asymptotic approach to the non-linearities, and focused attention on two questions, as follows: (a) Why are the experimentally observed buckling loads significantly lower than the predictions of the classical theory? (b) Why do the experimental observations have so much scatter?
>
> His explanation was in terms of *imperfection-sensitive buckling loads* and the *unavoidable presence of small imperfections*; and this has been abundantly fruitful in the general field of the mechanics of buckling and stability. But in Koiter's approach to the problem one asks, essentially, why the experimental observations do not agree with Eq. (4.10). This is actually a different question from the one I take to be more central, *viz*. Why does Eq. (4.11) have an exponent of 1.5 rather than 1.0? This, then, is my main task in the present paper.

Based upon experiments on small-scale, open-topped, silicone-rubber cylindrical shells loaded by self-weight (Calladine and Barber, 1970; Mandal and Calladine, 2000) Calladine and his co-workers advanced "a working hypothesis that the buckling phenomena in 'ordinary' buckling experiments are closely related to those of self-weight buckling," since Eq. (7.3) describes precisely the experimental self-weight compressive stress at the base of shells at the point of buckling. By means of finite-element computations, authors showed that "there is a post-buckling mode for which the load remains near a 'plateau' value. The mode involves the growth of a dimple near the base, which allows the upper part of the shell to deform inextensionally, falling outwards over the dimple region."

Moreover, analysis conducted by Guggenberger *et al.* (1999) "of dimples near the edge of locally-loaded shells shows that the dimple supports a more-or- less constant force proportioned to $Eh^{2.5}/R^{0.5}$, irrespective of the size of the dimple" a result a more-or-less consistent with a Pogorelov-type analysis of the dimple. "The experimental observations on self-weight post-buckling are consistent with the force holding the dimple in place being provided by the weight of the shell vertically above the dimple, so that the critical height depends on $h^{1.5}$, other things being equal." Calladine's experiments on small-scale silicone-rubber shells showed much less "scatter" than those on "ordinary" shells. He attributes this to the *static-determinacy* of open-topped shells, in contrast to the static indeterminacy of "ordinary" shells, which allows for "locked-in" stresses of unknown, but random, magnitude before the shell is loaded. This idea is supported by experiments reported by Calladine, Lancaster and Palmer (2000), on a cylindrical shell in which the shell wall was clamped to the end-discs by circumferential pre-stressed bands. Vertical stresses entering the shell were thus limited by friction — as the system was less statically indeterminate than shells with firmly connected end-discs; and the buckling loads measured in repeated tests had considerably less scatter than those for "ordinary" shells.

Calladine (2001) concluded his paper by the following thoughts:

> I hope that my arguments will not be assailed on the grounds that they lack absolute precision; for my "plateaux" are only flat "to first order", and my numerical constants are given only to about one significant figure... I take comfort from Francis Crick's remarks to the effect that some problems are so difficult that they can only be solved by *a process of over-simplification.* And indeed, my work is an example of what Robert May has called "the lie that tells the truth.

These intriguing conclusions resonate well with the statement in another paper (Lancaster *et al.*, 2000) that the aim of the authors was "investigating this putative 'missing ingredient' in current thinking about shell buckling." The authors noted that their proposed explanation of the conundrum was "qualitative and hypothetical. However, the hypothesis is testable."

Lower-bound estimates developed by Batista (1979), Batista and Croll (1979), Croll (1981, 1995), Ellinas and Cross (1981), Gavrilenko and Croll (2001a, 2001b), Gavrilenko (2003, 2007), Gavrilenko and Matsner (2007), Gavrilenko *et al.* (2000), Wang (2009), Sosa *et al.* (2006), and Wang and Croll (2008, 2013) appear to be of great interest. Croll (2006) writes:

One of the important objectives of much non-linear shell mechanics during the 1960s was the determination of a theoretical method for predicting the lower bounds to the imperfection-sensitive buckling of shells. After all, even then it was clearly recognized that such lower bounds existed. Increasingly vast collections of experimental buckling loads, when normalized with respect to theoretical classical critical loads and plotted against suitable geometric parameters, displayed clear evidence of lower bound plateaus being reached as the levels of imperfection were increased. Most design codes of the day reproduced these lower bounds in the form of one or other design envelopes distinguished from each other by the proportion of the reported buckling loads that fell above them. I was not immune from this goal of finding the theoretical basis for this lower limit to imperfection-sensitive buckling.

When the penny eventually dropped, the breakthrough was largely the result of luck and a good measure of laziness. Requiring a theoretical basis for the interpretation of some wind tunnel tests on the buckling of cantilevered cylinders and hyperboloids, the simplest set of available shell equations was adopted for the purpose of calculating critical eigenvalues. The resulting critical pressures showed remarkably good agreement with the observed test buckling pressures. Believing that the eventual report to the funding agency would require rather more sophisticated theoretical treatment these earlier comparisons were eventually checked against the solutions obtained for the set of full shell equations. It was with some surprise that the results from these more complete shell equations gave critical eigenvalues that were considerably higher than those obtained using the original, simpler, set of equations. Pondering this over the next few months it gradually dawned on me what was the source of this apparent paradox. Inadvertently, the original equations used were close to those adopted by Rayleigh for his investigations of the vibration characteristics of bells. These differed from the equations advocated by Love, and it might be recalled provided the substance of a fairly bitter disagreement between the two, in so much as Love included in his equations a more complete representation of the terms representation of the terms representing the membrane stretching of the shell's surface. Rayleigh's equations represented an effectively inextensional model of the shell vibrations; a not unreasonable model for bells having open boundaries. It was this recognition of the strong influence of membrane actions in controlling the critical eigenvalues of shells that was to provide my answer to the quest for lower bounds to the buckling of shells.

It is difficult to recall the exact sequence of events, but a fairly early stage it was realized that any significant non-linear behavior of structures was derived from changes to the initial membrane stiffness. This observation is as true of the transition from a membrane fundamental state to a purely bending state in say the imperfection-driven buckling of an axially loaded cylinder. It is equally true in say the buckling of plates where a largely inextensional critical buckling mode regains stiffness in the non-linear, post-buckling, regime as a result of an increase in the contribution of membrane stiffness. Virtually all non-linearity of any practical significance is the result of changes in the membrane stiffness. Or much more helpfully, any significant structural non-linearity is the result of changes to the membrane energy of resistance.

This then leads to the deceptively simple notion that if in shell buckling there is a loss of stiffness, or alternatively energy, then this energy (stiffness) contribution would have had to be present in the initial critical buckling mode. For post-buckling of bifurcating systems this in turn means that the distribution of energy within critical buckling mode should be capable of revealing which if any components are likely to be lost as a result of the modal couplings taking place in the post-buckling behavior. Put more simply, if

> it is not there to begin with it cannot be lost. As a lower bound to this loss of membrane energy would be a critical load within the particular mode from which the at risk membrane energy has been removed. The resulting critical load was initially referred to as the "quasi-inextensionnal" but later changed to "reduced stiffness" buckling load. This simple idea worked convincingly for many classes of shell buckling problem.

In his recent article Thompson (2013c) writes:

> "... Croll has developed an important analytical technique, see for example (Yamada and Croll, 1999; Croll and Batista, 1981), using his reduced-stiffness method which, by deleting certain energy terms in the formulation, aims to derive a lower bound to the experimental failure data."

1.2. *Local buckling*

Hutchinson *et al.* (1971) stress that "although imperfections distributions are likely to be random in nature, it is often observed that local dimples or shape imperfections are present in shell structures." Local buckling of shells was apparently pioneered by Rabotnov (1946). Further contributions include those by Ohira (1961, 1965), Shirshov (1962), Almroth (1966), Lipovtsev (1968), Vladimirov *et al.* (1969), Hutchinson *et al.* (1971), Amazigo and Budiansky (1972), Koiter (1974, 1978), Gristchak (1976), El Naschie (1975a, 1975b), Kaoulla (1977), Bauer (1978), Evkin *et al.* (1978a, 1978b, 1978c), Tovstik (1982, 1984, 1991, 2005), Mikhasev (1984), Krasovsky (1990), Murray (1997), de Vries (2006), Mamai (2011), Krasovsky *et al.* (2011), Khamlichi *et al.* (2004), Gavrilenko and Krasovsky (2004b, 2004c), Limam *et al.* (2011), and possibly others.

Specifically, Amazigo and Budiansky (1972) have derived a formula for the buckling load of an infinitely long cylindrical shell under axial compression containing a localized axisymmetric imperfection. As Hutchionson *et al.* (1971) mention, "their formula is an asymptotic one which is valid for sufficiently small imperfections in much the same way as is Koiter's (1945, 1963) formula for a sinusoidal axisymmetric imperfection." The theoretical and experimental investigation of the effect of local imprefections was conducted by Hutchinson *et al.* (1971). They demonstrated that

> localized dimples in constant thickness cylindrical shells have an effect which is somewhat less severe than a sinusoidal axisymmetric imperfection. Almroth (1966) has made the same observation on the basis of his studies of the effects of axisymmetric imperfections on cylindrical shells. Axisymmetric dimple studies of conical shells lead to similar conclusions (Schiffner, 1965; Arbocz, 1968b).

Koiter (1974) revisited this issue; he first quotes his classic equation

$$\left(1-\frac{\lambda^*}{\lambda_1}\right)=-\frac{6\cos\lambda^*}{\lambda_1}, \tag{7.4}$$

which is governing the buckling load of the imperfect structure with periodic imperfections in the form

$$w_0 = kh[\cos 2m\alpha + 4\cos(m\alpha)\cos(m\beta)]. \tag{7.5}$$

Then he considers the local imperfections of the form

$$w_0 = kh[\cos 2m\alpha + 4\cos(m\alpha)\cos(m\beta)]\exp\left[-\frac{1}{2}\mu^2(\alpha^2+\beta^2)\right] \tag{7.6}$$

and drives the following equation

$$\left(1-\frac{\lambda^*}{\lambda_1}\right) = -\frac{4\cos\lambda^*}{\lambda_1}. \tag{7.7}$$

Koiter makes following remarks: "The significant conclusion to be drawn from our first approximation is that more or less localized imperfections of the shape described by (7.5) and (7.6) are equally harmful as imperfections of a periodic type (7.5) with amplitude reduced by a factor of 2/3." Moreover, Koiter (1974) stresses:

> We emphasize that this result is only a first approximation, valid for sufficiently small values of μ^2/m^2. It is highly desirable to amplify this analysis by the evaluation of the terms of order μ^2/m^2, in order to ascertain whether the first approximation remains an adequate approximation when imperfections (7.5), (7.6) extend over only a few wave lengths.

Koiter's (1974) unpublished report appeared in 1978, in a slightly modified form. The task of investigation the second approximation was conducted by Gristchak (1976) conducted under guidance of Koiter. Gristchak (1976) concludes:

> A second approximation calculation of the effect of more or less localized short wave imperfections have been carried out. These imperfections are equally harmful as those of a periodic type with amplitude reduced by a factor of 2/3, as also was found in a first approximation study (Koiter, 1974). The influence of the factor μ^2/m^2 leads to some increase of the value of the parameter of buckling stress λ^*/λ_1, but not so much: for example, for the case $\mu^2/m^2 = 0.1$ at $kc = 0, \ldots$ the value of the factor λ^*/λ_1 in a second approximation increases only by 11.6 percent in comparison with the value of $\lambda^*/\lambda_1 = 1$ which corresponds to the buckling stress in a first approximation for the perfect shell.

In the subsequent publication, referring to Gristchak's (1976) result Koiter (1978) makes a comment to the effect that his first approximation is "adequate for most practical purposes."

Numerous topics associated with buckling of composite shells with imperfections were treated by Vanin and Semenyuk (1987), Semenyuk (1987a, 1987b), Semenyuk and Zhukova (2006), Sheinman and Goldfeld (2001, 2003, 2004), Tennyson (1995), Tennyson and Muggpridge (1973) and others.

1.3. *Localization*

Tvergaard and Needleman (1980, 1983, 2000) devoted their pioneering studies to the phenomenon of localization of the buckling patterns. In the earlier paper, the authors investigated the "possibility of localization of a buckling pattern... for a class of structures in which the initial buckling mode is periodic." They showed, on a simple model, that the "basic mechanism of localization involved a bifurcation at the maximum load point." In their later study, Tvergaard and Needleman (2000) studied localization in a cylindrical panel under axial compression. They established analogy of localization phenomenon with plastic flow localization in tensile test specimens. They concluded that "for the cylindrical panel... buckling localization develops shortly after a maximum load has been attained, and this occurs for elastic-plastic panels."

Recent revival of localized buckling pattern is due to three principal reasons. First two of these are associated with *linear* problems, whereas the third one manifests itself in *nonlinear* setting. The localization appearing in the linear problems is marked by the publication of the paper by Pierre and Plaut (1989), illustrating an interesting curve veering due to small misplacements in the structure. This model was further expanded in models by Zingales and Elishakoff (2000) and Challamel *et al.* (2006). Buckling and vibration mode localization was studied both in deterministic (Luongo, 1992; Nayfeh and Hawwa, 1994; Mikhasev, 1984; Elishakoff *et al.*, 1995; Li *et al.*, 1995a; Mikhasev and Tovstik, 2009) and probabilistic (Ariaratnam and Xie, 1995; Xie, 1995, 1997, 1998; Xie and Elishakoff, 2000) settings. Challamel *et al.* (2006) note

> Localization has been encountered in many engineering problems, especially when inelasticty occurs in a small part of a structure. Historically, the plastic hinge concept is probably the first appearance of such a phenomenon in structural mechanics. The phenomenon has been initiated by Coulumb (1773) in the 18th century, when studying earth retaining walls of military fortifications: Coulumb introduces the fundamental reasoning of the static approach by outside which is equivalent to a kinematic approach by a rigid block (Salencon,1990)... Nowadays, the localization phenomenon is encountered in many inelastic problems. This includes the onset of necking in a tensile test specimen, or the formation of shear bands, or a localization of a buckling pattern (see, for example, Tvergaard and Needleman, 1980 and Bažant and Cedolin, 2003)... It is quite a paradox that most researches dealing with localization in elastic structural mechanics have only begun in the 90s (see the bibliographical study; Luongo, 1993). The localization phenomenon has been found in many elastic stability (and dynamics) problems of continuous structures such as local buckling and overall buckling of thin-walled members (Luongo, 1993 or Coman, 2004) or buckling of repetitive structures in presence of irregularities (Pierre, 1988; Pierre and Plaut, 1989; Li *et al.*, 1995). Localization in a multi-span elastic column may indeed appear by slightly perturbing the length of each column in presence of additional external springs. Localization here means that one span buckles with larger amplitude than the other spans. Let us note that

localization in a multi-span elastic-plastic column has also been found by Needleman and Tvergaard (1982), by introducing an imperfection in the deflection.

This type of localization phenomenon is often referred to as Anderson localization, after the Nobel laureate physicist Phillip Warren Anderson (1958) predicted localization of electronic wave functions in disordered crystals, with attendad absence & diffusion of waves.

Benaroya (1996) and Xie (2000) edited special journal issues pertaining to the Anderson localization in various structures. Different facets of localization in stability setting have been elucidated by Kyriakides and Ju (1992), Ju and Kyriakides (1992), Ambartsumian and Belubekyan (1994), Belubekyan (2008), Belubekyan and Chil-Akobyan (2004), Coman and Houghton (2006), Banichuk and Barsuk (2008), and others.

The second reason for revival of the studies in localization is the series of papers conducted at St. Petersburg State University, Russia by Tovstik and his associates. Monographs by Tovstik (1995) and Tovstik and Smirnov (2001) provide appropriate theoretical background for the localization in the *linear* buckling context. Mikhasev and Tovstik (2009) define localized modes as those "modes which are concentrated in some small fixed regions (referred to as most weak) or moving lines or points on the middle surface, and exponentially decaying when distanced from the lines or points. One or both dimensions of localization region are smaller in comparison with the dimensions of the entire middle surface; hence for constructing localized forms of vibration asymptotic methods are used." The authors stress that "the localization is possible only when symmetry is violated (cylinder is noncircular and/or edges are inclined."

The third reason for renaissance of localization studies is constituted by the British school of applied mechanics. It deals with inherently *nonlinear* problems. Specifically, in the paper by Thompson and Virgin (1988), authors note: "It could be that... geometric concepts and techniques are ripe for a fruitful re-interpretation into the spatial domain, with the space coordinate s replacing the time t as the independent variable." They presented "an example of spatial chaos and localization in the planar deformations of an elastic rod..." They also treated "the response of an axially compressed beam on an elastic foundation, equivalent to the rotationally-symmetric deformations of a cylindrical shell" with softening cubic nonlinearity, with restoring force represented as $F = ky - ay^3$. The authors demonstrated that at values of compressive load P below its classic linear buckling counterpart P^c "the spatial localization corresponds to the homoclinic orbit leaving and returning to the phase-space origin." Hunt *et al.* (1989) (see also Bolt, 1989) reviewed localization phenomenon "from three complementary viewpoints: (a) from a modal perspective, (b) from a formulation which allows the amplitude to modulate in an asymptotically defined 'slow' space and (c) from a dynamical analogy in phase space suggested by the form of the underlying differential equation." Authors mention that "examples of localized buckling exist, railway tracks or pipelines on frictional beds for instance..." They stress:

> The restriction of a displacement pattern to a localized part of the available spatial domain is a practical feature of many buckling situations, and introduces an extra level of analytical complexity when compared with periodic buckle patterns (Potier-Ferry, 1983). Apparent localization may be caused simply by stress variations throughout a structure, because of, perhaps, the (modulating) presence of a second overall mode (Koiter, 1976); the buckle pattern then can apparently localize under continuously increasing load, albeit with a unique post-buckling solution. The pure localization of interest here, however, is associated only with a falling (unstable) equilibrium path, and is thoroughly multi-valued. As in the corresponding process of plastic necking (Tvergaard and Needleman, 1980), it can develop with equal likelihood in one of several regions, and a typical solution is then merely one of a number of competing possibilities; reminiscent of the dynamical phenomenon of chaos, the total topological description must accommodate many (strictly infinite) possible alternative equilibrium states.

The authors considered a beam on nonlinear elastic softening foundation, with restoring force $F = ky - ay^m$ under axial load P. The negative sign in the nonlinear term $-ay^m$ ensured that for even powers of m the foundation destabilizes with increasing y. Potier-Ferry (1983) and Thompson and Virgin (1988) treated only the symmetric case of $m = 3$. Hunt *et al.* (1989) considered the asymmetric case $m = 2$, "regarded as paradoxical by Arnold (1989)," as the authors note. In the paper by M.A. Wadee (2000) and M.K. Wadee and coauthors (Hunt and Wadee, 1991; Wadee and Bassom, 2000a, 2000b, 2011; Wadee *et al.*, 1997, 2000, 2003, 2004) studied buckle pattern localization in the context of it being a "fast emerging as an important area of study in structural mechanics." The method includes perturbation analysis, through partially numerical and partially analytical treatment (M.K. Wadee *et al.*, 2000), to fully numerical treatment (Champneys and Toland, 1993). In several studies, the authors consider what they refer to as "the archetycal case of an embedded strut" (M.K. Wadee *et al.*, 2000) or as "a model localized buckling problem" (M.K. Wadee, 1999).

Specifically, M.K. Wadee and Bassom (2000a) consider the case of a column on a quadratic-cubic foundation with restoring force given as $F = ky - c_1y^2 + c_2y^3$. For P just less than P^c, $P = 2 - \varepsilon^2$, with ε much less than unity, localized solutions have been obtained by utilizing a double-scale perturbation scheme. Localized buckling patters are portrayed in figures of the article, and the authors show that when approached value of 38/27 localized buckling pattern is no longer feasible.

M.K. Wadee *et al.* (2000) compared two versions of the Bubnov–Galerkin method, one in which the linear eigenvalue information is employed, wherein the latter version in the second, the shape functions are utilized as free variables. In the latter case "the assumed solution need not be a linear superposition of modes" turning to be preferable to the standard version. M.K. Wadee *et al.* (2004) revisited the column on quadratic-cubic foundation and compared the buckling solutions derived via using the numerical code AUTO97 with the Rayleigh–Ritz method.

M.K. Wadee and Bassom (2011) studied the cubic-quintic foundation and again found localized solutions near the value of classical critical buckling load. Interested readers can also consult papers by Coman (2004a, 2004b, 2010) and by Coman and Houghton (2006).

These studies perhaps could be generalized by incorporating initial imperfections. Indeed, in the paper by Thompson and Virgin (1988), the authors do mention initial imperfections. According to them, at P less than P^c, "the spatial localization corresponds to the homoclinic orbit leaving and returning to the phase-space origin. So we can conjecture that the amplitude of any local buckling mode that might be induced in a beam by small imperfections or dynamic disturbances will tend to zero as P is increased to P^c." Lagrange and Averbuch (2012) do take into account initial imperfections in the study of localization phenomenon. They note the existence of "a maximum imperfection size which leads to a limit point in the equilibrium curve of the system. The existence of this limit point is very important since it governs the appearance of localized phenomena."

We should also mention papers by Hunt *et al.* (1993, 2003), Lord *et al.* (1997), and Hunt (2006, 2011). Hunt (2006) argues that at it is the "*minimum energy density periodic solution*, picked out at the Maxwell load that governs the final buckle pattern, Maxwell load being defined as the load at which the energy levels in the unbuckled and buckled periodic states are the same." To digress, this author learned first about Maxwell load from the definitive paper by Chater *et al.* (1983) who studied buckle propagation on a beam on a nonlinear foundation. It is interesting to note that Maxwell's thermodynamic criterion was apparently first used to the shell buckling problems by Tsien (1942) but was rejected (Fig. 7.1). The idea was resurrected by Chater, Hutchinson and Neale (1983) and later by Hunt and Neto (1991, 1993), although in different contexts. In the latter paper, the authors write

> For long structures, where the response may localize over a portion of the length, a time-like interpretation of the spatial dimension has a great appeal... With the focus exclusively on localized response, it is then possible to predict maximum amplitudes directly from energy considerations. For problems like the axially loaded cylindrical shell, which first destabilize and then restabilize in the post-buckling range, such amplitudes are apparently only attainable for loads greater than the classical Maxwell critical load, where the energy levels on the unbuckled fundamental path and restabilized post-buckled path are equal. Maxwell load should thus represent a lower bound on the appearance of localized response.

The authors chose, "as the first check on this new theory, to compare with the careful experiments of Yamaki (1984) on axially loaded cylindrical shell. The agreement

Fig. 7.1. Hsue-Shen Tsien's (1911–2009) bust at the Museum of the National University of Defence Technology, Changsha, People's Republic of China (courtesy of Dr. Yuan Li).

with minimum experimental load, and the corresponding wavelength, is found to be excellent." Hunt (2011) notes that "the Yamaki (1984) cylinders are not really long enough to see whether localization occurs along the length, but other experiments with longer samples (Esslinger and Geier, 1972, e.g.) show quite clearly that it does."

Hunt *et al.* (2003) and Hunt (2006, 2011) suggest the following mechanism for the selection of the number of buckling waves in circumferential direction, depending on the length of the shell:

— For relatively short shells, the critical buckling mode is that which occurs on the Koiter's circle with a single half wave over the length L of the cylinder. This is the suggestion made by Yamada and Croll (1999).
— For longer shells, the mode likewise occurs on the Koiter's circle, but comprises two half waves over the length.
— For even longer shells, the mode may theoretically span three or more half waves over the length L; however, the available experimental data does not appear to contain shells that are sufficiently long for this to occur.

The latter part of the above quote implies the need of additional experimental investigations. The following natural questions arise: "Why not to look for localized solutions... in the imperfect structure? Why is it not possible to use asymptotic methods or numerical methods in the (realistic) imperfect structure — in conjunction with localization phenomenon? Why such an approach ought to yield the lower possible load for the imperfect structure (as treated by Yamaki, 1984) that was not introduced into the analysis? Also, Yamaki (1984) cylinders' length to radius ratio was about unity. Can such a shell be treated as a long one? It is possible though, that such seemingly short shells still exhibit the behavior of the long ones? How do the Maxwell loads relate to the formulas recommended in NASA SP-8007 (Anonymous, 1968)? Is it practical to use Maxwell loads for design purposes?"

It was refreshing to read the recent paper by Hutchinson (2013). He does take into account initial imperfections and the localization phenomenon in the study of the role of substrate nonlinearity in the stability of wrinkling of thin films bonded on compliant substrate, investigating the initial post-bifurcational range. It turned out that the "localization phenomena usually set in at compressive strains well above the bifurcation strain and are not captured by the initial post-bifurcational approach."

According to Hunt (2011), "the quest for the minimum load that the buckled shell can support (Hutchinson and Koiter, 1970) has refused to lie down and die." Indeed, numerous investigators, as described in this chapter, have pursued it. Likewise, even when one takes into account uncertainty, albeit non-probabilistically, one can determine the minimum load that the shell can support, as described in Chapter 6. Recently, NASA has shown a renewed interest in this topic (Haynie and Hilburger, 2010, 2012; Mispoli *et al.*, 2012). Edlund (2007) notes: "In a number of papers on buckling, especially of compressed cylinders, by Hunt and co-workers..., interesting approaches dealing with localized buckling, cellular buckling, Maxwell critical loads, homoclinic and heteroclinic bifurcation are presented."

Professor Thompson (2013) writes:

> "I would like to add some personal comments about the localization of post-buckling paths and the Maxwell energy criterion that has attracted a burst of activity in recent years in engineering mathematics groups in Bristol and Bath. These researchers have employed, with great success, the concepts and techniques of nonlinear dynamics using the static dynamic analogy.
>
> Applied mathematicians study physical problems in mechanics for many reasons. Sometimes just out of curiosity, sometimes to explore quite fundamental issues in mechanics — providing insights that might spill over into other areas. Helping engineers in their design work may well be the last thing on their mind. This can be seen in fluid mechanics, in the topic of turbulence. An enormous amount of work has gone into examining fundamental models, such as the Taylor–Couette flow between counter-rotating cylinders, with no conscious thought that it might be of immediate help in aircraft design. This type of background activity is very important, and does often lead to useful results and advances in the general subject of mechanics. Meanwhile,

> between the applied mathematician and the practicing engineer is a very broad spectrum of activity, some of it very difficult to classify.
>
> To the applied mathematicians, the Maxwell 'energy criterion' load has certainly generated a lot of ideas about the organisation of equilibrium paths in the advanced post-buckling of long-thin structures which exhibit severe shell-like imperfection sensitivity, including in particular those paths that involve spatial localization. This work could be important in very many areas, far away from shell buckling, such as pattern formation, etc.
>
> My interest in the Maxwell load is certainly not to predict a lower bound to the scattered experimental shell buckling results. Rather, it is the possibility of escape from the unbuckled meta-stable trivial solution due to lateral dynamic or static disturbances. These disturbances can be very important in engineering environments, so they should certainly be deeply discussed and examined.
>
> The first thing anyone realizes is that the depth of the alternative (post-buckled) potential well has no fundamental relevance to escape, which is entirely dependent on the height of the intervening energy barrier. Curiously, within the recent explosion of interest in the Maxwell load, I find that the height this barrier has only once been estimated, in the excellent paper by Horak, Lord and Peletier (2006).
>
> Working with Gert van der Heijden at University College London, I have now shown (Thompson and van der Heijden, 2014) that for a spectrum of post-buckling problems (including twisted rods and beams on a nonlinear elastic foundation) there is a very significant and instantaneous lowering of the energy barrier once the Maxwell load is exceeded. We are calling this 'shock sensitivity', and the reason for its occurence is easily described.
>
> It is now firmly established that localizing sub-critical post-buckling paths emerging from the classical critical buckling state are destroyed by what is called a heteroclinic collision at the Maxwell load. Now we find, in all cases studied, that the energy barrier of the localizing solution is quite close to the energy of a *single* periodic wave. But a single such wave is not kinematically admissible, and the corresponding periodic barrier must be calculated for all the waves in the long structure, N. Now in practice, N will be large, and does indeed tend to infinity with the length of the structure. Thus the shock sensitivity increases by a large factor of N as the Maxwell load is exceeded.
>
> This finding applies (even) to perfect shells with no geometrical imperfections. Of course it would seem sensible to check out the findings for imperfections shells. The reluctance of applied mathematicians to put in imperfections can, to some extent, be explained. The post-buckling of cylinders is already very complicated — the addition of one more complication will likely obscure some of the findings and reduce the elegance of the solution. It might even make the problem so complicated that no simple ideas can emerge. Nevertheless, it does seem important to address this issue.
>
> Indeed, the extreme sensitivity to localized side loads must surely be reflected in an extreme sensitivity to localized imperfections.

To summarize this brief section, we can note that the jury is still out on the deterministic understanding of the cylindrical shell buckling phenomenon. Additional theoretical studies are needed to see if the deterministic studies would be sufficient to resolve the perplexing behavior of cylindrical shells. Moreover, Thompson (2013b) associates Maxwell loads with "the enhanced sensitivity to *localized imperfections* in the shells that would surely accompany the demonstrated super-sensitivity to lateral disturbance. This is something that shall certainly be addressed by our mathematical colleagues, embracing, perhaps, the probabilistic viewpoint of Elishakoff (1983)." It appears that the above statement could be strengthened stressing that researchers, interested in deterministic and/or stochastic analysis are highly recommended to follow the above advice, for we do not want to overlook anything that may prove useful.

At the same time, it is rather hoped that the deterministic analysts will delve into the probabilistic and/or convex models of uncertainty. It appears that by doing so they would follow Karl Menger's (1724–1804) injuction: "It is vain to do with fewer what requires more." According to Max Born (1949) "The conception of chance enters in the very first steps of scientific activity in virtue of the fact that no observation is absolutely correct. I think chance is a more fundamental conception than causality." One cannot dismiss Nate Silver's (2012) sage advise: "We must become more comfortable with probability and uncertainty."

1.4. *Shell testing process*

Another currently abandoned avenue of research is a thorough investigation of the shell testing process itself. In series of papers Hoff (1951, 1954) and Hoff *et al.* (1952, 1955) dealt with dynamic investigation of the "ordinary column test" despite the fact that "no controversy has yet arisen on the behavior of a perfectly elastic column." Still, Hoff (1954, p. 33) notes that "the situation is not clear-cut when a short column is compressed in a very rigid testing machine." Note that the probabilistic analysis of Hoff's problem was undertaken by Elishakoff (1980). It appears advisable to study the buckling of the shell in the shell buckling test as a dynamic process. The analogous investigation of the shell buckling test in the dynamic setting may reveal additional insights into the imperfection sensitivity of shells. For additional work in this direction the interested reader may consult the papers by Babcock (1967) and Agamirov (1968).

1.5. *Modal density of eigenvalues*

Studies on mode interaction are numerous and well known and will not be listed here. However, there are some studies that appear to be beyond reach of Western authors. Kornev (1969, 1972a, 1972b, 1975a, 1975b, 1976), Bendich and Kornev (1971), and Kornev and Ermolenko (1980) advocate a connection between the drastic imperfection sensitivity and the high density of shell eigenvalues. Novikov (1988) and Goltzer (1986) argue that the deterministic instability represents a manifestation of internal resonance. These papers raise subtle points that deserve closer investigation.

1.6. *Imperfection-insensitive structures*

One of the most interesting recent topics in structural stability appears to be a theme on conversion of the imperfection-sensitive structure into an imperfection insensitive one. I recall discussing various topics with Professor Hans Besseling of the Delft University of Technology. He expressed the view that engineers ought to devote much effort to transforming imperfection-sensitive structures into imperfection-insensitive ones by some structural modifications. I asked him if he had a chance to discuss this topic with Professor Koiter. Professor Besseling replied that he never posed this question to Professor Koiter. This intriguing subject was investigated very recently. This very topic is extensively studied in the recent decade by Professor Herbert A. Mang of the Vienna University of Technology, Austria, and his group. The associated publications include the papers by Mang *et al.* (2006), Schranz *et al.* (2006), Steinboeck and Mang (2008), Steinböck *et al.* (2009), Hoefinger and Mang (2009), Jia and Mang (2010), and Jia (2010). Professor Mang (2012) informs: "In my talk at the Conference in the Georgian National Academy of Sciences I have compared loss of stability of an imperfection-sensitive structure to a malignant tumor. By means of sensitivity analysis of the initial post-buckling behavior, imperfection-sensitive structures may be converted into imperfection-insensitive ones. (Admittedly, architectural and/or functional constraints may render such conversions unfeasible.)

I have then compared loss of stability of an imperfection-insensitive structure to a benign tumor. In contrast to buckling of structures where the conversion from a malignant to a benign situation is frequently possible, it is up to now, to the best of my knowledge, impossible in oncology.

An arch bridge may serve as a practical example for what is meant by conversion of an originally imperfection-sensitive structure into an imperfection-insensitive one. The configuration of the bridge (span: 40 m, width: 6 m, maximum height of arch: 5 m) is shown in part (a) of Fig. 7.2.

A uniformly distributed vertical surface load $\lambda\bar{p}$ is applied on the deck, which is the geometrical form of a shallow cylindrical shell. The reference value $\bar{p}$ of the load is 0.004 kN/cm^2; λ is a dimensionless load factor. Bifurcation buckling of the desk is symmetric with respect to the drawing plane. Hence, II_0, in part (b) of the Fig. 7.2, is the projection of the secondary path onto this plane. The buckling mode as such is antisymmetric with respect to the transverse symmetry line of the deck. The monotonically decreasing curve II_0 shows that this structure is imperfection-sensitive. Applying hangers does not only result in an increase of the stability limit from S_0 to S_1 but also in a monotonically increasing projection II_1 of the secondary path onto the drawing plane, indicating a successful conversion of the originally imperfection-sensitive arch into an imperfection-insensitive structure."

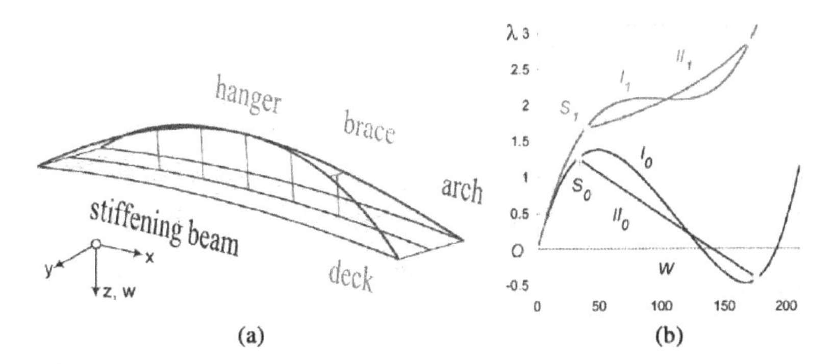

Fig. 7.2. Arch bridge: (a) configuration, (b) load–displacement paths of the center of the deck (S_0, I_0, and II_0 denote the stability limit, the primary path, and the secondary ("postbuckling") path, respectively, of a bridge without hangers: S_1, I_1, II_1 denote these items for hangers with a diameter of 8 mm.

A concept for imperfection-insensitive composite structures is identified as a robust design in a definitive study by Lee *et al.* (2010). Indeed, a robust design is defined by Phadke (1989) as one that is able "to make product's performance insensitive to variations in material, geometry, manufacture and operating environment."

Closely related paper is that by Obrecht *et al.* (2006). They write:

> . . . the work on foam-filled circular cylindrical shells reported in [paper by] Karam and Gibson (1995a, 1995b) suggests — on the basis of judiciously simplified analytical model analyses as well as experiments on silicon rubber cylinders filled with rubber foam — that compliant foam cores having comparatively high porosities and correspondingly low moduli may lead to remarkable increases in the load-carrying capacity of cylindrical shells while at the same time significantly reducing their intrinsic imperfection-sensitivity.

Some unintuitive results were reported by the above authors:

> . . . the addition of a fairly small amount of highly compliant material may stabilize the respective shell to such an extent that for an imperfection amplitude of the order of shell thickness the buckling load will exceed that of the imperfect unfilled shell by about a factor of 10, and the classical bifurcation stress of the hollow perfect shell — which normally is considered unattainable — is exceeded by a factor of 2.5.

It is impossible to disagree with the above authors when they state that "improvements of this magnitude may be considered rather unusual." One should note earlier papers on this topic by Almroth and Brush (1963), Goree and Nash (1962), Seide (1962), Evans *et al.* (1999), and Hutchinson and He (2000).

These papers answer, in a certain way, questions that were posed in an electronic message, to Arbocz by Budiansky (see, for example, Arbocz and Singer, 2000, p. 42):

> "One other question that has been nagging at me for a long time is this: 'Are we necessarily doomed to accept forever the unhappy coexistence of efficient shell design and

imperfection-sensitivity? Or are there some structural design secrets to be discovered that will retain minimum weight and rid us of the curse of imperfection sensitivity?'

Still, deterministic analysis of shell buckling does not appear to be closed either. Bodner and Rubin (2005) write:

> Despite enormous attention that has been given to the problem of the buckling of thin-walled elastic cylindrical shells under axial compression, there is still some uncertainty with the proposed treatments. This is evidenced by the occasional notes and comments on the subject that still appear in the literature. One of the reasons is that the main features of geometric imperfections that influence post-buckling response have not been fully identified. Recent approaches to buckling load calculations include high-fidelity analyses that emphasize the need for sophisticated numerical methods and an international data bank of shell imperfections (Arbocz and Starnes, 2002; Singer *et al.*, 2002).

Recent papers by Lopanitsyn and Matveyev (2011a) is tellingly titled as "*The Possibility of a Theoretical Confirmation of the Experimental Values of External Pressure of Thin-walled Shells.*" Another paper of the same authors (2011b) informs that Lopanitsyn and Frolov (2008) also studied the circular cylindrical shells under axial compression. We were unable to obtain the copy of the latter study. For recent comprehensive review of deterministic approaches to shell buckling the reader can consult with paper by Edlund (2007).

It must be borne in mind that the stochasticity-based research ought to utilize the most adequate deterministic theories that are available. In this sense stochastic analysis *is not detachable* from the deterministic one.

2. Comments on Some Later Non-deterministic Works

> *This tends to encourage the hope that quantitative predictions of the buckling strengths of imperfection-sensitive structures eventually may be possible on the basic of the knowledge of a few statistical parameters descriptive of the imperfections.*
>
> B. Budiansky and J. Hutchinsan (1966)

> *There are known knowns; there are things we know we know. We also know there are known unknowns; that is to say we know there are some things we do not know. But there are also unknown unknowns — there are things we do not know we don't know.*
>
> D. Ramsfeld (2002)

It appears instructive to start with a quote from the paper by Schenk and Schuëller (2003) commenting on the study by Elishakoff and Arbocz (1982) as follows:

> Elishakoff and Arbocz (1982) treated the Fourier coefficients of measured imperfections as random variables, whereby the histogram of the limit loads has been obtained either by Monte Carlo simulation or approximate methods. Common to these approaches is the determination of the limit load by means of analytical or semi-analytical procedures. This may also be the reason for the stochastic representation of geometric imperfections

> by means of a two-dimensional Fourier series with random Fourier coefficients, since an analytical buckling analysis of cylindrical shells yields a two-dimensional Fourier series representation of the critical modes. However, it is well known that severe shortcomings of analytical or semi-analytical approaches, respectively, for predicting buckling, are on one hand their limited capability of modeling more complex shell structures and on the other hand the implementation of a more general type of imperfections.

Schenk and Schuëller (2003) utilized Karhunen–Loève expansion "replacing the traditional Fourier series representation." However, the authors overlooked that "traditional Fourier series representation" and Karhunen–Loève expansion yield identical results for failure probability (Bourinet, 2012), although the Kahrunen–Loéve expansion may lead, in the paper by Schenk and Schuëller (2003), to reduction of a number of dominant random variables. Indeed, Bourinet (2012) informs:

> The goal of K.L. [Kahrunen-Loeve] series expansion is to coincide with the mode expansion, at a given accuracy due to truncated form of the K.L. series. The K.L. series expansion has the main advantage of reducing the number of random variables required for modeling the imperfection field. The results obtained so far show no difference between the two stochastic models in terms of failure probability.

It also appears puzzling to read (Schenk and Schuëller, 2003) about the so-called "well known . . . severe shortcomings of analytical or semi-analytical approaches," for there is no need to employ, as proverbial wisdom maintains, tanks in order to fight the flies. If a problem can be solved analytically or semi-analytically, why should we resort to the FEM? Don't we expose in our strength of materials courses, beam bending problems, for example, by analytical means? Had above authors (2003, p. 1120) addressed the problem of "either reinforced or unreinforced large cutouts, which are in most cases not amenable for analytical solutions," one could fully understand and justify the preference to the FEM. Still, they could have chosen to use the universal technique, namely, FEM, without apparent criticism of the "analytical or semi-analytical approaches" and compared them with the results yielded by the non-FEM numerical codes developed by Arbocz (1987). It appears that there is a sufficient place under the Sun both for analytical methods and FEM. The former can at least be utilized for the verification of the latter, otherwise one would follow an anonymous "advisor," who quipped: "Don't think, use computer." Naturally, we all share the view on the extreme importance of the FEM. Indeed, in the paper devoted to Dr. Starnes (see Elishakoff, 2006, p. 153) it is stated: "It appears that FE method based general program is needed for the attendant reliability analysis."

According to de Borst (2002)

> . . . Arbocz emphasizes a hierarchical approach for shell design. His philosophy: to get a good view on a shell's behavior one should apply programs based on simplified and well known theories. This allows designer to understand the behavior of the shell and improve the accuracy of the solution gradually without reverting to lengthy computer simulation ("sledgehammer approach").

Along these lines Eli Sternberg (1985) said in his Timoshenko Medal acceptance speech:

> . . . no one would earnestly question the enormous value of computing to applied mechanics and indeed to all of applied science. Yet computing without a proper theoretical background can be hazardous to the public health. Thus there are grounds to worry about the rising traffic in finite-element codes for stress analysis, which are often secret codes (in the sense that their theoretical basis remains a closely guarded secret) and the authors of which are occasionally uninhibited by a more than cursory acquaintance with the theory governing the problems they purport to solve.

In an analogous vein, Oden and Bathe (1978) write about "overconfidence in modern computing capabilities" and "computer overkill, which refers to the use of existing large sophisticated codes to study systems which could be better understood, at considerably less expense and effort, by the use of much simpler models in the hands of those with deeper comprehension of the physical phenomena and the basis of the methods available to analyze them."

According to Bažant (2009), "brute-force computer simulations, of course, cannot provide full understanding. But, if carefully calibrated, they can extend the experimental evidence and reveal the essential trend. Thus, one can get a clue for an analytical model — the ultimate prize." It should be noted that in their other papers, Schenk and Schuëller (2002, 2007) dealt with shells with cutouts; in these cases resorting to the FEM was fully justified due to the absence of "semi-analytic" approaches. Semi-analytic approach was utilized recently by Kriegesmann *et al.* (2012) to deal with axially compressed stiffened composite shells.

The paper by Cederbaum and Arbocz (1996) is close, in its spirit, to the study by Elishakoff *et al.* (1987). Both of these studies deal with the Hasofer–Lind (1974) method, which is based on the linearization of the nonlinear performance function in the vicinity of the design point. On the other hand, it appears that the study by Cederbaum and Arbocz (1996) represents, in a certain sense, a step back, in comparison with the investigation conducted by Elishakoff *et al.* (1987).

Indeed, the former uses the asymptotic formula by Koiter (1945) whereas the latter employs the advanced numerical code. Additionally, with Koiter's (1945) asymptotic relationship, one can obtain an exact expression for reliability, as was done by Roorda (1980). Hence, the study by Cederbaum and Arbocz (1996) ought to be considered as a comparison between Roorda's (1980) (Fig. 7.3) exact solution and the result that is obtainable by the Hasofer–Lind (1974) method. For comparison of the Hasofer–Lind method with exact solutions the interested reader can consult with the paper by Elishakoff and Hasofer (1985).

In another paper, Cederbaum and Arbocz (1996) utilize Koiter's (1945) asymptotic relationship:

$$(1 - \lambda_s)^{3/2} - Q\lambda_s|\xi| = 0 \tag{7.8}$$

Fig. 7.3. Professor John Roorda (1939–1999) of the University of Waterloo was an active contributor to the random imperfection sensitivity of structures (obtained through courtesy of Prof. Arde Guran).

where $\lambda_s = P_s/P_{cl}$ is the ratio between the applied load, which causes buckling and the classical buckling load of the perfect shell, Q is a constant, and ξ is the non-dimensional initial imperfection amplitude. The absolute value of ξ is used, since either positive or negative imperfection causes the same reduction of the buckling load. The actual initial imperfections are modeled as the following function:

$$\bar{w}(x, y) = \xi h \sin \frac{k\pi x}{L} \cos \frac{ny}{R} \tag{7.9}$$

where h is the shell thickness, L and R are shell length and radius, respectively, k and n are half wave and full wave numbers along the cylinder length and circumference, respectively. With formula (7.8) Cederbaum and Arbocz (1996) derive some simple reliability expressions. In other cases, when exact solution is unavailable, the authors resort to the Hasofer–Lind (1974) method. Within this method, the involved random variables are transformed into standard normal variables. This can be done by utilizing the Rosenblatt transformation in the general case of correlated or independent nonnormal variables, or by employing the Cholesky decomposition for correlated random variables. Then, in the space of normalized variables the distance β from the origin to the nonlinear failure surface allows the determination of reliability:

$$R = \Phi(\beta) \tag{7.10}$$

where $\Phi(x)$ is the cumulative normal distribution function. The authors conclude that "the loading randomness has a major effect on reliability." Such an effect obviously should have been expected.

The loading randomness, in addition to random imperfections, was addressed in the subsequent paper by Cederbaum and Arbocz (1997). Referring to treating initial imperfections as a random variable by Bolotin (1958, 1962) and Roorda (1969, 1971, 1980) the authors wrote:

> The enthusiasm for the new random variable was so big that, to our best knowledge, all other possible random variables received a deterministic treatment and thus, the reliability of such structures was considered as a univariant problem — that of the initial imperfection. The additional effect of the loading randomness on the reliability prediction of shells within the Koiter's formulae was recently investigated by the authors of the present work (Cederbaum and Arbocz, 1996).

Cederbaum and Arbocz (1997) were apparently unaware of this author's paper (1983b) as well as the paper by Li *et al.* (1995), which dealt with the effect of load imperfections. It was shown that the load randomness can have *either beneficial or detrimental effect* on structural reliability: it may lead, as was shown in Chapter 3, to either increase or decrease of reliability that was estimated by treating only initial imperfections as random quantities. Analogous conclusion was arrived at in the extensive study by Elishakoff *et al.* (2010).

In the paper by Dubourg *et al.* (2009) the authors utilize asymptotic numerical method by Baguet (2001) and a shell finite element model within the EVE code. The stochastic model takes into account the initial imperfections in the form of random combination of the most critical modes, in addition to randomly varying shell material properties and thickness. According to Bourinet (2012), the analysis

> results in 4 design points of equal weights in the reliability analysis, which represent worst case scenario at each corner of the shell (small values of the Young's modulus, yield strength and thickness fields, high values of the geometric imperfection). Combining the 4 corresponding linear limit-states of the FORM analysis in series system analysis gives a failure probability which is roughly the same as the one obtained by simulations.

(see also the paper by Bourinet *et al.*, 2000).

In the beginning of 1990s, Dr. W. Jefferson Stroud of the NASA Langley Research Center spent his sabbatical at the Florida Atlantic University within the prestigious fellowship program. We studied reliability of panels with random imperfections (Stroud *et al.*, 1992, 1993).

At this juncture it appears most appropriate to cite the definitive work conduced by the Greek School of applied mechanics. These comprise two groups of works. One is the series of works by Palassopoulos (1991, 1997), dealing with reliability concept for imperfection-sensitive structures. The other group starts with the work by Argyris *et al.* (2002) who developed the stochastic finite element method in a buckling context. To digress, I will note that I spent one month at the Argyris department at the University of Stuttgart in 1976. Since then he invariably referred to me not less than "Isaac — King of Israel." Later I learned from Professor Ted Belytschko that he would confer extremely hyperbolic titles upon other colleagues.

The work by Argyris *et al.* (2002) was extended by Stefanou and Papadrakakis (2004),

> taking into account random variations in Young's modulus and Poisson's ratio as well as the thickness of the shell. These random parameters are assumed to be uncorrelated. They are described by independent two-dimensional univariate (2D-1V) homogeneous Gaussian stochastic fields, which are represented via the spectral representation method... Under the assumption of prescribed power spectral density function of stochastic fields, it is possible to compute the response (displacement and stress) variability of the shell using the direct MCS [Monte Carlo simulation] technique.

It would be nice to combine this technique with some real data that is available in the Depository of the Delft University of Technology online. Papadopoulos and Papadrakakis (2005) extended the above work, noting (p. 1406) that "in the majority of studies these influencing parameters [variability of thickness, material properties, boundary conditions and misalignment of loading — I.E.] have not been treated as stochastic variables in a rational manner." The authors concluded (p. 1425) that "the investigation of the buckling behavior of the axially compressed cylinder showed that the incorporation of material and thickness imperfections to the model of the non-homogeneous initial geometric imperfections resulted in close predictions of the distribution of the buckling loads of the cylinder with respect to the experimental results." It was pleasing to read that the authors dispensed with the assumption of the spatial homogeneity of initial geometric imperfections.

Optimization of shells with random parameters was addressed by Lagaros and Papadopoulos (2006) and Baitsch and Hartmann (2006). Nonuniformity of axial loading on the buckling behavior in the contest of random imperfection was investigated by Papadopoulos and Iglesis (2007). The paper by Papadopoulos *et al.* (2009) abandons the assumption of Gaussianity of material and thickness properties.

Alibrandi *et al.* (2010) conducted a definitive study where the response surface was represented as a ratio of polynomials. Tootkaboni *et al.* (2009) resorted to asymptotic spectral analysis to nonlinear behavior of a stochastic structure with random material properties. Brar *et al.* (2008) dealt with axially compressed shells with axisymmetric initial imperfections resorting to Fourier series. Bayer and Roos (2008) utilized non-parametric structural reliability analysis using random fields and robustness evaluation. Salerno *et al.* (1996) resorted to the FEM for stochastic imperfection sensitivity. Kala (2007) dealt with stochastic and fuzzy uncertainty in stability problems. Schillinger (2004), Du *et al.* (2005), and Schillinger *et al.* (2010) resorted to stochastic FEM to deal with imperfect structures, de Paor *et al.* (2012) presented extensive experimental data permitting a full probabilistic characterization of imperfections.

Another novel way of deriving knockdown factors for anisotropic shells was derived recently by Takano (2012). He suggests the "A- and B- basis values of the knockdown factors." These are associated with "specified reliability and are not 'ultra-conservative.'" Takano (2012) notes: "From a practical viewpoint, using NASA's design criteria are extrapolations of the lower-bound curve of experimental data on an isotropic cylindrical shell . . . , and this leads to an extremely low value." The author also notes that "the accuracy of the knockdown factors cannot be increased without evaluating more experimental data."

The work that was performed at the Weimar University, Federal Republic of Germany by Bucher and his associates ought to be mentioned (Bucher, 2006; Bucher and Ebert, 2000; Schorling, 1997; Schorling, Bucher, 1999; Schorling *et al.*, 1998).

Here a comment of general nature appears to be instructive. Hunt (2002) concludes his review of our book (Elishakoff, Li and Starnes, 2001) as follows: ". . . authors . . . suggest a new twist to old, favourite problems for workers in stability theory — the axially compressed cylindrical shell and the compressed strut on elastic foundation to name two examples. This much is true, yet the deeper underlying debate, between determinism and stochastic modelling, is one that reverberates across many disciplines." Indeed, the question arises if it is absolutely necessary to invoke probabilistic analysis, and in general, uncertainty analysis of non-probabilistic nature (put forward in Chapters 4 and 6) to resolve the 20th century conundrum in elastic stability. Naturally, the purely deterministic design criterion without resorting to notion of uncertainty would be preferable to stochastic or anti-optimization techniques, since it would not invoke additional concepts, be it probability or boundedness. Hopefully, such a criterion would distinguish different manufacturing processes with attendant different imperfection patterns, and not penalize carefully designed shells.

3. Topical Personal Reminiscences

Fond memory brings the light of other days around me.

Thomas Moore (1779–1852)

"While memory holds a seat
In this distracted globe
Remember thee!"

William Shakespeare (1564–1616), *Hamlet*, Act 1, v. 96

Naturally, many anecdotes can be told about personal interactions with illustrious scientists like Bolotin, Koiter, Budiansky, and others, some of them alive and well.

Since Professor Koiter's postal address contained the words "Laboratory of Technical Mechanics," I asked him where the laboratory was located, as I intended to visit it. He replied: "I am the laboratory!"

This statement may appear to be not humble, at the first glance. But, alas, Koiter was telling the truth. He was a walking encyclopedia, not merely a laboratory, of theoretical and applied mechanics.

Upon the birth of our daughter Orly at Delft's Bethel hospital, I ordered printing a festive announcement (Fig. 7.4), as was customary in Holland. We decided to have the announcement in the Dutch language, for the benefit of the local audience. The Koiters were very pleased with our decision, and Mrs. Lous presented us with her own handmade embroidery commemorating this historic (for our family) event.

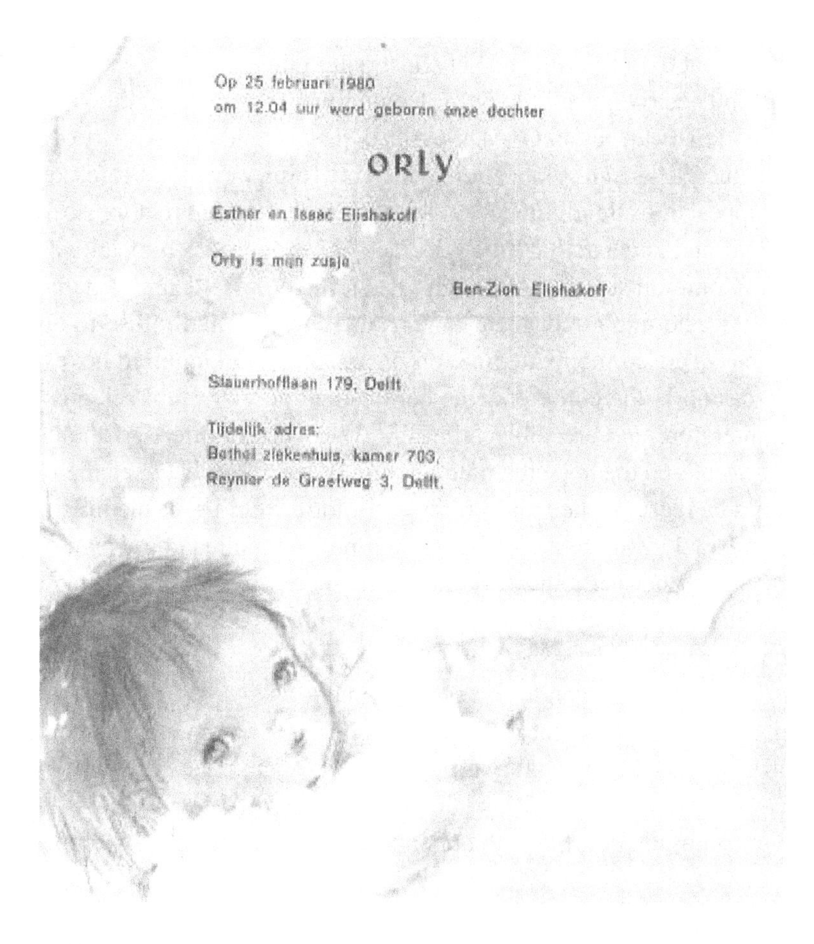

Fig. 7.4. Birth announcement of our daughter Orly, in the Dutch language.

Koiter was critical of extremely active researches, citing as an example Bolotin, who at the IUTAM Congress at Stresa, Italy, undertook to deliver, as Koiter mentioned, not less than five talks. Now, I keep meeting at ASME or AIAA meetings colleagues who give not less than 10 presentations: technology can be partially credited with this fact, for it is extremely easy to prepare PowerPoint presentations. This begs the question: how many new and original ideas can one have, which in turn reminds me of a story about Leopold Infeld. He once visited Albert Einstein at Princeton, and asked the host if he takes midday naps. Einstein replied in the negative, to which Infeld responded: "As far as I am concerned, I'm getting my good ideas after I wake up from the midday nap." Einstein countered by saying that he had been blessed with two ideas, and was skeptical about any further new ones; for this reason, he dispenses with naps.

Twice in my academic lifetime, until now, I have danced in the street. The first occasion was when Koiter asked me to expand my lecture notes, written in Delft, into a book. The "dance" took place on my way from Koiter's residence on Charlotte de Bourbonstraat to our home on Rietveldstraat, in the historic center of Delft.

The second occasion was at Lyngby, Denmark, two years later, in 1982. Its cause was an invitation by Professor Ove Ditlevsen of the Technical University of Denmark to lecture at the Euromech symposium, which my former mentor, Professor Bolotin, was also supposed to attend. In view of the extremely strained relations between the former USSR and Israel at that time I decided not to approach him, a loyal member of the State Communist Party. Although I had left the Soviet Union legally 10 years earlier in 1972, he might still regard me as a "renegate" and refuse to talk to me. However, he surprised me: patted me on the shoulder and inquired how I was. On another occasion at the conference he complained how Western scientists did not know the literature and are reinventing the wheel yet so often. At the end of the meeting, I expressed the hope that in the future the relations between the USSR and Israel would be reestablished and Professor Bolotin would visit me in Israel.

"Only if Israel institutes a communist regime!" he replied to my own surprise as well as of some other participants at the conference, who were standing close by.

"We already have communism, in our communes, called kibutz. So no need to establish it," I responded.

"This is Zionist propaganda" was his comment.

I said, "You see, you do not believe, in essence, that communism is possible. But our kibutzes do have it, and some of them even have never dealt with money!"

Professor Bolotin modified his statement: "Okay, if Meir Wilner — the head of Israel's Communist Party — is appointed a minister in the government, I will come!"

I replied to postpone his visit indefinitely, in such a case. Indeed, he never did, though in 1991 the Soviet Union ceased to exist — peacefully, not unlike a cylindrical shell under axial compression. Indeed, like shells, the Soviet Union had much too many imperfections, not that other super-powers lack them. Still, he was not uncritical of his country. I was told by Professor Hans Besseling that in 1979, when he met Bolotin upon his arrival to Delft to celebrate Koiter's 65th birth anniversary, he asked him why he came without his wife; Bolotin replied: "They will not allow her to travel with me until I will reach the age of 90 and on top of that, she will be hardly able to walk!" But after 1991 he and his wife, Professor Kira Sergeevna Bolotina, traveled extensively abroad, twice visiting Florida Atlantic University.

In 1984 Koiter was invited to Israel to give the Theodore von Kármán lecture at the Israeli Annual Conference on Aeronautics and Astronautics. Professor Josef Singer asked me to meet him and his wife at the airport, as well as take to care of the various arrangements. After Koiter's lecture, Professor Singer instructed me to ask him whether he would prepare a paper for the conference proceedings. Koiter declined: "All I have to say is published already. It is not my habit to repeat."

Later in our conversation he mentioned Stepan Prokofievitch Timoshenko, of whom he was critical. However, as he was speaking in whispers, I could not make out the *spirit* of his criticism, and regrettably it did not occur to me to verify what I did hear with Professor Singer.

I told Koiter the stories about Timoshenko that I heard from Ing. Yehoshua Borishansky of the Technion, who had attended Timoshenko's Strength of Materials course before the 1917 Revolution, and went to Israel in 1924. He recalled that Timoshenko was an inspiring lecturer and an extremely handsome and charismatic man. Accordingly his lectures, given in a large hall, would attract tens or even hundreds of young female students, even though they had not registered for his course. This was a stark contrast to the lectures of Professor Zhukovsky, who attracted not more than six students.

By contrast, it turns end that Timoshenko was almost a sadistic examiner. He would design for each student a personal exam, which would be taken several times, until the final passing grade would be given. Every subsequent test was built upon the mistake(s) the student had made in the previous try. Borishansky had been lucky: by some miracle he had passed the test on the first attempt. Since before the revolution of 1917 there was a talk — largely due to Timoshenko — circulating in Russia: "One cannot afford to have a girlfriend or get married until one has passed the exam in Strength of Materials!"

This author's work became intertwined with Koiter's on another topic besides imperfection sensitivity: the so-called follower forces. In fact, the two papers preceding Koiter's last one deal with our joint studies with Dr. Yivei Li and Dr. James Starnes. Already in 1966 Koiter criticized the notion of purely statically applied follower forces, and in the 1985 Ludwig Prandtl lecture he voiced even harder criticism. Since this paper was published in a relatively inaccessible journal, I advised

Fig. 7.5. The cover page of the report of the Koiter Institute Delft, for years 1998–1999.

him to propagate his idea more intensively by writing to the journal that published numerous papers on this subject. He prepared such an article titled "Beware of Follower Forces" and then changed the title to "Unrealistic Follower Forces," mentioning my name as well. I felt that there was no need to acknowledge my modest input, but Koiter insisted. My own work on follower forces was inspired by the paradoxical result obtained by Smith and Herrmann (1972), who demonstrated that a clamped-free column, subjected to a follower-concentrated load at its free end, does not "feel" an elastic foundation however stiff, and loses its stability at the same load as found by Beck (1952) for a column without such a foundation. As one can see, stability problems are rife with paradoxes and conundrums. For an extensive review of this subject, the interested reader can consult some studies (Elishakoff, 2005) and Ruta and Elishakoff. Let me add also that for about 10 years now I have been occupied, whenever some time becomes available, writing a monograph titled *Follower Forces: Fact or Fantasy?* The full story, and my interactions with Koiter, Bolotin, Sugiyama, Herrmann, and others, on that topic, are beyond the scope of this monograph, and will have to await the completion of that book.

Once at dinner at their home, the Koiters spoke of the need for timely recognition of deserving people. He told the story of Max Tailleur (1909–1990), a famous Dutch humorist writer, who was knighted on his 70th birth anniversary. Asked by the media to comment, he recounted a joke: "A man was going out with a lady. After ten years, he proposed marriage. The lady said: 'This proposition is so sudden, so unexpected, I need time before I respond!'" Tailleur added that he too had to reflect on how to comment on his knighthood. Lous remarked that Professor Koiter would never be knighted, but her husband responded that this was fine with him. By the way, Tailleur had a daily joke recorded at a specially designated pay-phone. I told the Koiters that the landlady of the apartment where I lived would dial the number several times a day. "But you always hear the same joke, do you not?" I inquired once. She replied: "Yes, this is the same joke throughout the 24 h period, and it evokes the same laughter from me every time that I dial!"

In 1994 Koiter invited me to a small symposium in honor of his 80th birthday but I was unable to attend. Our last discussion on the follower-forces topic took place in 1996, during my visit to Delft University of Technology at the kind invitation by Prof. Johann Arbocz. On this occasion I learned that he was ill and had decided not to undergo an operation. I told him about our preliminary results obtained for a fluid-conveying pipe on an elastic foundation, which did not exhibit the curious behavior of the Smith and Herrmann (1972) column. He was very happy to hear this news (the appropriate paper, by Elishakoff and Impollonia, 2001, would appear several years after Koiter's passing away in 1997). Thus, although we had different points of departure — his emphasis was an experimental verification of follower-force results, whereas I was busy resolving the Smith and Herrmann paradox — we were of a like mind on which problems were worthy of attack to resolve the above paradox.

During Professor Budiansky's sabbatical stay at the Technion, we invited him and his wife Nancy to our home and served a dinner composed of a diversity of courses, including one called "Satsivi" in the Georgian language. We were surprised to learn that they were familiar with it, having had served to them in Tbilisi at the home of Professor Ilya Nesterovich Vekua (1907–1977). Professor Budiansky opined that in Vekua's wife's rendering this particular item was tastier. This taught my wife that this dish should be prepared two days prior to serving. (On that occasion, in order to save time, she had prepared it on the same day it was served.) Some years later, already in Delft, Mrs. Luis Koiter gave us the recipe for another Georgian dish called "Satsebeli," written for her by the wife of Professor Niko Muskhelishvili.

During our stay at TU Delft in 1996, I brought him reprints of some papers on worst initial imperfections, written at the University of Michigan, by Professors Nicolas Triantafyllidis and Ralf Peek. He started to write an article on this topic but feared that he would not be able to complete it. Some months later, I wrote to Professor Hans Besseling suggesting that TU Delft could assign to him an assistant, just as Leonhard Euler had in his final years. In reply, I was informed that Professor Koiter was gravely ill and unable even to dictate. However, he was proud that his grandkid who was living in France did extremely well in the mathematics tests.

In the course of his sabbatical stay, Budiansky undertook to give a lecture at the Technion, for which he chose the theme of initial imperfection sensitivity. At the question time, Dr. Itzhak Lottati asked why a certain orthogonalization procedure was used for obtaining the results. Budiansky answered: "God told me so."

As secretary of the structures group at the Technion's Faculty of Aerospace Engineering, I invited Professor Budiansky to give an additional lecture on the first anniversary of passing away of the great Israeli engineer Markus Reiner (1886–1976). For the topic, I suggested a review of researches he had performed at

Harvard. After some hesitation, he agreed. Three years later, when I served as a visiting scientist at M.I.T. [courtesy of Professor Stephen H. Crandall (1920–2013)], the Budianskys invited me and Esther for an evening reception at their house in Lexington, Massachusetts. Our then five-year old son Benzion was specifically not invited, as he was bound to damage the home computer. We managed to find a baby-sitter for our infant daughter, but were forced to take our son with us — with the foreseen consequences to Bernie's computer.

We met repeatedly at the ASME Annual Conferences, and he would ask my opinion of certain Russian and American scientists. Once he told me that he had learned that Bolotin was working on two books at the same time, and wondered how this could be accomplished: Was Bolotin alternating between one in the mornings and the other in the evenings? I responded that Bolotin, in addition to his unusual talent, was extremely hardworking, and demanded the same from his associates, graduate students and close relatives. He was also surprisingly familiar with literature, emphasizing the need to know nearly everything done on the subject in question. Every week we had a departmental get-together where a faculty member belonging to the Department of Dynamics and Strength of Machines, or a graduate student, or a visitor presented a seminar. The penetrating questions Bolotin would pose could only be matched by Koiter's at the seminar on Technical Mechanics, which I invariably attended during the academic year 1979–80.

At the Weibull Anniversary Conference in Stockholm in 1985, Bolotin asked me if I felt a certain nostalgia for Russia. I responded that except for many fond memories, I did not. Then I quoted to him, from Pushkin's (1799–1837) novel in verse *Eugene Onegin*, where the author — then on the staff of the Russian Foreign Ministry — refers to his punitive transfer in 1820 from St. Petersburg to the provinces. He says [my own free translation from the Russian original]: "I was there for a while, but found the North was not my style." "But you were not exiled, were you?" Bolotin retorted. I explained that my family's desire to immigrate to Israel stemmed exclusively from our religiosity. He was shocked. "Pursuit of a better life would not surprise me, but religion?!" I responded that this indeed was the second reason. By contrast, Warner Koiter, probably also a secularist, lamented the decline of religion and its negative influence on the morals of people.

Professor Bolotin would argue with his host at the Center for Applied Stochastics Research at Florida Atlantic University that fatigue was not a diffusion process and thus stochastic differential equations and their elegant solutions were irrelevant to reality. Brutal truth, however, is not much welcome (even) in scientific circles.

Most unfortunately, many recognized American scientists, laureates of prizes, and members of national academies, could not find an opportunity to write a book and be judged by the substantial reviews that would follow its publication. Naturally, one who writes historic papers that revolutionize the field is "exempt" from the prerequisite of book writing. What better (and extreme) example than Einstein, whose five revolutionizing papers were not published in a book format?

Still, there is a strong feeling that much could be improved in the recognition process in the United States; one researcher wrote to me about lots of "politics"; another claimed that since university ranking depends on the number of faculty memberships in academies, universities encourage nominations from one's own institution; still another told me that the process of "scratch my back and I will scratch yours" is common in the United States, and possibly everywhere else in the world. This inevitably fosters mutual-admiration relationship where "Cuckoo praises the Rooster for what he praises the Cuckoo," to quote from Ivan Andreevich Krylov (1769–1844), the Russian Jean La Fontaine (or more exactly, the Russian Aesop).

Maybe I have digressed too far, but the need for reform in the American academic establishment cannot be questioned. Had I not been hopeful that change is possible (despite the apparent failure of the "Yes, we can!" slogan) I would have refrained from mentioning this topic here.

Another relevant problem is the nearly lifelong tenure of much too many program managers or directors in the US research establishments. They become, as it were, the academic "aristocracy." It is all too natural that in the process (not unlike the Soviet and Chinese political leadership of the past) they develop tastes and preferences and seldom entertain new researches in their nearly hermetically sealed "clubs." The situation is exacerbated even further because faculty (nearly invariably members of academies) is not forbidden to nominate program directors (from when they get funds) for academic awards. Something appears to be rotten, to paraphrase Shakespeare's immortal *Hamlet*, in the academic world of the United States.

4. Establishment of the ASME Warner T. Koiter Medal

I have a vivid recollection of New Year's Eve in December 1942 when my wife and I sat in front of a very modest fire and discussed, in addition to the most acute war — time problems, what would happen to my work which I considered to be significant but questioned whether it would be recognized as such.

W.T. Koiter (1979)

I have been enormously influenced, instructed and encouraged by Warner Koiter, the sage of Delft.

B. Budiansky (1989)

A transition period is identified between 1935 and 1960, in which several attempts were made to reconcile theory and experiments. But it was not until the work of W.T. Koiter became known and understood that a new paradigm was established in the sense of Kuhn . . . Koiter developed a new conceptual system and introduced new methodologies for the analysis.

L.A. Godoy (2002)

It appears instructive to cite Herakovich (1997), the then-chair of the ASME's Applied Mechanics Division:

> As I pause to reflect on the past year of the Applied Mechanics Division, there are several highlights that come to mind. The most significant development is undoubtedly the establishment of the ASME Warner T. Koiter Medal to be bestowed "in recognition of distinguished contributions to the field of solid mechanics with special emphasis on the effective blending of theoretical and applied elements of the discipline, and on a high degree of leadership in the international solid mechanics community." This medal was established by the Applied Mechanics Division which will be responsible for reviewing nominations and making recommendations to the ASME Committee on Honors. A Koiter Medal Committee, consisting of the five most recent Koiter Medalists, and five current members of the Executive Committee, has been established for this purpose. The medal honors Warner T. Koiter for his seminal contributions to solid mechanics, in particular for his fundamental work in nonlinear stability of structures. It is also pleasure for me to report that upon recommendation of the Koiter Medal Committee and with the approval of the ASME Committee on Honors, the first Koiter Medal has been awarded to Professor Koiter. That is as it should be! The final approval awarding the first medal to Koiter came shortly after the IMECE in Atlanta. John Hutchinson, former Chair of the Division, presented Professor Koiter with a certificate and honorarium during a ceremony at Delft University on January 22, 1997.
>
> The decision of the Executive Committee to recommend that this medal be established was not made lightly . . . the Koiter medal provides the opportunity to recognize those individuals who have made significant contributions in solid mechanics and international leadership. Koiter's contributions to mechanics are of the highest caliber and having a medal named after him will only add honor to the Division . . . All ASME medals must be endowed and we are grateful to Delft University for providing the full endowment for the Koiter Medal.

In the summer of 2000, as TU Delft's inaugural holder of the W.T. Koiter professorship, I visited the Koiter family twice. Mrs. Koiter complained that there was no building named Koiter Institute. I explained both to Mrs. Lous and to their son Klaas that the Koiter Institute was alive in the hearts and minds of every buckling specialist of the world, and that the time will come when such a building will be consecrated. Indeed, it would be nice if TU Delft establishes a museum in his house on 14 Charlotte de Bourbon Straat, or at least dedicates his former office at the university to his seminal contributions. Klaas reminded me that in 1996, when we last met with Professor Koiter alive, the latter walked me out for quite a distance from their home. Klaas noted that Warner Koiter had not done this with anyone else. Perhaps he felt that this would be our last meeting, since he mentioned that he decided against an operation to remove a cancer. We had only one more occasion of the telephone conversation, when he thanked me for initiating the Koiter medal, mentioning that I was his true friend.

The decision to include this section in the history of the establishment of the Koiter medal was dictated by the desire of the present writer to keep the story

FLORIDA ATLANTIC UNIVERSITY
P.O. BOX 3091
BOCA RATON, FLORIDA 33431-0991

COLLEGE OF ENGINEERING
Department of Mechanical Engineering
(407) 367-3420

December 13, 1994

Professor John Hutchinson
Chairman, ASME
Applied Mechanics Division
Gordon MacKay Professor of Applied Mechanics
313 Pierce Hall
Harvard University
Cambridge, MA 02138

Dear John:

I am writing to you on the following matter. Every applied mechanitian, who attends the ASME winter Annual Meeting, feels pleasantly obliged to attend the Applied Mechanics Dinner, where, amongst others, and most importantly, the S. P. Timoshenko Medal's winner is announced and the winner gives an address, reflecting the winner's own experiences as well as talks on pertinent issues of general interest.

It appears to me that the Applied Mechanics Division may want to consider a possibility of establishing an additional medal, which will become even more prestigious than that bearing the name of Timoshenko, a great pedagogue, but I don't know if the greatest scientist, although he appears to have had a profound influence on American mechanics.

Amongst our contemporaries the name of Warner Tjrdus Koiter comes immediately and invariably in mind. I must not introduce Professor Koiter to you, since you yourself had cooperated with Professor Koiter and participated so vividly in forming the modern architecture of buckling and postbuckling.

It appears to me that it will be not an exaggeration to say that since Leonard Euler no scientist had contributed to the science of instability more than Warner Tjrdus Koiter.

Additional medal's come into mind too: why not have medals named after Leonhard Euler or Caughey?

I am assuming that there must be sufficient funds available in ASME to sponsor the proposed three medals.

Looking forward to hearing from you at your earliest convenience,

Yours sincerely,

Isaac Elishakoff

Fig. 7.6. Letter of Professor J.W. Hutchinson dated 13 December 1994.

available for the future generations of applied mechanicians, in case they will be interested. Perhaps, however, it was partially dictated by the keen observation of Marcel Reich-Ranicki, Germany's, and perhaps world's greatest literary critic: "I never met an author who wasn't vain and egocentric — unless you count very bad authors" (Hickley, 2013). Above all, this topic appears to fit extremely well the subject of the book. In elastic stability Koiter played a most central role in the 20th century (see also Appendix B). It appeared appropriate therefore to include this material in the book, as a tribute to the man who explained to us what was going on with the perplexing behavior of cylindrical shells.

Fig. 7.7. Professor John Hutchinson — foremost authority in stability.

HARVARD UNIVERSITY
PIERCE HALL, CAMBRIDGE, MASSACHUSETTS 02138
(617) 495-2848
HUTCHINSON @ HUSM.HARVARD.EDU

DIVISION OF APPLIED SCIENCES
JOHN W. HUTCHINSON
GORDON MCKAY PROFESSOR OF APPLIED MECHANICS

January 30, 1995

Professor Isaac Elishakoff

Dear Isaac,

There is some discussion underway about a new medal, but the name Koiter has not been mentioned. Of course, I like the idea of a Koiter Medal very much. It will only fly if someone (like you) gets behind it and pushes hard. You should also be aware, that ASME does not have funds for this purpose. Indeed, part of the work--perhaps the biggest part-- you would be taking on is the solicitation of $25,000 to endow any new ASME medal.

I'm passing the buck back to you, as they say. Happy new year to you too.

With best wishes,

John W. Hutchinson

Fig. 7.8. Letter of Professor J.W. Hutchinson dated 30 January 1995 (reproduced by permission of Prof. Hutchinson).

FLORIDA ATLANTIC UNIVERSITY
P.O. BOX 3091
BOCA RATON, FLORIDA 33431-0991

COLLEGE OF ENGINEERING
Department of Mechanical Engineering
(407) 367-3420

July 27, 1995

Professor John Hutchinson
Chairman, ASME
Applied Mechanics Division
Gordon MacKay Professor of Applied Mechanics
313 Pierce Hall
Harvard University
Cambridge, MA 02138

Dear John:

Last week I was in Paris and then in Delft, where I have met with Warner Koiter, Johann Arbocz and Edward Riks. After my lecture on "Variational Principles for Stochastic Structures & Associated Finite Element Methods" I have met with Professor Besseling. We have touched, inter alia, the topic of the Koiter medal. I have asked if he could check if the Dutch Academy of Science could contribute funds to ASME for the medal. He was excited with the idea, and will communicate to me the results of his discussions.

Let us wait.

Cheers,

Isaac Elishakoff
Professor

IE/tp

Fig. 7.9. Letter of Professor J.W. Hutchinson dated 27 July 1995 (reproduced by permission of Prof. Hutchinson).

5. Some Remarks about Priority

A man must either resolve to put out nothing new, or to become a slave to defend it.

Sir Isaac Newton (1676)

It is evident that the era of dusty libraries, meditative readings, acrimonious controversies about the priority of discovery, have immediately declined.

P. Villaggio (2013)

In the opinion of Hoff (1966), "The question of priority in the discovery of scientific information is always an interesting one, although perhaps not an important one."

By the end of my stay at the Delft University of Technology during the academic year 1979/80, it was clear to all of us that Professor Arbocz had fully subscribed to the probabilistic methodology. Our Report LR-206 appeared in November 1980,

HARVARD UNIVERSITY
PIERCE HALL, CAMBRIDGE, MASSACHUSETTS 02138
(617) 495-2848
HUTCHINSON @ HUSM.HARVARD.EDU

DIVISION OF APPLIED SCIENCES
JOHN W. HUTCHINSON
GORDON MCKAY PROFESSOR OF APPLIED MECHANICS

September 23, 1995

Professor Isaac Elishakoff
Department of Mechanical Engineering
Florida Atlantic University
P.O. Box 3091
Boca Raton, FL 33431-0991

Dear Issac,

Many thanks for your letter of July 27th, and sorry for taking so long to respond. To be honest, I am way behind in my correspondence, but now that I am no longer on the Executive Committee of AMD, I have less of an excuse. It is very good news indeed that there might be a possibility to obtain some endowment funds for a Koiter Medal from the Dutch Academy. Will you be at the winter meeting of ASME in San Francisco? If so, let's try to get together to talk about how we might try to go forward with this. I'll be there on the Monday and Tuesday. Since I am no longer on the AMD committee, I am going to forward you letter and a copy of this one to Tom Cruise, the current Chair of AMD. I'll have a word with him about your efforts. I do think it would be marvelous to have a Koiter Medal.

Best wishes,

John W. Hutchinson

Fig. 7.10. Letter of Professor J.W. Hutchinson dated 23 September 1995 (reproduced by permission of Prof. Hutchinson).

and was published in paper form in 1982. However, the excitement was so great that one could not wait until our joint paper will be out of the press to tell the rest of the researchers that this author's methodology was working for cylindrical shells — albeit for axisymmetric imperfections only at that stage — that Arbocz submitted on 14 May 1981 (following its presentation at the Annual Meeting of Deutsche Gesellschaft für Luft- und Raumfahrt in Aachen, earlier that year) to the *Zeitschrift für Flugwissenschaften und Weltraumforschung*. The paper had a telling title: "Past, Present and Future of Shell Stability Analysis." In it Arbocz wrote: "Several investigators (Bolotin, (1958); Amazigo (1969), Roorda and Hansen (1972)) have studied the static and dynamic buckling of imperfection sensitive structures with small random imperfections. For an authoritative review the reader should consult Amazigo's paper from 1974."

Arbocz then continued: "However, it was not until 1979, when Elishakoff published his reliability study of the buckling of a stochastically imperfect finite column on nonlinear elastic foundation, that a method has been proposed, which made it

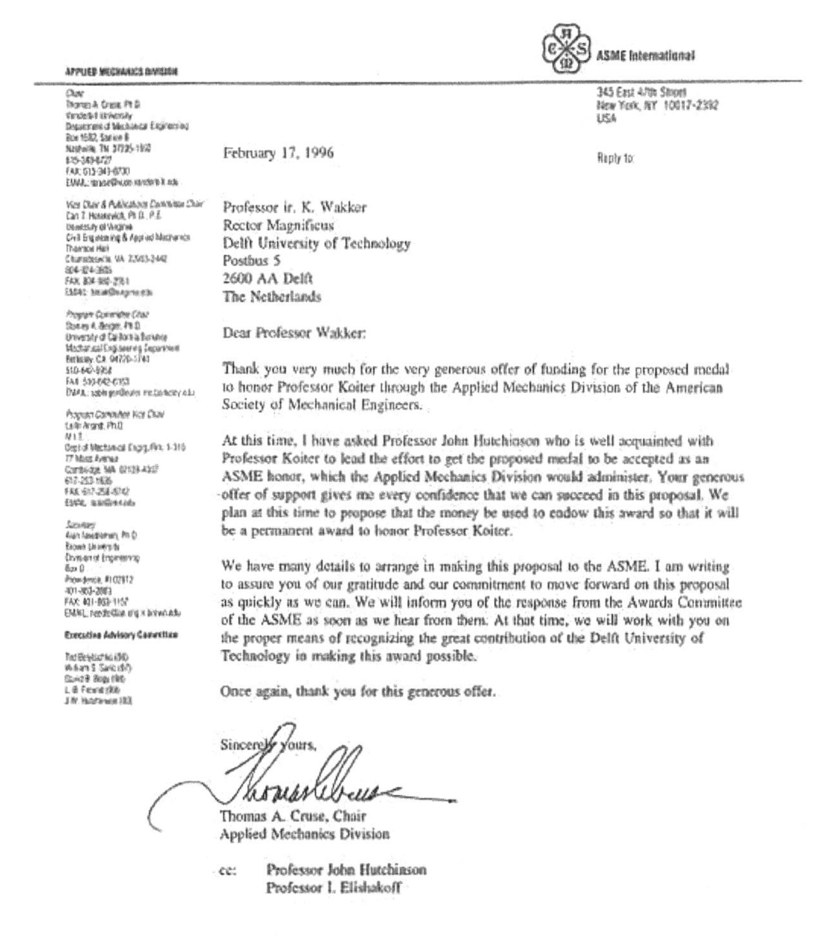

APPLIED MECHANICS DIVISION

ASME International

345 East 47th Street
New York, NY 10017-2392
USA

Reply to:

Chair
Thomas A. Cruse, Ph.D.

Executive Advisory Committee

February 17, 1996

Professor ir. K. Wakker
Rector Magnificus
Delft University of Technology
Postbus 5
2600 AA Delft
The Netherlands

Dear Professor Wakker:

Thank you very much for the very generous offer of funding for the proposed medal to honor Professor Koiter through the Applied Mechanics Division of the American Society of Mechanical Engineers.

At this time, I have asked Professor John Hutchinson who is well acquainted with Professor Koiter to lead the effort to get the proposed medal to be accepted as an ASME honor, which the Applied Mechanics Division would administer. Your generous offer of support gives me every confidence that we can succeed in this proposal. We plan at this time to propose that the money be used to endow this award so that it will be a permanent award to honor Professor Koiter.

We have many details to arrange in making this proposal to the ASME. I am writing to assure you of our gratitude and our commitment to move forward on this proposal as quickly as we can. We will inform you of the response from the Awards Committee of the ASME as soon as we hear from them. At that time, we will work with you on the proper means of recognizing the great contribution of the Delft University of Technology in making this award possible.

Once again, thank you for this generous offer.

Sincerely yours,

Thomas A. Cruse, Chair
Applied Mechanics Division

cc: Professor John Hutchinson
Professor I. Elishakoff

The American Society of Mechanical Engineers

Fig. 7.11. Letter of Professor T.A. Cruse, dated 16 February 1996.

possible to introduce the results of the initial imperfection surveys routinely into the analysis."

Finally, Arbocz (1981) outlined the methodology of our report, reproducing some figures from the future paper.

I, although slightly uncomfortable with the preliminary publication of the joint results in another journal, still felt elated about the "future of shell stability analysis" being fully based upon my methodology. The latter quotation appeared, *verbatim*, in at least 10 publications by Arbocz and his other co-authors, including the probably most recent *ESA Handbook on Structural Stability* (Arbocz *et al.*, 2010).

ASME International

December 18, 1996

Professor Wawner T. Koiter
Delft University
Charlotte de Bourbonstraat 14
Delft, The Netherlands

Dear Professor Koiter:

It is with the greatest of pleasure and honor that I write to you on behalf of the Applied Mechanics Division of the American Society of Mechanical Engineers (ASME) to inform you and congratulate you on the occasion of two very proud and wondrous events for you and the Division. The first event is that upon the recommendation of the Applied Mechanics Division, ASME has established a new medal to be named the Warner T. Koiter Medal. In recommending this new medal to ASME, the following statement was given:

> *The Warner T. Koiter Medal is bestowed in recognition of distinguished contributions to the field of solid mechanics with special emphasis on the effective blending of theoretical and applied elements of the discipline, and on a high degree of leadership in the international solid mechanics community.*
>
> *To honor Warner T. Koiter for his foundational works in nonlinear stability of structures in the most general sense, for his diligence in the effective application of these theories, his international leadership in mechanics, and his effectiveness as a teacher and researcher.*

The significance of this new medal is exemplified by the fact that it is only the second medal to be sponsored by the Applied Mechanics Division, the other being the Timoshenko Medal which was first awarded in 1957. The fact that forty years have passed between the establishment of these two medals clearly indicates that it is not done lightly and that the medals are considered to be of the highest caliber and most prestigious.

We congratulate you on this wonderful recognition of your achievements in solid mechanics and your standing in the international community. Your name will be forever enshrined in the mechanics community and the Applied Mechanics Division of ASME.

You should know that Professors Isaac Elishakoff, John Hutchinson and Tom Cruse have been instrumental in bringing about the establishment of the Koiter Medal and Delft University has made a significant financial contribution to endow the Medal.

The second event of which I have the honor of informing you is that you have been selected to receive the first Koiter Medal. The citation recommending you for the Medal reads:

> *For his foundational works in nonlinear stability of structures in the most general sense, for his diligence in the effective application of these theories, his international leadership in mechanics, and his effectiveness as a teacher and researcher.*

We are delighted that you will be the first recipient of the Koiter Medal. It is as it should be! I'm sure you will be interested to know that letters of recommendation nominating you as the first Koiter Medalist were written by Bernard Budiansky, John Hutchinson, Pierre Ladeveze, Frithiof Niordson and Jim Simmonds.

ASME is having the first Koiter Medal struck and we will get one to you at the earliest possible time. We would be delighted to make a formal presentation of the Medal to you at an ASME Meeting. We meet this summer, June 29 - July 2, 1997 at Northwestern University and in the fall, Nov. 16-21, 1997 in Dallas, Texas. If it would be possible for you to attend either of these meetings, please let me know and I will make the appropriate arrangements. However, in the meantime, John

Fig. 7.12. Letter of Professor C.T. Herakovich, dated 18 December 1996 (reproduced by permission of Prof. Herakovich with a comment: "Please add a note to the effect that there was a typo in Warner's name. I didn't catch it until the letter had gone out. Koiter never mentioned it to me. He was a gentleman.").

Wawner T. Loiter
Page 2

Hutchinson is now in Denmark on sabbatical and will be contacting you so that he can present you with an official recognition of the award.

In closing, on behalf of the Applied Mechanics Division of ASME, I want to tell you how delighted we are with the establishment of the Koiter Medal and the fact that you are the first recipient. A standard has been established that will be difficult to maintain, but we will do everything we can to ensure that only those of the highest caliber will be awarded the Koiter Medal.

All of us on the Executive Committee, as I am sure, all the members of the Applied Mechanics Division look forward to the time that we might personally convey our congratulations and warmest wishes to you.

With best personal regards,

Sincerely yours,

Carl T. Herakovich

CTH/bm

cc: ASME Committee on Honors
AMD Executive Committee
Rector Magnificus Professor ir. K. Wakker, Technical University of Delft
Professor Bernard Budiansky, Harvard University
Professor Thomas A. Cruse, Vanderbilt University
✓Professor Isaac Elishakoff, Florida Atlantic University
Professor John Hutchinson, Harvard University
Professor Pierre Ladeveze, Ecole Normale Superior de Cachan
Frithiof Niordson
Professor James G. Simmonds, University of Virginia

Fig. 7.12. *(Continued)*

Likewise, our work's main results were reproduced in a paper (Arbocz, 1982) aptly titled "The Imperfection Data Bank, A Means to Obtain Realistic Buckling Loads" that also appeared prior to journal publication (Elishakoff and Arbocz, 1982), again demonstrating Professor Arbocz's extremely positive evaluation of my method and our joint results. The above title, however, seemed to be imprecise. Imperfections have been measured before my method was proposed, but these measurements have not been shaped into a "realistic" methodology. The main point was our stochastic analysis of the initial imperfection data bank reproduced in detail in the last-named paper, where (p. 353) Arbocz again stressed the importance of this author's methodology: "Looking into the future, it is to be expected that the existence of extensive data on characteristic initial imperfection distributions classified according to fabrication processes, the availability of improved versions of the present generation nonlinear structural analysis codes (Almroth *et al.*, 1973), the stochastic stability approach via the reliability functions and the greatly increased computational speed

Fig. 7.13. Prof. John Hutchinson presents the ASME Koiter Medal certificate to its first recipient, Prof. W. T. Koiter. (Courtesy of Professors Dick van Campen and Paul van Woekrum)

offered by the so-called super-computers will finally result in a series of improved design recommendations which will incorporate the latest theoretical findings and make them routinely accessible to the designers."

Nine years later, Arbocz (1991) in his paper titled "Towards an Improved Design Procedure for Buckling Critical Structures," made an even stronger statement:

> ...I have been looking in the past 15 years or 20 for ways to define an improved lower bound curve for a specific set of nominally identical shells... Please notice, that "improved" means a "higher" knockdown factor.
>
> The turning point in my research came in 1978 [*Note*: the correct year is 1979/80 instead –I.E.] when Isaac Elishakoff spent his sabbatical with me in Delft and introduced me, among other things, to the concept of reliability function $R(\lambda)$.

It was pleasing to read the dedication in the special issue of the *International Journal of Solids and Structures* (de Borst, 2002), dedicated to Professor Arbocz on the occasion of his retirement: "... In the early 1980s cooperative efforts with Elishakoff resulted in work on a statistical analysis approach of geometrically imperfect thin walled shells. The paper was presented in a invited lecture at the IUTAM-Symposium on collapse of structures in London in 1983."

Rene de Borst then indicates the results of the above cooperation: "In the period thereafter the flavor of the activities changed from deterministic toward probabilistic techniques..."

Koiter (1979, p. 244), in his reminiscences write, referring to his course at Brown University in the academic year 1961/62: "Interest in this field also mushroomed at that time, in particular at Harvard University where I fondly believe my work had some influence..."

Likewise, I can state that I felt pleased with my influence on the research at Delft University of Technology, and elsewhere. Koiter, although did not write about my general methodology in his review paper (1985) published after my work has appeared, endorsed it during my public lectures that he attended; specifically these were seminars given over the years at the Delft University of Technology; conference organized by him in Delft in March 1980; my lecture at the IUTAM Congress in Toronto, in August 1980; my lecture at the 1982 conference organized by Professors J. Michael Thompson and Giles Hunt, at the University College London; at the IUTAM Congress in Grenoble, France in 1988.

The latter two lectures were attended *inter alia* by Professor Bernard Budiansky, who also endorsed the methodology in conversations with me. The lectures given in 1980 and 1982 were also attended, amongst others, by Professor John Hutchinson.

Simitses (1986) notes:

> Comparison between theoretical predictions (critical loads) and experimental results (buckling loads) revealed discrepancy of unacceptable magnitude. A tremendous effort was made in order to explain the discrepancy both analytically and experimentally ... Elishakoff (1979) proposes a method for incorporating experimentally obtained imperfection distributions into a statistical imperfection-sensitivity analysis. The feasibility of this approach has been demonstrated (Elishakoff and Arbocz, 1982, 1985) for axially loaded imperfect cylindrical shells.

Ikeda and Murota (2002, pp. 115–117) partially reproduced the results of our study (Elishakoff, 1979) in their monograph. Additionally, they wrote (p. 250):

> The effect of random axisymmetric imperfections on the buckling of circular cylindrical shells under axial compression was investigated by Elishakoff and Arbocz (1982). The statistical properties of the shells were evaluated based on the measured data, and the reliability function of the shells was computed by means of the Monte Carlo method. Later on, on the basis of an assumption that the initial imperfections are represented by normally distributed random variables, the first-order second-moment method was employed to replace the Monte Carlo method and, in turn, to greatly reduce computational costs (e.g., Elishakoff *et al.*, 1987; Arbocz and Hol, 1991).

6. Conclusion

> *. . . I sometimes wondered what a wonderful opportunity it must have been when beautiful facts, such as the critical load of an elastic column, still awaited discovery. But similar opportunities exist today and are actually more numerous. The growing body of human knowledge may be imagined as the growing volume of a sphere. The unknown is the infinite exterior, but that is currently knowable is only what is in contact with the surface of the sphere. As the surface grows, the knowable unknown grows with it, representing the problems ripe to tackle.*
>
> Z.P. Bažant (2009)

> *Another issue that must be addressed to obtain a new set of useful and practical design monographs is design uncertainties. A significant contribution to this area can be made by providing guidelines for determining which shell stability issues are more adequately handled in a deterministic rather than in a probabilistic manner. From a practical viewpoint, this information indicates approximately the number of experiments and analyses needed to establish meaningful design recommendations and reliable, but not overly conservative, knockdown factors.*
>
> M.P. Nemeth and J.H. Starnes, Jr. (1998)

It appears instructive to quote from Arbocz's (1974) paper published in the *Festschrift* volume to Professor Sechler, where he comments on the state of the art of shell stability at that time:

> Despite considerable research effort in the past *50 years* [author's emphasis — I.E.] the behavior of thin circular cylindrical shells under axial compression is still a subject for controversy. In-plane boundary conditions (Ohira, 1961; Hoff, 1961) and prebuckling deformation caused by edge restraints (Almroth, 1961; Cohen, 1966) have been shown to affect buckling appreciably. However, after *half a century* [author's emphasis — I.E.] of intensive investigations the initial geometric imperfections are accepted as the main cause for the wide experimental scatter and the poor correlation between the predictions of the linearized small-deflection theory and the experimental results. Despite this recognition the incorporation of imperfection sensitivity into engineering practice has not been accomplished.

Arbocz (1974) mentions here the period characterized with the lack of bridge between theory and practice in two manners: "50 years" and "half a century."

Five years later, Budiansky and Hutchinson (1979) again reviewed the state of buckling, starting with "the simple but confident introductory assertion that great progress in the understanding of buckling phenomena has been achieved in the past four decades." They did not explicitly mention the chasm between theory and experiment, but made the following comment on stochastic buckling problems:

> Given the importance of small imperfections in imperfection-sensitive structures, and the uncertainty of their magnitudes and shapes, it has long been believed that a rational approach to the design of such structures should relate failure probability under a given

> loading to appropriate statistical information concerning imperfections. Such correlations are straightforward, in principle, when single-mode buckling is pertinent, for then the probability of failure may be related directly to the probability distribution function of the amplitude of the relevant imperfection. This procedure may be extended when a few modes interact, but becomes cumbersome for many modes.

This particular difficulty was overcome by the present writer (Elishakoff, 1979). Four years later, Babcock (1983) reviewed the state of shell stability on the occasion of the 50th anniversary of the *Journal of Applied Mechanics*. He posed the following question: "What has the research community produced in understanding and useful results over the last 20 years or so?", and responded as follows, "with some trepidation": "Equally apparent is the fact that this knowledge and understanding has become a *self-perpetuating monster* [author's emphasis — I.E.] which awaits a modern-day Timoshenko to tame it and place it in the service of the practicing engineer."

Bažant and Cedolin (1991, p. 474) wrote:

> ... the design at present still needs to be based on the classical critical loads obtained by linear analysis and corrected on the basis of extensive experimental data. The need to use these empirical corrections is nevertheless *deplorable* [author's emphasis — I.E.] for the theorists since it makes generalizations outside the range of the existing experiments, and especially to different types of shells, uncertain.

Babcock (1983) stressed: "The great volume of work has firmly established the imperfection sensitivity of some shell/load combinations and developed analysis procedures to examine other problems. This certainly has been one of the great successes in mechanics research, but unfortunately one that had little effect on engineering practices." And finally:

> Elishakoff (1983) pursues the concept of reliability using the Monte Carlo method. Again the problem of suitable initial imperfections arises. Arbocz and Babcock (1969, 1976, 1978) have engaged in a lengthy program to use laboratory scale shells to establish the validity of methods to predict 'knockdown' factors.

It was gratifying that Babcock (1983) connected my reliability analysis with the experimental and numerical work of Arbocz and himself.

Nine years after the Sechler (1976) *Festschrift*, in the proceedings of the IUTAM Symposium that took place at University College, London (see volume edited by Thompson and Hunt, 1983), Arbocz's tone vastly differed from his 1976 paper. Mentioning the works that preceded Elishakoff's paper of 1979, he wrote:

> In the absence of experimental evidence about the type of imperfections that occur in practice, and in order to reduce the mathematical complexity of the problem, all these investigators have worked with some form of idealized imperfection distribution. However, it is not obvious how the methods employed can be extended to the general

imperfection distributions observed in practice. Thus it was not until 1979 that a method was proposed by Elishakoff (1979), which made it possible to introduce the results of the general initial imperfection distributions obtained experimentally into the analysis.

Further Arbocz (1983, pp. 67, 68) wrote

> Basically, Elishakoff had suggested the utilization of the Monte Carlo method to solve the stability problem of axially compressed cylindrical shells with random initial imperfections. The feasibility of this approach has been demonstrated for axisymmetric (Elishakoff and Arbocz, 1980a) as well as general asymmetric imperfections (Elishakoff and Arbocz, 1980b).

In the same proceedings, I wrote (see also the slightly modified figure below):

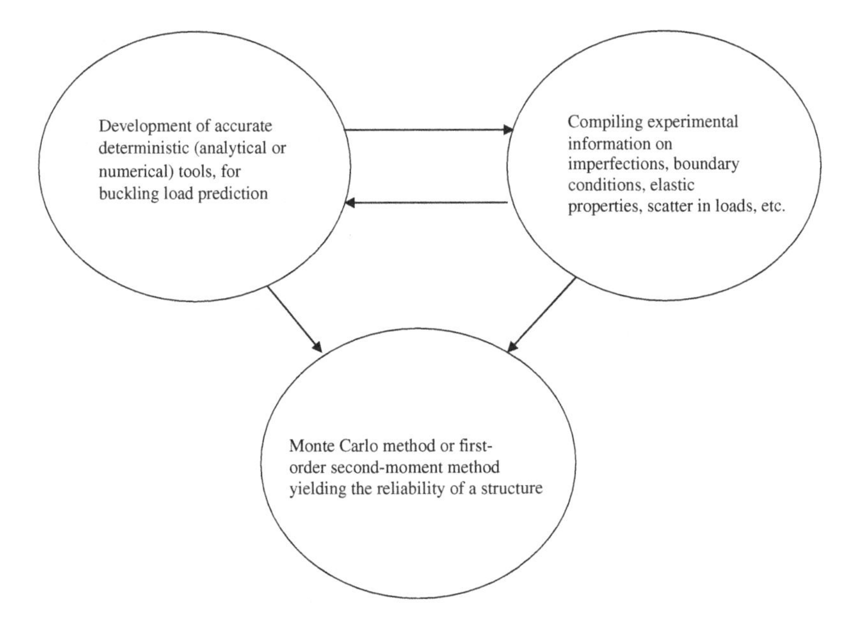

Fig. 7.14. Steps for introduction of the imperfection-sensitivity concept into design.

To conclude, introduction of the initial-imperfection concept into design calls for:

(1) Further development of combined analytical, numerical and experimental tools for predicting the buckling loads under deterministic analysis being one of the cornerstones of the Monte Carlo method, with appropriate establishment of the accuracy, so vital for the reliability prediction. Alternatively, one can resort to the first-order second-moment method which is computationally *vastly less* extensive than the Monte Carlo technique.
(2) Compilation of extensive experimental information on imperfection sensitive structures, classified according to the manufacturing process, with a view to finding their mean and autocovariance functions.

With the above developments in analytical, numerical and experimental tools and compilation of experimental information, the Monte Carlo method or its alternative, first-order second-moment method represent a promising means for narrowing the present chasm between theory and practice.

Arbocz (1983, pp. 67–68) had fully identified himself with the above philosophy of my studies (Elishakoff, 1979, 1983):

> Based on these preliminary results by [Elishakoff, 1979; Elishakoff and Arbocz, 1980a, 1980b], one can envisage the following procedure to derive improved lower bounds for the buckling loads of axially compressed cylindrical shells:
>
> (1) Compile extensive experimental information on initial imperfections classified according to the manufacturing process used.
> (2) Expand the measured initial imperfections in terms of the buckling modes of the corresponding perfect structures. The Fourier coefficients are then considered as random variables.
> (2) Compute the required probabilistic properties (the mean vector and the variance–covariance matrix) of the initial imperfections which are characteristic of a given fabrication process. This is accomplished by taking appropriate "ensemble averages" of the experimentally measured Fourier coefficients of a small sample of shells.
> (4) Simulate the random initial imperfections, that is, "create" a large number of shells with generally different imperfection profiles. This is done by using the simulation procedure for space-random fields proposed by Elishakoff (1979a).
> (5) Obtain the histogram of buckling loads for the large number of simulated shells. This involves a deterministic buckling load calculation of the "created" imperfect shells.
> (6) Calculate the reliability function associated with the given manufacturing process.

Thus Arbocz's (1976) original complaint about the persisting chasm between theory and practice, and Babcock's observation that "the great successes in mechanics research . . . had little effect on engineering practices" were replaced by sound optimism (Arbocz, 1983, p. 68):

> This new method is intended to replace, step by step, the old rigid Lower Bound Design Philosophy.

Jones (2006, p. 682) stresses in his definitive monograph:

> Various attempts have been made to develop a probabilistic approach to buckling — critical shell design — see, for example, the papers by Arbocz *et al.* (2001) and Elishakoff and Starnes (1999) and the book by Elishakoff *et al.* (2001). That approach will take some time to mature and come to the point where it is both appreciated and accepted by designers.

It appears that the most important steps in this direction have already been taken. The methodology that I lucked out to develop was accepted and adopted by the NASA and ESA researchers (Anonymous, 2010) and the German Aerospace

Center (Kriegesmann *et al.*, 2010, 2011; Kriegesmann *et al.*, 2010; Elishakoff, Kriegesmann, Rolfes, Hühne and King, 2012).

Thompson and van der Heijden (2014) write: "Elishakoff continues to argue with conviction (Elishakoff, 2012) that from an engineering standpoint the final target for the analyst must be to study imperfect shells (a challenge for many colleagues) and do express the results in probabilistic terms."

The fact that some researchers may not know how the 20th-century conundrum in elastic stability was resolved may perhaps be ascribed to human frailty. It is hoped that this book will fill this "void."

Ramm and Wall (2002) refer to cylindrical shells as *the prima donna of structures.* They explain: "*Prima donna* — because the performance of shells very much depends on how it is treated ... Hence, the analysis of shells is at least as delicate as their construction, and persists to be one of the most challenging tasks in structural mechanics."

In Babcock's (1983) words

> ... It is apparent that there have been great advances in the understanding of buckling problems — Equally apparent is the fact that this knowledge and understanding has become a self-perpetuating monster that awaits a modern-day Timoshenko to tame it and place it in the service *of the practicing engineer.*

It can be claimed with a certain amount of gratification and humility that the capricious, prima-donna behavior of shells, namely, their hyper-sensitivity to imperfections, has been tamed.

Appendix A

Analytical Analysis of the Nonsymmetric Version of the Budiansky–Hutchinson Model

This appendix concentrates on the simple model structure proposed by Budiansky and Hutchinson (1964) for illustrating Koiter's deterministic imperfection-sensitivity notion. This section closely follows this author's paper (1980). Such a model of an idealized column is shown in Fig. A.1. The three-hinge, rigid-rod systems constrained laterally by a nonlinear spring with the mass concentrated at the hinge joining the two rods.

We suppose that the restoring force F is related to the end shortening (or elongation) x of the spring by

$$F = k_1\xi + k_2\xi^2 + k_3\xi^3 \tag{A.1}$$

where $\xi = x/L$. Refraining temporarily from a restriction as to the sign of k_2 or k_3, we assume $k_1 > 0$. Fig. A.1(1) shows the system in its straight (undisplaced) state, and Fig. A.1(2) in its displaced state. The horizontal reactions at the hinges equal $F/2$. Furthermore, the moment about the middle hinge must vanish. This requirement leads to the equilibrium equation:

$$\lambda\xi = \frac{1}{2}F\sqrt{1-\xi^2} = \frac{1}{2}(k_1\xi + k_2\xi^2 + k_3\xi^3)\sqrt{1-\xi^2}. \tag{A.2}$$

For small values of ξ we obtain the following asymptotic result:

$$\lambda\xi = \lambda_c(\xi + a\xi^2 + b\xi^3 + \cdots) \tag{A.3}$$

$$\lambda_c = \frac{1}{2}k_1 \tag{A.4}$$

$$a = \frac{k_2}{k_1}, \quad b = \frac{k_3}{k_1} - \frac{1}{2}. \tag{A.5}$$

Equation (A.3) represents the load–displacement relationship. Now the trivial solution

$$\xi = 0 \tag{A.6}$$

satisfies this equation for all values of the load λ, which merely confirms that the straight, vertical state is one of equilibrium. However, when ξ is not zero, we can

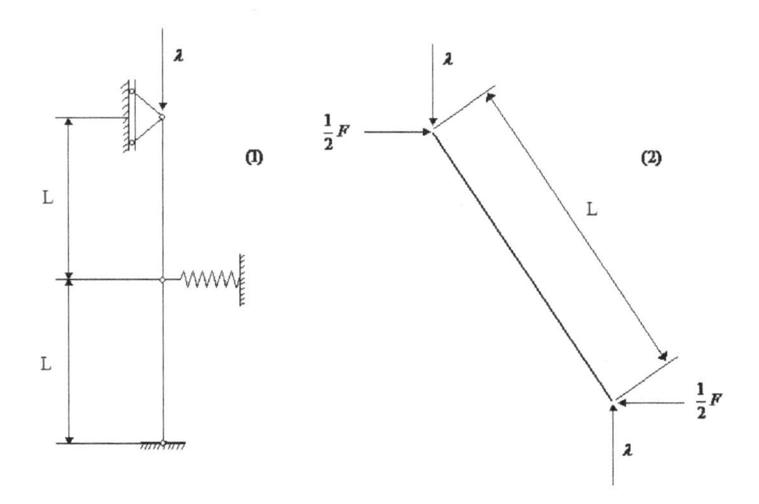

Fig. A.1. (1) Idealized column, (2) equilibrium of single bar.

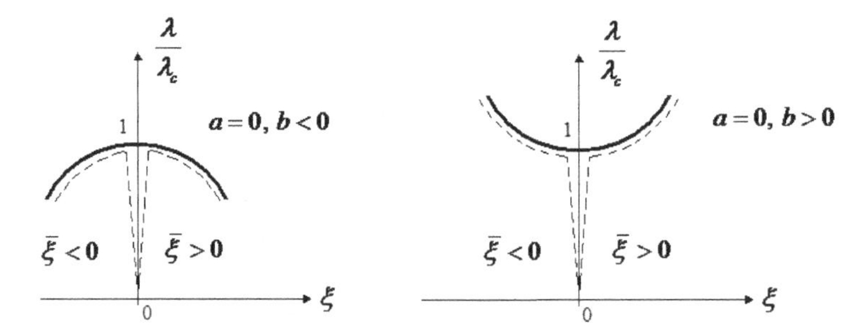

Fig. A.2. Nondimensional load versus additional displacement relationship for symmetric structure.

still satisfy this equation, that is, have equilibrium, if

$$\lambda = \lambda_c(1 + a\xi + b\xi^2 + \cdots). \tag{A.7}$$

The solid curves in Fig. A.2 show the equilibrium load λ plotted against the displacement ξ. At point $\xi = 0$, $\lambda = \lambda_c$, there is a *branching (bifurcation)* of the curve, one branch continuing along the straight line $\xi = 0$ and the others moving out along a parabola (A.7). This splitting into branches indicates the onset of instability; therefore the load λ_c at the branching point is called the *classical bifurcation buckling* load.

The structure is designated as *nonsymmetric* in the general case $a \neq 0$, $b \neq 0$, as *symmetric* if $a = 0$, $b \neq 0$, and as *asymmetric* if $a \neq 0$, $b = 0$. In the latter case the parabola degenerates into a straight line:

$$\frac{\lambda}{\lambda_c} = a\xi + 1. \tag{A.8}$$

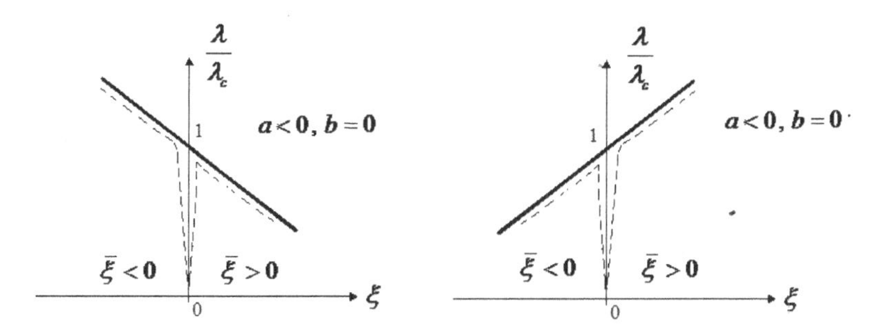

Fig. A.3. Nondimensional load versus additional displacement relationship for asymmetric structure.

We now proceed to the realistic, *imperfect* structure. Assuming that an unloaded structure has an initial displacement $\bar{x} = L\bar{\xi}$, then equilibrium dictates, instead of Eq. (A.3), the following relationship:

$$\lambda(\xi + \bar{\xi}) = \frac{1}{2}F[1 - (\xi + \bar{\xi})^2]^{1/2} = \frac{1}{2}(k_1\xi + k_2\xi^2 + k_3\xi^3)[1 - (\xi + \bar{\xi})^2]^{1/2}. \quad \text{(A.9)}$$

where ξ is an additionnal displacement and $\xi + \bar{\xi}$ a total displacement. For small values of ξ, we arrive at the following asymptotic result:

$$\lambda(\xi + \bar{\xi}) = \lambda_c[\xi + a\xi^2 + b\xi^3 + O(\xi^2\bar{\xi})]. \quad \text{(A.10)}$$

Equation (A.10) indicates that ξ and $\bar{\xi}$ have the same sign (i.e., additional displacement ξ of the system is such that the total displacement $\xi + \bar{\xi}$ is increased by its abolute value). Otherwise, the assumption $\xi\bar{\xi} < 0$ would imply $\lambda < 0$ for $0 < |\xi| < \bar{\xi}$, the presence of tension, which is contrary to our formulation of the problem. Note also that the graph λ/λ_c vs. ξ for an imperfect structure issues from the origin of the coordinates. Additional zeroes of λ/λ_c coincide with the zero points $(-a \pm \sqrt{a^2 - 4ab})/2b$ of the parabola (A.7) representing the behavior of a perfect structure. The dashed curves in Figs. A.4 and A.5 show the equilibrium load λ plotted against the additional displacement ξ, in the imperfect structure.

We now seek the *static buckling load* λ_s, which is defined as the maximum of λ on the branch of the solution $\lambda - \xi$ originating at zero load, for specified $\bar{\xi}$.

We refer to a structure as *imperfection-sensitive* if an imperfection results in *reduced* values of the maximum load the structure is able to support; otherwise, we designate the structure *imperfection-insensitive.*

To be able to conclude whether or not a structure is sensitive to initial imperfections, we have to find whether the first derivative of λ with respect to ξ:

$$\frac{d\lambda}{d\xi} = \frac{\lambda_c}{(\xi + \bar{\xi})^2}[2b\xi^3 + (a + 3b\bar{\xi})\xi^2 + 2a\bar{\xi}\xi + \bar{\xi}]. \quad \text{(A.11)}$$

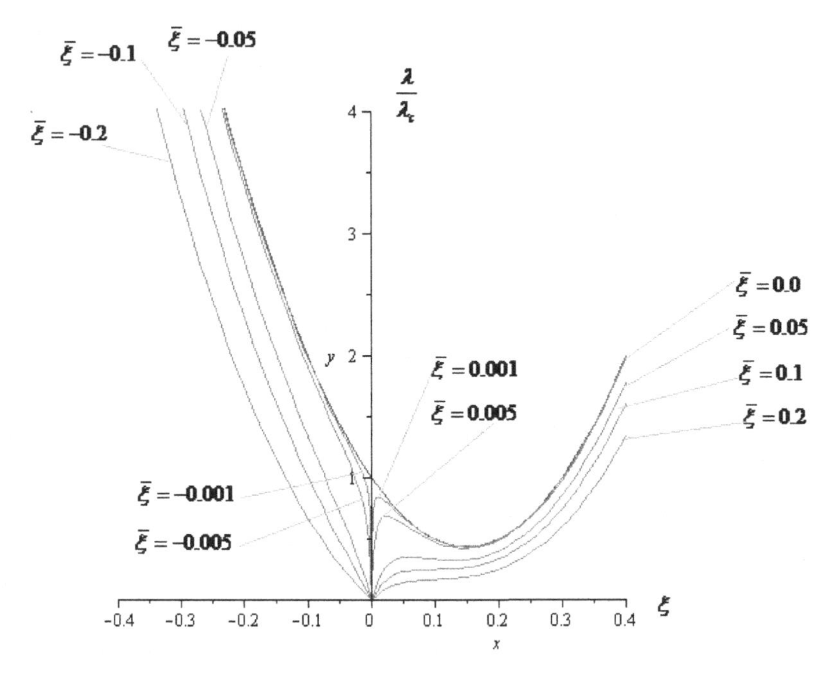

Fig. A.4. Nondimensional load versus additional displacement relationship for nonsymmetric structure ($a = -7.5, b = 25$).

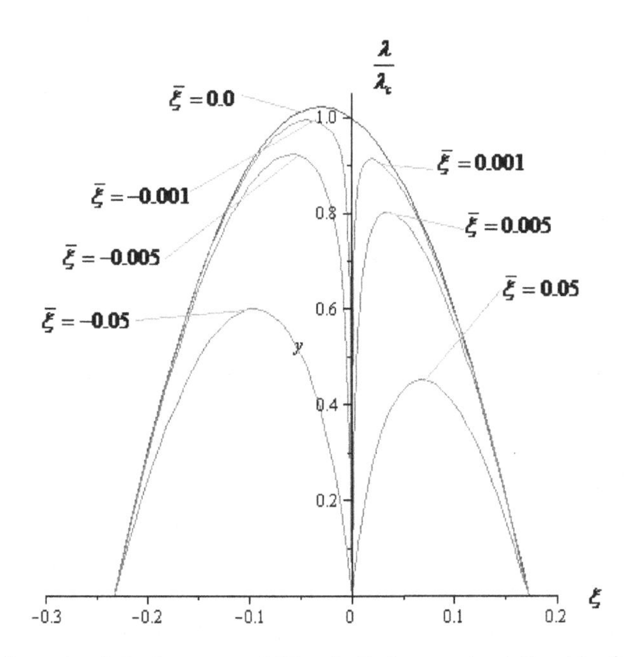

Fig. A.5. Nondimensional load versus additional displacement relationship for nonsymmetric structure ($a = -1.5, b = -25$).

has at least one real root. For our purpose, it suffices to examine the numerator

$$\varphi(\xi) = 2b\xi^3 + (a + 3b\bar{\xi})\xi^2 + 2a\bar{\xi}\xi + \bar{\xi}. \tag{A.12}$$

The structure buckles if the equation

$$\varphi(\xi) = 0. \tag{A.13}$$

has at least one real positive root for $\bar{\xi} > 0$, or at least one real negative root for $\bar{\xi} < 0$.

In some cases, Descartes's rule of signs provides an answer on buckling of the structure. This rule states that if the coefficients $a_0, a_1, a_2, \ldots, a_n$ of the polynomial

$$\varphi(\xi) = a_0\xi^n + a_1\xi^{n-1} + \cdots + a_{n-1}\xi + a_n$$

have v variations of sign, the number of positive roots of the polynomial equation, $\varphi(\xi) = 0$ does not exceed v and is of the same parity. The number of negative roots of this equation equals that of positive roots of equation $\varphi(-\xi) = 0$.

Consider the case $b < 0, \bar{\xi} > 0$. We wish to know whether there is at least one positive root between those of $\varphi(\xi) = 0$. In the subcase $a < 0$, we have

$$a_0 = 2b < 0 \quad a_1 = a + 3b\bar{\xi} < 0 \quad a_2 = 2a\bar{\xi} < 0 \quad a_3 = \bar{\xi} > 0 \tag{A.14}$$

so there is a single change in sign, indicating the occurrence of buckling. In the subcase $a > 0$, we have

$$a_0 = 2b < 0 \quad a_2 = 2a\bar{\xi} > 0 \quad a_3 > 0. \tag{A.15}$$

So, irrespective of the sign of $a_1 = a + 3b\bar{\xi}$, there is again a single change in sign. Thus the structure has a buckling load if $b < 0, \bar{\xi} > 0$.

Consider now the case $b < 0, \bar{\xi} < 0$. The question is now whether $\varphi(\xi) = 0$ has at least one negative root. Then

$$\varphi(-\xi) = -2b\xi^3 + (a + 3b\bar{\xi})\xi^2 - 2a\bar{\xi}\xi + \bar{\xi} \tag{A.16}$$

and for $a > 0$ we have

$$a_0 = -2b > 0 \quad a_1 = a + 3b\bar{\xi} > 0 \quad a_2 = -2a\bar{\xi} > 0 \quad a_3 = \bar{\xi} < 0 \tag{A.17}$$

that is, a single change in sign. For $a < 0$, we have

$$a_0 = -2b > 0 \quad a_2 = -2a\bar{\xi} < 0 \quad a_3 = \bar{\xi} < 0 \tag{A.18}$$

and irrespective of the sign of a_1 we again have a single change in sign, the conclusion being that for $b < 0$ (irrespective of the signs of a or ξ) the structure carries a finite maximum load. In complete analogy, it can be shown that the structure is imperfection-insensitive for $b > 0$ and $a\bar{\xi} > 0$.

For $b > 0$ and $a\bar{\xi} < 0$, neither Descartes's rule nor the Routh–Hurwitz criterion for the number of roots with a positive real part (used in conjunction with the fact that $\varphi(\xi) = 0$ always has one real root) suffice for a conclusion. This case, however, can be treated by Evans's root-locus method (see, e.g., Ogata, 1970), frequently used in control theory.

Let us consider first the particular subcase $b > 0$, $a < 0$ and $\bar{\xi} < 0$. The formal substitution $\xi \to s$, where $s = \text{Re}(s) + i\text{Im}(s)$ is a complex variable, in Eq. ()

$$1 + \bar{\xi}\psi(s) = 0 \quad \psi(s) = \frac{3bs^2 + 2as + 1}{s^2(2bs + a)}. \tag{A.19}$$

We now construct the root-locus plot with $\bar{\xi}$ varying from zero to infinity (obviously, only $\xi\square 1$ has physical significance). For $\bar{\xi}$ approaching zero, the roots of Eq. () are the poles of $\psi(s)$, marked with crosses (X's):

$$s_1 = s_2 = 0 \quad s_3 = -\frac{a}{2b}. \tag{A.20}$$

The $\bar{\xi} \to \infty$ points of the root loci appraoch the zeroes of $\psi(s)$, marked with circles (O's):

$$s_{1,2} = \frac{1}{3b}(-a \pm \sqrt{a^2 - 3b}). \tag{A.21}$$

$\psi(s)$ has three poles: one double at zero, and another at $(-a/2b) > 0$. A root locus issues from each pole as $\bar{\xi}$ increases above zero; a root locus arrives at each zero of $\psi(s)$ or at infinity as $\bar{\xi}$ approaches infinity. For the case $a^2 = 3b$, the two "circles" coincide. As seen from Fig. A.6, Eq. (A.19) has two real positive roots, and the structure buckles for any $\bar{\xi} > 0$. For $a^2 < 3b$, both "circles" are complex (Fig. A.7); for a certain value of $\bar{\xi}$ (called the critical value, $\bar{\xi}_{\text{cr,s}}$ a pair of loci break away from the real axis. For $\bar{\xi} > \bar{\xi}_{\text{cr,s}}$, Eq. (A.19) has no real positive root, and consequently the structure is imperfection-insensitive.

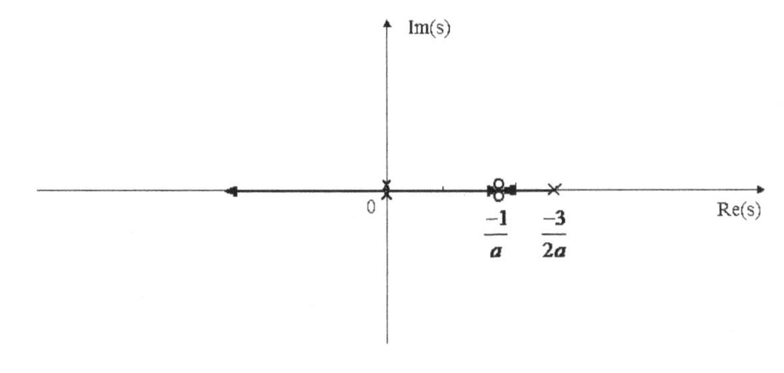

Fig. A.6. Root-locus plot ($a < 0, b > 0, \bar{\xi} > 0, a^2 = 3b$).

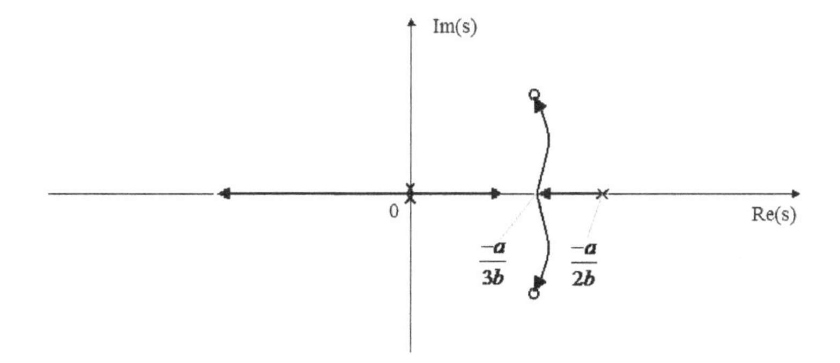

Fig. A.7. Root-locus plot ($a < 0, b > 0, \bar{\xi} > 0, a^2 < 3b$).

The breakaway point is found as the root of the equation

$$\frac{\mathrm{d}C}{\mathrm{d}s} = 0 \quad C(s) = \frac{1}{\psi(s)} = \frac{s^2(2bs + a)}{3bs^2 + 2as + 1}. \tag{A.22}$$

This equation has only one real root, $s_0 = -a/3b$. The appropriate value of $\bar{\xi}$ equals $-C(s_0)$:

$$\bar{\xi}_{\mathrm{cr,s}} = -\frac{a^3}{9b}\frac{1}{3b - a^2}. \tag{A.23}$$

The static buckling load associated with $\xi = s_0$ and $\bar{\xi} = \bar{\xi}_{\mathrm{cr,s}}$ is

$$\frac{\lambda_{\mathrm{s}}}{\lambda_{\mathrm{c}}} = \frac{a}{3b}\left(\frac{2}{9}\frac{a^2}{b} - 1\right)\left(\bar{\xi}_{\mathrm{cr,s}} - \frac{a}{3b}\right)^{-1}. \tag{A.24}$$

For example, for

$$\frac{b}{a^2} = \frac{2}{3}, \quad \bar{\xi}_{\mathrm{cr,s}} = -\frac{1}{6a}, \quad \frac{\lambda_{\mathrm{s}}}{\lambda_{\mathrm{c}}} = \frac{1}{2}, \tag{A.25}$$

and for $\bar{\xi} > \bar{\xi}_{\mathrm{cr,s}}$ static buckling does not occur.

For the case $a^2 > 3b$ (see Fig. A.8), there are always two real positive roots to $\varphi(\xi) = 0$, and the structure buckles. Consequently, the structure turns out to buckle for $a^2 \geq 3b$. In the range $a^2 < 3b$ the structure buckles if $\bar{\xi} \leq \bar{\xi}_{\mathrm{cr,s}}$.

We next consider the case $a > 0, b > 0, \bar{\xi} < 0$. It is readily shown that the *inverse root loci* for $-\infty < \bar{\xi} \leq 0$ are the mirror images of the original root loci for $0 \leq \bar{\xi} < -\infty$ with respect to the imaginary axis. The system buckles for $a^2 \geq 3b$ and also for $a^2 < 3b$ if $\bar{\xi} \geq \bar{\xi}_{\mathrm{cr,s}}$, as indicated by Eq. ().

For the particular case $a = 0$ the structure buckles if $b < 0$ (for both $\bar{\xi} > 0$ and $\bar{\xi} < 0$) and is insensitive if $b > 0$. For $b = 0$, the structure buckles if $a\bar{\xi} < 0$ and does not buckle in the opposite case.

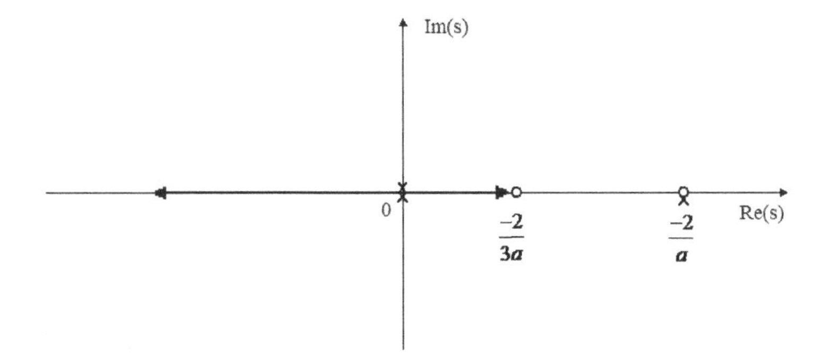

Fig. A.8. Root-locus plot ($a < 0, b > 0, \bar{\xi} > 0, a^2 = 4b$).

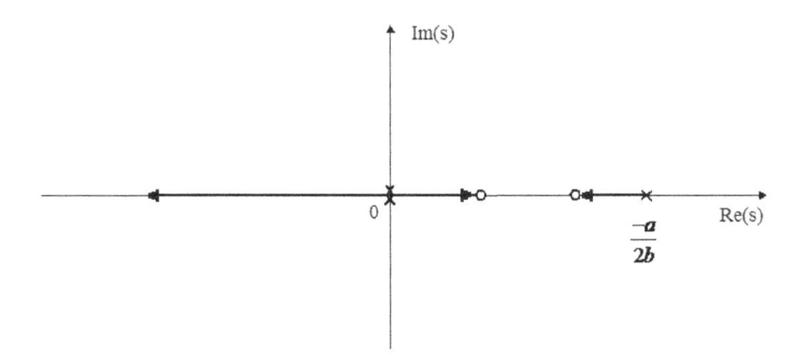

Fig. A.9. Root-locus plot ($a < 0, b > 0, \bar{\xi} > 0, 3b < a^2 < 4b$).

As for the structure that buckles, differentiating Eq. () with respect to ξ and setting (for $b \neq 0$)

$$\frac{d\lambda}{d\xi} = 0 \quad \lambda = \lambda_s \tag{A.26}$$

we obtain the relation between the buckling load λ and initial imperfection amplitude $\bar{\xi}$:

$$\left(1 - \frac{\lambda_s}{\lambda_c} - \frac{a^2}{3b}\right)^3 = -\frac{27}{4}b\left[\frac{a}{3b}\left(1 - \frac{\lambda_s}{\lambda_c}\right) - \frac{2}{27}\frac{a^3}{b^2} + \frac{\lambda_s}{\lambda_c}\bar{\xi}\right]^2. \tag{A.27}$$

Note that Budiansky (1967) dealt with the quadratic-cubic nonlinear model but did not provide the closed-form expression (A.27) from which buckling load can be determined. However, he did detect (p. 95), via numerical analysis that the value $b/a^2 = 1/3$ played an important role.

For $b < 0$ and $a = 0$, we get from Eq. (A.27)

$$\left(1 - \frac{\lambda_s}{\lambda_c}\right)^{3/2} - \frac{3\sqrt{3}}{2}|\bar{\xi}|\sqrt{-b}\frac{\lambda_s}{\lambda_c} = 0. \tag{A.28}$$

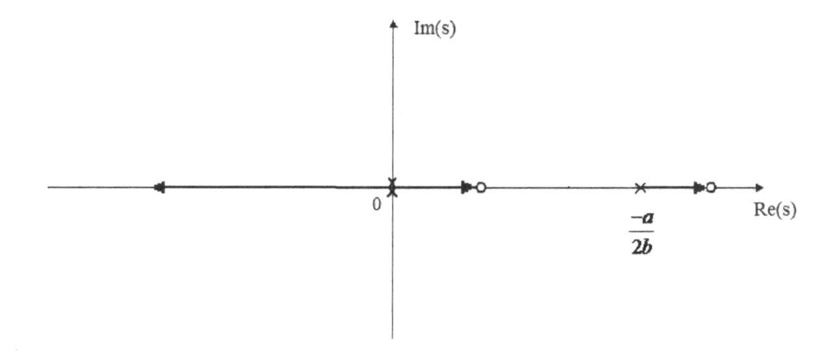

Fig. A.10. Root-locus plot ($a < 0, b > 0, \bar{\xi} > 0, a^2 > 4b$).

Note that the displacement ξ corresponding λ_s/λ_c is given by

$$\xi_{1,2} = \frac{1}{3b}\left[-a \pm \sqrt{a^2 - 3b\left(1 - \frac{\lambda_s}{\lambda_c}\right)}\right] \tag{A.29}$$

where $\xi_{1,2}$ depends upon $\bar{\xi}$ via λ_s/λ_c. The static buckling load can be obtained from Eq. (A.29), given the initial imperfection $\bar{\xi}$. The meaningful root λ_s/λ_c of Eq. (A.29) is the greatest of those that meet the requirement $\xi\bar{\xi} > 0$.

The case $b = 0$ has to be considered separately. Equation () reduces to

$$\varphi(\xi) = a\xi^2 + 2a\bar{\xi}\xi + \bar{\xi} \tag{A.30}$$

Descartes's rule then immediately yields the conclusion that the structure is imperfection-sensitive if $a\bar{\xi} < 0$. Equation () then leaves us with

$$\left(1 - \frac{\lambda_s}{\lambda_c}\right)^2 + 4a\bar{\xi}\frac{\lambda_s}{\lambda_c} = 0. \tag{A.31}$$

Note that expressions (A.28) and (A.31) for symmetric and asymmetric structures, respectively, were obtained by Budiansky and Hutchinson (1964), whereas Eq. (A.27) was obtained by this author (1980). The latter study includes also the dynamic counterpart of Eq. (A.27).

A modified version of the Budiansky–Hutchinson model was proposed by Bodner and Rubin (2005). The modification consisted in treating the bars to be elastic in compression and rigid in bending in order "to predict more realistic load–deflection response and to allow for the possible transfer of energy in compression to energy in the lateral spring." Specifically, Bodner and Rubin (2005) derived the following formula, generalizing Eq. (A.31):

$$\left(1 - \frac{\lambda_s}{\lambda_c}\right)^2 = \frac{8}{3}\left[1 - \frac{\lambda_c}{E_b A_b}\right]\frac{\lambda_s}{\lambda_c}\frac{\theta_0}{\beta} \tag{A.32}$$

where E_b is Young's modulus of each bar, A_b is the cross-sectional area of each bar, θ_0 in an angular measure of geometric imperfection and β relates to the spring force N. Authors concluded that "determining the lateral spring characteristic of the model β in terms of the basic shell geometry $\kappa(=h/R)$ [based on Calladine's (2001) empirical formula

$$\frac{\sigma_s}{\sigma_c} = 5\sqrt{3(l-\nu^2)}\kappa e^{\frac{1}{2}} = 8.3\kappa e^{\frac{1}{2}} \tag{A.33}$$

for $\nu = 0.3$] leads to good agreement with details of the shell buckling behavior." In Eq. (A.33) σ_c is the classical value of the axial stress at bifurcation, σ_s is the value of axial stress at peak load.

Bodner and Rubin (2005) also stressed: "it might be useful to consider buckling as a local instability problem with the boundaries determined as part of the solution. This calls for detailed examination of the localized imperfections and the deformation modes from the onset of loading.... Also, attention needs to be given to manufacturing variabilities."

Other definitive works including simple models include those by Augusti (1964), Thompson and Hunt (1973), Calladine (1983), Raftoyiannis and Kounadis (1988), Yithak *et al.* (1988), Simitses (1990), Kounadis *et al.* (1991), Kounadis (1993), Elishakoff *et al.* (1996), Augusti *et al.* (1998), and many others.

Appendix B

Elastic Stability: From Musschenbroek & Euler to Koiter — There Was None Like Koiter

Here is a family announcement dated 2 September 1997: "Verdrietig, maar opgelucht dat hem een langere lijdensweg bespaard is gebleven, delen wij u mede dat thuis is overleden Warner Tjardus Koiter Geboren 16 juni 1914te Amsterdam"

The giant of the modern theoretical and applied mechanics, and uniformly recognized father of the modern theory of stability, is gone. We will not listen to his excellent lectures, we will not hear reports on his correspondences with contemporaries about their misconceptions, we will not be able to witness tough remarks like "On your entire figure only one point is correct, belonging to another author," or "It will be no catastrophe if instead of the catastrophy theory you will research the chaos." We will not have a privilege to have his razor-sharp questions that can change direction of one's research. We will not be able to hear about the uncompromising positions in the modern world where everyone, so it seems, is advised to keep quiet, to be politically correct and not to be judgmental. Many of his propositions were acceptable since they did not have a flavor of "I told you so," but rather a desire to repair the scientific world, and the scientists themselves, as researchers and humans as well. Warner Tjardus Koiter dealt with the subject of stability. We encounter with this notion in our everyday lives. Its ordinary, dictionary meaning is "steadiness of firmness of character, resolution of purpose, constancy, steadfastness." The meaning of the word stability underwent a transformation over the centuries. Its first description perhaps goes to the Bible, where the Tower of Babel lost its stability under its own weight, and the weight of too high self-esteem of people who decided to reach the unreachable in the place where unreachable cannot be attained. The term itself was introduced by Plinius referring to *Stabilitas dentium* (stability of teeth) (Villaggio, 2001). Moiseev (1949) ascribes it to the Carmelite monk Paolo Foscarini in his book under the title *De Mobilitate Terrae et Stabilitate Solis* (in English *On Motion of Earth and Immovability of the Sun*). As we see, the Latin word *stabilitate* was used for denoting immovability, not related to the disturbance of the equilibrium state. In words of Piero Villaggio (2001),

"there can be no doubt that the axioms of Archimedes (287–212 B.C.E.) on the principle of the lever contain an almost modern meaning of infinitesimal stability, and that his treatment of hydrostatics helps to solve complicated stability problems for floating segments of parabolords of revolution. The notion of infinitesimal stability is also present in the work of Jordanus Nemorarius (1237), who is credited with the invention of the virtual work principle. Heron of Alexandria (c. 100 B.C.E.), in the course of a "long dull work on statics" (Truesdell, 1960) endeavored to explain why the strength of a piece of wood reduces as its length increases. Leonardo da Vinci (1452–1519) provided two empirical rules for the strength of columns in compression. The Jesuit M. Mercenne (1588–1648), in his *Reflexions* on the causes of resistance in solids, observed that "iron, copper and other metals, even single bodies, when subject to a force or weight, curve and bend to the form of an arch before breaking (Benvenuto, 1990)."

Mercenne's conclusions were unexpectedly confirmed by the consistent program of experiments conducted by Musschenbroek (1692–1761), the inventor of testing machines designed to allow systematic variations in experimental parameters. Musschenbroek even proposed a quantitative law for the failure in compression of a parallelepiped composed of wood. According to Galileo Galilei (1638), "...a small oblisk or column or other solid figure can certainly be laid down or set up without danger of breaking, while very large ones will fall apart at the slightest provocation, and that purely on different account of their own weight" (see also Godoy, 2010).

In the famous encyclopedia by Diderot and d'Alembert (1778), the word is used to denote the property of what is fixed, immobile: they say unchangeability of the contract, character, spirit, views, etc. The concept was long associated with inertia. The tendency was to think of stability as an integral part of the notion of equilibrium of the material body. In his letter dated 11 February 1605 to Fabricius, Johannes Kepler notes: "Any material body is made so by the nature to rest in place, where it is located. Indeed, the rest, as darkness, is sort of the neutrality."

The concept of lack of stability is vividly illustrated in the short story *A Sound of Thunder* (highly recommended all those who deal with mechanics) by a famous science-fiction writer Ray Bradburry (1952). The plot is set in the year 2055, immediately after the U.S. presidential election has been won by the progressive candidate. Time Safari Inc., an outfit specializing in guided tours of the past, has organized a hunting trip to a jungle 60 millions years back. Participants are warned to keep to a specially provided floating path and shoot only marked animals, but otherwise refrain from tampering in any way with the environment, as even the slightest error may multiply over the ages. For example, killing a mouse would mean no progeny, a fox would starve for want of mice; a lion would starve for want of foxes; and in some later era a man would starve for want of game with the attendant effect of future demography. Accordingly, the hunt is confined to animals preselected and marked in advance as "safe," namely, Tyrannosaurus Rex destined to be killed anyway by a falling tree. The hero of the story, however, disobeys these instructions, steps off the path, and inadvertently treads on a butterfly. Back in the future, in the year 2055, he finds to his dismay that the "anti-everything" candidate became a President, and, if this were not bad enough, the sign on the wall of "Time Safari Inc." is grotesquely misspelt, since the spelling has been changed over the centuries, due to the different development humankind underwent due to the (small) butterfly effect. Thus a presumably small disturbance leads to drastic consequences in the future development. This is lack of stability.

Experimentally (see Truesdell, 1968) the topic of *elastic* stability was pioneered by Pieter van Musschenbroeck (1729). *Mathematically* the theory of *elastic* stability was initiated by a Swiss mathematician Leonhard Euler (1744). This theory found applications only in the 19th century, when the transportation revolution demanded lightweight structures, in the form of very thin plates or shells. Around 1910, Euler's theory was applied to thin shells (like the outer surface of an ordinary egg, or a bamboo stick) by the engineers in three different countries: Timoshenko (1910, 1914) from Russia, Southwell (1913, 1915) from England, and Lorenz (1908, 1911) in Germany (see an unusually extensive review by Grigoliuk and Kabanov, 1967). Since then there was a sort of a "quarrel": Who was the first amongst them? Yet, this question is a non-problem. The reason is very simple: their theoretical predictions were *not* substantiated by experiments. As we all know, Its Excellency the Experiment is the final judge, and in Huxley's terminology, it is a "tragedy of science: a beautiful theory killed by the ugly facts."

These shells were not needed in the kitchen for eggs or in the forest for the bamboos. Shells and plates are integral parts of the airplanes, cars, ships, space shuttles, and what not. The theory of stability of columns, plates, and shells is needed to design the vital elements of modern transportation. Engineers suggested a weird thing: Since the experiment does not match the theory, first make theoretical calculations, then divide the result by some large correcting, "fudge" knockdown factor, say, 5, 6, or even 7, and utilize that number for the design. At this crucial time young Warner Tjardus Koiter (1945) presented to the Delft University of Technology his Ph.D. dissertation titled "Over de stabiliteit van het elastisch evenwicht." The thesis was done during the Nazi occupation of the Netherlands. As Professor Koiter recalled at his retirement lecture (1979):

> I have a vivid recollection of New Year's eve in December 1942 when my wife and I sat in front of a very modest fire and discussed, in addition to the most acute war time problems, what would happen to my work which I considered to be significant but questioned whether it would be recognized as such.

In September 1945 he presented the thesis to the Senate of the Delft University of Technology. In his work he explained why the "ugly" experiments were not "dancing" in accordance with the theory: shells possess unavoidable small imperfections, deviations from the desired ideal form. They, as small butterflies in Ray Bradbury's story, have a dramatic influence on the stability. This influence is so big that it sometimes reduces the ideal theoretical results by a factor of 10 (though more of ten by a factor of 2.5) and thus previous "classical" results are totally invalidated.

As Godoy (2011) notes, "the identification of the origins of what we now call the theory of elastic stability in not an easy task. Most authors trace the origins to the pioneering work of Leonhard Euler in 1744, and some shift this origin to the experimental works of Petrus van Musschenbroeck 1729. However, other contemporary authors, interested in the history of the discipline postulate that the works of medeaval and Renaissance scholars should be considered as the true sources of the buckling studies performed in the XVIII century."

Yet young educator of TH Delft did not extensively publish in the 1940s the results of his Ph.D. dissertation and the scientific papers on this subject, as it is so widely done nowadays, when an idea is "dissected" and then published and republished in numerous papers. He recalled:

> ...I considered I had already published my basic thinking on elastic stability, and that it would be improper to reiterate on the same topic, a curious mixture of modesty and immodesty, the latter because I took it for granted that my published work was accessible in principle to anyone actively interested in the field.

It took however two decades before the country that was uncovered by Christopher Columbus uncovers Koiter's work. Then this dissertation was enthusiastically translated two times, independently, into English, in 1967, as a translation by NASA, and the second one performed at the Stanford University. Then the stability research around the world flourishes, due to Koiter's breakthroughs, both in the West and in the East. Koiter's work was performed nearly 200 years after Euler's. But we can safely say that in the subject of elastic stability "from Euler to Koiter there was none like Koiter." Professor Koiter always paid attention to the mathematical part of the subject, but always stressed that engineers need to come up with just appropriate theoretical assumptions to catch the essence of the physical phenomenon. I recall that it was in 1966 when, an upper-level undergraduate, I first met him in Moscow. At the scientific congress he said at the panel discussion on the future of shell research: "We cannot excessively occupy ourselves with second order theoretical effects. We are engineers, and must come up with workable engineering solutions."

In 1979/1980 I and my family spent a sabbatical year at Delft as a guest of the Aerospace Department, through a kind invitation of Professor Johann Arbocz and C. Vakgroep. At that time, I also had a privilege to get invitations from several other prestigious universities but chose to come to a compact and scientifically and otherwise beautiful city of Dutch "nuchters" (sober-minded people). One scientist, whose University was not chosen, asked me: "Why did you go to Delft? Where is it? What do they do?" The reply was simple: "Delft is a city where Koiter is working." Indeed, during that academic year, the already retired mastermind of stability gave no less than five scientific seminars on various topics.

Scientists live in this world and cannot and shouldn't escape from its problems. Professor Koiter actively expressed his worldly views. In 1976, during the international mechanics congress in Delft, the then Soviet Union did not allow some of its out-of-favor scientists to attend. Koiter stood firm; he announced from the podium that he disagreed with such a treatment; that the time slots allotted to Soviet scientists would not be filled with other lectures: the empty room would serve as a reminder of the silent protest. Many were deeply impressed with this uncompromising stand. All, it so seemed, shared a human pride for this prominent scientist did not yield to the pressure. This was one of the highlights in the lives of the Congress participants: such moments fill one's heart with optimistic spirit and human pride that the truth would be eventually found. Was this because Hugo Grothius (1583–1645), whose statue firmly stands in the very center of Delft, somehow continues to inject part of himself in every Delftenaar? One of the ingredients of the chair Koiter occupied was "stiffness." He vividly illustrated this stiffness of human mind at the world congress. But he was flexible too. In his writings he would sometimes insert very flowery and elucidating statements. Here are two examples: (a) "Flexible bodies like thin shells require a flexible method." (b) "Any two-dimensional theory of thin shells is necessarily of an approximate character. An exact two-dimensional theory of shells cannot exist, because the actual body we have to deal with, thin as it may be, is always three-dimensional. We may perhaps illustrate this point in a somewhat facetious way: even if it appears to be the fashion for ladies to be as thin as possible, fortunately, in our view, they remain essentially three-dimensional."

He endorsed our developments of probabilistic analysis (Elishakoff, 1979, 1985; Elishakoff and Arbocz, 1982, 1985) of initial imperfections pioneered earlier by Bolotin (1958) and encouraged several of my stochastic studies. Yet, his deep grasp of every possible disadvantage of the theory of probability together with lucky interaction with several "non-probabilistic uncertainty" scientists (Professors Rudolf Drenick, Yakov Ben-Haim, Ezra Zeheb) led to an enrichment of my research scopes (Ben-Haim and Elishakoff, 1990; Elishakoff and Ohsaki, 2010). I know others who had such a blessing too. I was also very happy; several years go, to cooperate with him on a scientific project posed by the NASA Langley Research Center. His hand-evaluated elegant analytical solutions, based on legendary physical insights could only be matched and generalized by using the latest, suitably modified, symbolic computer packages (Koiter, Elishakoff, Li and Starnes, 1994a, 1994b). His last published paper (1996) (see also his review, 1985, that he planned to extend for the *Applied Mechanics Reviews* upon my invitation; yet he could not complete this work due to illness) will also have a lasting effect: it warned engineers about dangers of using the concepts of unrealistic, fictitious, *static* "follower" forces, until convincing experimental validation of Beck's column is found. He did not mind mathematicians to be engaged with these forces. We humbly submit that his admonition was timely, since many hundreds of papers were written up to now about pseudo-phenomena, which could be only simulated but which neither could be directly reproduced experimentally, nor can be considered useful, since one can directly deal with the *real* phenomena themselves rather than with oversimplified models. These ideas are not shared, as anticipated, by all. This paper led to the response by Sugiyama *et al.* (1979) and numerous attempts that are underway to realize follower forces experimentally, some maintaining that they have succeeded so (Bigoni and Noselli, 2011; Ingerle, 2013).

His and Mevr. Lous's house in Delft always was a place for many foreign scientists who came to Delft to refine their understanding of stability. His research was about the effect of imperfections, and in life too, I recall him complaining once about the decline of the positive influences of religion on the modern society, and the attendant imperfections.

Ms. Lous and Professor Koiter's hospitality was an invariable element of the Delft Atmosphere. I know of children of at least two foreign scientists for whom Mevr. Lous commemorated their birth in Delft by handmade embroidery. Professor Dr. Ir. Warner Tjardus Koiter left this world much more perfect than he found it, since only the identification of and the action upon both human and engineering imperfections can repair the world. One thing is above a disputation: from Musschenbroek & Euler to Koiter there was none like Koiter, in the field they both shared, namely "theory of elastic stability." It is our privilege that his monumental scientific work and most fond memories remain with us.

Appendix C

1. Data Basis

The shells that are considered in Chapter 6 have been manufactured, measured, and tested at the German Aerospace Centre in Braunschweig (see Degenhardt *et al.*, 2010).

The shell surfaces were measured using the optical measurement system ATOS. For the determination of the wall thickness, ultrasonic measurements have been executed. Coupon tests have been executed for determining the material parameters. A loading imperfection occurred within the buckling tests, which has a significant influence on the test results. This imperfection has not been measured, but because its influence is too significant to be neglected, these imperfections are determined indirectly (see Sec. C.1.d). Additionally, inaccuracies of the fiber angle, fiber waviness and void inclusions can occur for fiber composite shells, which have not been measured for the given set of shells.

(a) *Geometric imperfections*

The geometric imperfections $\overline{W}$ are defined as the radial deviation of the measured surface from the perfect cylinder (see Fig. C.1). For the uncertainty analyses it is necessary to describe geometric imperfections by a preferably small number of parameters that capture the characteristics of imperfection patterns. This is usually done with double Fourier series (see e.g. Arbocz and Hol (1991) and Arbocz and Hilburger (2005)). For the considered set of shells, the representation (C.1) is used:

$$\overline{W}(x, y) = t \sum_{k=0}^{n_1} \sum_{l=0}^{n_2} \bar{\xi}_{kl} \cos \frac{k\pi x}{L} \cos \left(\frac{ly}{R} - \bar{\varphi}_{kl} \right). \tag{C.1}$$

L, R and t are the length, the radius and the wall-thickness of the shell, x and y are the coordinates in axial and circumferential direction and n_1 and n_2 are the number of considered Fourier coefficients. $\bar{\xi}_{kl}$ and $\overline{\varphi}_{kl}$ are the amplitude and the circumferential phase shift angle of an imperfection mode with k axial half-waves and l circumferential full waves. For the axial direction, it is also possible to use the half-wave sine-approach, which affords the representation of initial rotations at

Table C.1. Characteristics of the regarded cylindrical shells.

Nominal dimensions of all shells
Radius: 250 mm, length: 510 mm, wall-thickness: 0.5 mm
Laminate setup [±24°, ±41°]

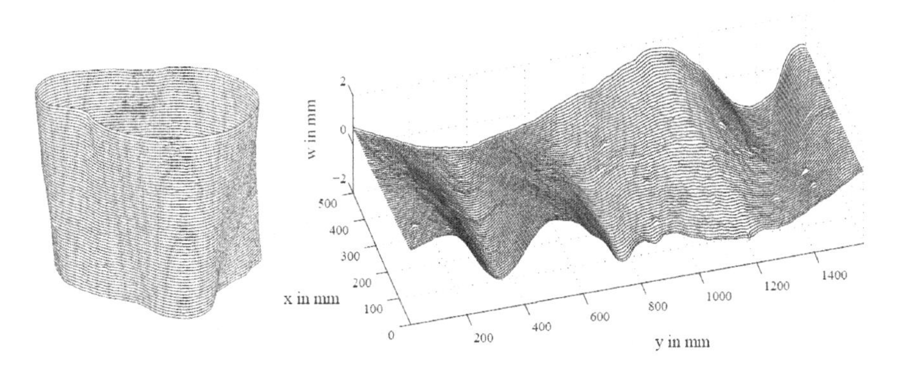

Fig. C.1. 3D-model (left) and unwounded measured imperfection pattern (right) of one test shell with scaling factor.

the edges, but without deflections. The half-wave cosine-approach is more suitable for clamped edges and thus it is used subsequently. For a detailed discussion on the choice of the Fourier representation the reader should refer to Kriegesmann *et al.* (2010).

For the considered set of shells, the double Fourier series (C.1) describes the imperfection patterns sufficiently accurate for $n_1 = 10$ and $n_2 = 20$ and hence $11 \cdot 21 \cdot 2 = 462$ coefficients are taken into account. The amplitudes $\overline{\xi}_{kl}$ and the phase angles $\overline{\varphi}_{kl}$ describe the scattering of geometric imperfections and are regarded as random parameters within the probabilistic analyses.

In order to reduce the number of parameters, a Mahalanobis transformation (see Härdle and Simar 2007) is executed. The Mahalanobis transformation describes the relation between a general random vector $\mathbf{X}$ and a random vector $\mathbf{Z}$ with independent entries, each with mean value 0 and variance. Hence, each realization $\mathbf{x}$ can be transformed to a realization $\mathbf{z}$ of $\mathbf{Z}$ using the following expression:

$$\mathbf{x} = \mathbf{\Sigma}^{\frac{1}{2}}\mathbf{z} + \boldsymbol{\mu} \quad \text{and} \quad \mathbf{z} = \mathbf{\Sigma}^{-\frac{1}{2}}(\mathbf{x} - \boldsymbol{\mu}). \tag{C.2}$$

$\mathbf{\Sigma}$ is symmetric and positive semidefinite and hence, $\mathbf{\Sigma}$ may be singular, the eigendecomposition of $\mathbf{\Sigma}$ is used to find the root of $\mathbf{\Sigma}$ and its inverse

$$\mathbf{\Sigma}^{\alpha} = \mathbf{Q}\mathbf{D}^{\alpha}\mathbf{Q}^{\mathrm{T}} \tag{C.3}$$

where $\mathbf{Q} = (\mathbf{q}_1, \ldots, \mathbf{q}_{rn})$ is the matrix with eigenvectors of $\mathbf{\Sigma}$ and $\mathbf{D} = \text{diag}(\lambda_1, \ldots, \lambda_{rn})$ is a diagonal matrix with the eigenvalues of $\mathbf{\Sigma}$. The scalar nm is the number of measurements; moreover rank($\mathbf{\Sigma}$) $= rn$ cannot be greater than nm. Thus, if the number of parameters np exceeds the number of measurements nm, $\mathbf{\Sigma}$ must be singular. The matrix $\mathbf{B}$ shall be defined as

$$\mathbf{B} = \mathbf{Q}\mathbf{D}^{\frac{1}{2}} = \mathbf{\Sigma}^{\frac{1}{2}} \quad \text{and} \quad \mathbf{B}^{-1} = \mathbf{D}^{-\frac{1}{2}}\mathbf{Q}^{\mathrm{T}} = \mathbf{\Sigma}^{-\frac{1}{2}}. \tag{C.4}$$

since $\mathbf{Q} \in \mathrm{R}^{np \times rn}$ and $\mathbf{D} \in \mathrm{R}^{rn \times rn}$, $\mathbf{B} \in \mathrm{R}^{np \times rn}$ and so $\mathbf{z} \in \mathrm{R}^{rn}$. This means that using the reducing Mahalanobis transformation (C.5) the number of parameters, which describe the scattering of geometry, is reduced form np to rn

$$\mathbf{x} = \mathbf{B}\,\mathbf{z} + \boldsymbol{\mu} \quad \text{and} \quad \mathbf{z} = \mathbf{B}^{-1}(\mathbf{x} - \boldsymbol{\mu}). \tag{C.5}$$

In the given example $rn = 9$ and thus, the 462 Fourier coefficients can be substituted by 9 parameters $(z_1, \ldots, z_9)$.

(b) *Scattering of material parameter*

The material parameters E_{11}, E_{22}, and G_{12} have been determined experimentally and their stochastic distribution has been determined. Due to the test procedure it was not possible to detect correlations and hence material properties are assumed to be uncorrelated even though it seems reasonable that, for example, the shear modulus G_{12} increases at the same time as the Young's modulus E_{22} increases. However, it is shown in Kriegesmann *et al.* (2011) that the influence of the scattering of material properties on the distribution of buckling load has minor influence. Hence, the nominal values of the material properties, which are given in Degenhardt *et al.* (2007) and Table C.2, have been used in the calculations.

(c) *Wall-thickness deviations*

The wall thickness varies for different shells, as well as within the surface of a single shell. Orf (2008) compared finite element buckling analyses of shells with measured wall-thickness patterns and shells with a constant thickness. Orf concluded that for the given ensemble of shells there is no significant difference of buckling load as long as the average wall thickness is equal. Thus, a constant wall thickness is regarded as random parameter for which the average wall thickness is the database. If the distribution of the wall thickness over the shell surface has to be considered, this two-dimensional function can be represented by Fourier coefficients as well.

Table C.2. Material parameters of the regarded cylindrical shells.

E_{11}	E_{22}	G_{12}	G_{23}	ν_{12}
157362 MPa	10095 MPa	5321 MPa	4000 MPa	0.277 MPa

(d) *Loading imperfection*

During the tests the shells are subjected to small unwanted inclination. It is caused by the test setup, but has not been measured. Because of the significant influence of the resulting bending moment on the buckling load, the inclination should be considered in the analysis. It is determined indirectly. As demonstrated in C.2, in FE simulations the shell is bended with the bending angle θ about an axis, which is described by the circumferential variation angle ω. Within these simulations the measured geometric imperfection pattern, the measured average wall thickness and the mean values of the material properties are applied.

Figure shows the numerically determined buckling load of one test shell with respect to the inclination. For the imperfect shell the position of the bending axis has an influence on the buckling load; however, the position during test is unknown. In order to approach the applied bending angle, the mean buckling load for fixed θ and varying ω is fitted to the experimental result. (For example in Fig. , a good approximation is given for $\theta = 0.009°$.) This procedure is applied for the 10 shells and 10 bending angles are obtained.

Because the value of ω is unknown during the tests, the estimation of θ includes measurable uncertainties. For the example given in Fig. the estimated bending angle θ equals 0.009°. If the circumferential variation angle ω had a value of 45° during the test, θ must equal 0.01°. If ω was equal 135°, a bending angle of 0.008° would fit the test result. This shows that the estimation of θ has an accuracy of about 0.001°.

2. Minimum Volume Enclosing Hyper-Rectangle

A hyper-rectangle in the d-dimensional space is defined by the lengths of its edges and by its orientation, which can be described with $d - 1$ rotation angles $(\alpha_1, \ldots, \alpha_{d-1})$. In order to find the minimum volume enclosing hyper-rectangle, the measurement points are rotated to the coordinate system parallel to the edges of the

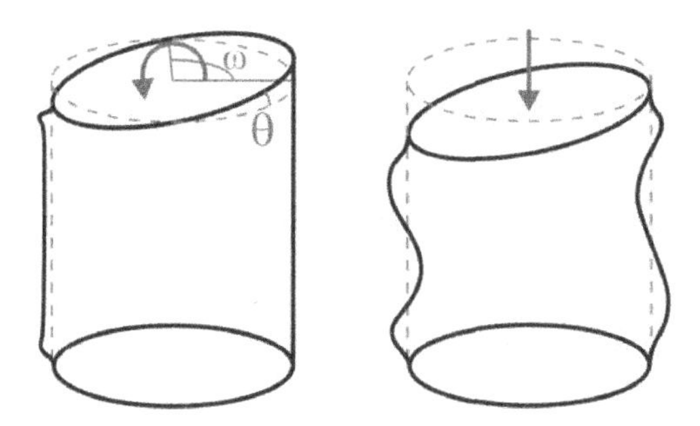

Fig. C.2. Illustration of the FE-simulation for the determination of bending angles.

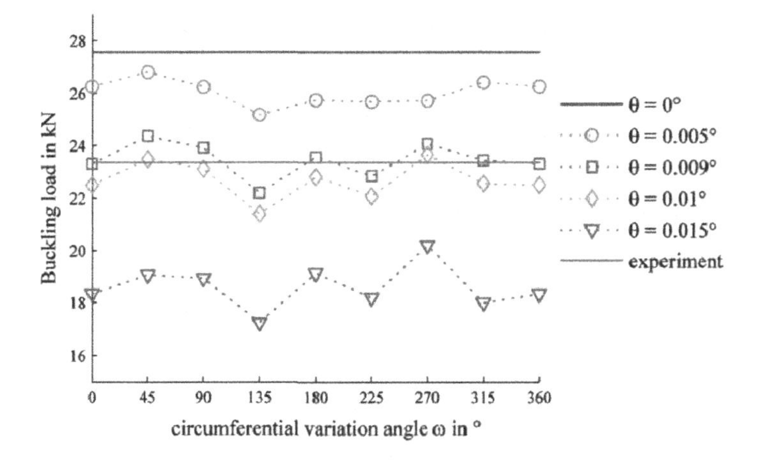

Fig. C.3. Buckling loads for different inclinations of one imperfect shell.

hyper-rectangle. The rotation of the measurement points is given by

$$\xi = \mathbf{T}(\alpha_1, \ldots, \alpha_{d-1}) \cdot \mathbf{x} \tag{C.6}$$

with

$$\mathbf{T}(\alpha_1, \ldots, \alpha_{d-1}) = \mathbf{T}_{d-1}(\alpha_{d-1}) \cdot \ldots \cdot \mathbf{T}_1(\alpha_1) \tag{C.7}$$

and

$$\mathbf{T}_i(\alpha_i) = \begin{pmatrix} \cos(\alpha_i) & 0 & \cdots & 0 & -\sin(\alpha_i) & 0 & \cdots \\ 0 & 1 & & & 0 & & \\ \vdots & & \ddots & & \vdots & & \\ 0 & & & 1 & 0 & & \\ \sin(\alpha_i) & 0 & \cdots & 0 & \cos(\alpha_i) & 0 & \cdots \\ 0 & & & & 0 & 1 & \\ \vdots & & & & \vdots & & \ddots \end{pmatrix} \begin{matrix} \\ \\ \\ \\ \{\text{row i+1}. \\ \\ \\ \end{matrix} \tag{C.8}$$

$$\underbrace{}_{\text{column } i+1}$$

The volume of the enclosing hyper-rectangle is given by

$$V_R = \prod_{i=1}^{d} [\max(\xi_i) - \min(\xi_i)]. \tag{C.9}$$

The orientations describing the minimum volume enclosing hyper-rectangle can be found numerically by increasing each α_i stepwise in the interval [0°, 90°]. The computation time of this procedure increases exponentially with the dimension, which means that for certain dimensions it may be cumbersome to find the minimum volume enclosing hyper-rectangle this way. This shall be demonstrated with the following calculation of computing time. Rotation of all measurement points and determining the current minimum volume enclosing hyper-rectangle for one specific orientation takes approximately 0.001 s with MATLAB R2008a. When each rotation angle is varied in 50 steps in the interval $I_\alpha = [0°, 90°]$, leading to a step size of 1.8°, the determination of the minimum volume enclosing hyper-rectangle requires 50^{d-1} calculations. Regarding 11 parameters, this would lead to a total calculation time of

$$50^{10} \cdot 0.001\text{s.} = 9.7656 \cdot 10^{13}\ \text{s} = 3.1 \cdot 10^{6}\ \text{years.} \tag{C.10}$$

Obviously, the search for the minimum volume enclosing hyper-rectangle cannot be executed this way. However, this consideration can be used in order to determine a step size that leads to an acceptable calculation time. For example, if seven steps are used, the calculation requires 3.3 calculation days, for eight steps 12.4 days are necessary for evaluation. Varying the rotation angle in seven steps in the interval $[0°, 90°]$ leads to a step size of $\Delta\alpha = 15°$, which leads to a relatively coarse approximation of the minimum volume enclosing hyper-rectangle. In this work, each of the 10 angles is varied with a step size of $\Delta\alpha = 15°$, since this allows the calculation in realistic time. Then the procedure is repeated, and the interval I_α in which the angles are varied is reduced to

$$I_\alpha = [\max(\alpha_{\text{opt}} - \Delta\alpha, 0°), \min(\alpha_{\text{opt}} + \Delta\alpha, 90°)]. \tag{C.11}$$

For the search in the new interval a smaller step size is used. This way, the search for the optimal angles can be refined efficiently. It should be noted that this procedure has a disadvantage, namely that the global minimum may be missed. If, for example, the global minimum can be obtained for an angle $\alpha_{\text{opt}} = 53°$, but in the first iteration the smallest volume is determined for $\alpha = 15°$ the global minimum will not be found. However, the described procedure delivers at least a local minimum with acceptable computational cost. Once the minimum is determined, additional random search is recommended to be conducted to check selected points, in view of locating the global minimum, in case it was missed by the above procedure.

3. Minimum Volume Enclosing Hyper-Ellipsoid

A hyper-ellipsoid with semiaxes g_i parallel to the coordinate system is given by

$$\sum_{i=1}^{d} \frac{(x_i - x_{c,i})^2}{g_i^2} = 1. \tag{C.12}$$

If the semiaxes are not parallel to the coordinate system, an ellipsoid can be described by

$$(\mathbf{x}-\mathbf{x}_c)^T \mathbf{W}(\mathbf{x}-\mathbf{x}_c) = 1 \tag{C.13}$$

with the weight matrix $\mathbf{W}$. Then, the points can be transformed to the coordinate system parallel to the semiaxes and the ellipsoid can be written as

$$\sum_{i=1}^{d} \frac{(\xi_i - \xi_{c,i})^2}{g_i^2} = 1, \quad \xi = \mathbf{T}(\alpha_1, \ldots, \alpha_{d-1}) \cdot \mathbf{x}. \tag{C.14}$$

If an ellipsoid is given by (C.14), the weight matrix $\mathbf{W}$ and the center point $\mathbf{x}_c$ can be determined by

$$\mathbf{W} = \mathbf{T}^T \Omega \mathbf{T} \quad \text{with} \quad \Omega = \text{diag}(g_1^{-2}, \ldots, g_d^{-2}) \quad \text{and} \quad \mathbf{x}_c = \mathbf{T}^{-1} \cdot \xi_c, \tag{C.15}$$

(see Ben-Haim and Elishakoff, 1990).

If an ellipsoid is given by Eq. (C.14), the semiaxes g_i, the center point ξ_c, and the transformation matrix $\mathbf{T}$ can be determined as follows. By spectral decomposition of $\mathbf{W}$, $\mathbf{T}$, and Ω can be found and ξ_c is given by

$$\xi_c = \mathbf{T} \cdot \mathbf{x}_c. \tag{C.16}$$

It is noted that the angles $(\alpha_1, \ldots, \alpha_{d-1})$ are not yet known. They can be obtained with the following observation. The columns of $\mathbf{T}$ are the eigenvectors of $\mathbf{W}$, which are pointing to the directions of the semiaxes. Each eigenvector $\mathbf{t}_i$ can be projected on the x_1–x_i-plane and the projection $\mathbf{s}_i$ is given by $\mathbf{s}_i = (t_{i,1}, 0, \ldots, 0, t_{i,I}, 0, \ldots, 0)^T$. The angle between the projection $\mathbf{s}_i$ and the unit vector $\mathrm{e}_1 = (1, 0, \ldots, 0)^T \in \mathrm{R}^d$ equals the angle α_i, which can be obtained from

$$\cos(\alpha_i) = \frac{\mathbf{s}_i^T \cdot \mathbf{e}_1}{|\mathbf{s}_i| \cdot |\mathbf{e}_1|} = \frac{s_{i,1}}{|\mathbf{s}_i|} = \frac{t_{i,1}}{\sqrt{t_{i,1}^2 + t_{i,i}^2}}. \tag{C.17}$$

The volume of a hyper-ellipsoid is given by

$$V_E = C_d \prod_{i=1}^{d} g_i = C_d \det(\mathbf{W})^{-\frac{1}{2}} \tag{C.18}$$

where C_d is volume of the d-dimensional hyper unit sphere

$$C_d = \frac{\pi^{\frac{d}{2}}}{\Gamma(\frac{d}{2}+1)}. \tag{C.19}$$

4. Numerical Determination of Buckling Load

For calculating buckling loads a multitude of analytical, semianalytical, and numerical codes has been developed in the past. For a broad overview the reader should

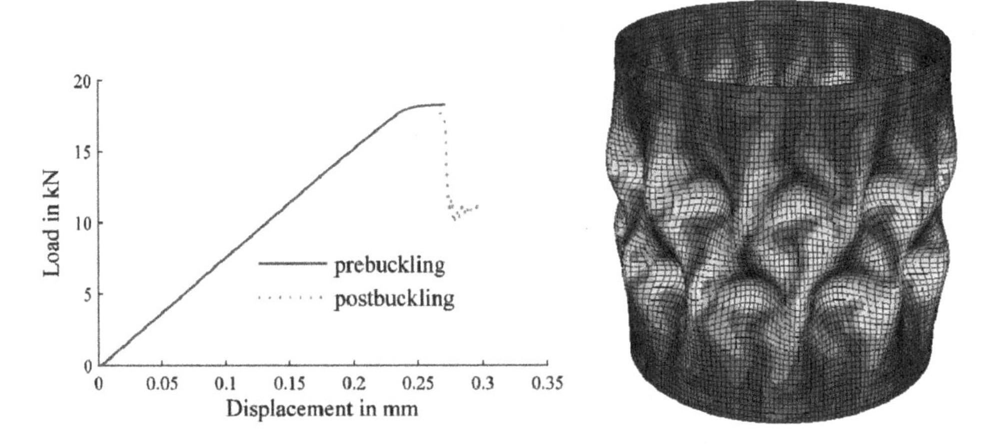

Fig. C.4. Load-displacement curve (left) and example of postbuckling deformation (right).

consult Singer *et al.* (1997) and Bushnell (1985). The numerical calculations of buckling loads within the probabilistic analysis are executed using the finite element code ABAQUS. A four-node shell element with reduced integration is used, which allows fast simulations. Because the shell edges were clamped in the experiments, this boundary condition has also been applied in the numerical simulation.

Numerical determination of buckling loads with ABAQUS is possible in several ways (see Hühne and Rolfes, 2002). The buckling load can be obtained from linear eigenvalue analysis, where the lowest eigenvalue multiplied with applied load equals the buckling load and the associated eigenvector describes the buckling mode. This is the fastest way to determine buckling loads, but it turned out to be too inaccurate for the considered set of composite shells, because some shells showed a significantly nonlinear prebuckling behavior (compare Fig., left). To capture this effect, nonlinear quasi-static analysis using the Newton–Raphson method can be performed. It has the disadvantage that it is not possible to determine postbuckling behavior of compressed shells. Within the convex analyses the postbuckling patterns are not of interest and hence the Newton–Raphson method is used since it predicts the buckling loads sufficiently accurate with acceptable numerical cost. If postbuckling deformations (like Fig., right) shall be determined, path following algorithms, like Riks (1984) method, or explicit time integration have to be used, which require more computing time than Newton–Raphson method.

5. Numerical Derivatives of Buckling Load

For the following considerations the coordinates x_i are assumed be parallel to the semiaxes, which can easily be archived by transformation.

In order to approximate the gradient φ and the Hessian Ξ of the buckling load function λ, the following derivatives have to be estimated at the center point $\mathbf{x}_c$:

$$\frac{\partial\lambda}{\partial x_i}, \frac{\partial^2\lambda}{\partial x_i^2} \quad \text{and} \quad \frac{\partial^2\lambda}{\partial x_i\,\partial x_j} \quad \text{for } i = 1, \ldots, d. \tag{C.19}$$

The first derivative can be estimated by

$$\frac{\partial\lambda(\mathbf{x}_c)}{\partial x_i} \approx \frac{\lambda(x_{c,1}, \ldots, x_{c,i} + \Delta x_i, \ldots, x_{c,d}) - \lambda(\mathbf{x}_c)}{\Delta x_i}. \tag{C.20}$$

A better approximation can be achieved by using the following expression:

$$\frac{\partial\lambda(\mathbf{x}_c)}{\partial x_i} \approx \frac{\lambda(x_{c,1}, \ldots, x_{c,i} + \Delta x_i, \ldots, x_{c,d}) - \lambda(x_{c,1}, \ldots, x_{c,i} - \Delta x_i, \ldots, x_{c,d})}{2\,\Delta x_i}. \tag{C.21}$$

Equation (C.21) requires twice as much calculation of bucking loads as Eq. (C.20), but for the second order approach these calculations have to be executed anyway, because the second derivative is approximated by

$$\frac{\partial^2\lambda(\mathbf{x}_c)}{\partial x_i^2} \approx \frac{\lambda(x_{c,i} + \Delta x_i) - 2\,\lambda(\mathbf{x}_c) + \lambda(x_{c,i} - \Delta x_i)}{\Delta x_i^2}. \tag{C.22}$$

The second derivative with respect to two different variables can be estimated by

$$\frac{\partial^2\lambda(\mathbf{x}_c)}{\partial x_i\,\partial x_j} \approx \frac{\lambda(x_{c,i} + \Delta x_i, x_{c,j} + \Delta x_j) - \lambda(x_{c,i} + \Delta x_i) - \lambda(x_{c,j} + \Delta x_j) + \lambda(\mathbf{x}_c)}{\Delta x_i\,\Delta x_j}. \tag{C.23}$$

With the same number of calculations it is also possible to determine the coefficients of the function

$$\lambda \approx \mathbf{x}^T \mathbf{A}\,\mathbf{x} + \mathbf{b}^T \mathbf{x} + c \tag{C.24}$$

with the coefficients $\mathbf{A} \in \mathrm{R}^{d\times d}(\mathbf{A} = \mathbf{A}^T)$, $\mathbf{b} \in \mathrm{R}^d$ and $c \in \mathrm{R}$ Gradient and Hessian are then given by

$$\varphi \approx 2\mathbf{A}\,\mathbf{x} + \mathbf{b} \quad \Xi \approx 2\,\mathbf{A}, \tag{C.25}$$

respectively. If the firstorder approach is executed and the derivatives are determined using (C.20), $1+d$ buckling load calculations are required; one calculation for $\lambda(\mathbf{x}_c)$ and d calculations for each $\lambda(x_{c,i} + \Delta x_i)$. For the second order approach $1 + 2d + \frac{1}{2}(d^2 - d)$ buckling load calculations have to be executed; d additional calculations for each $\lambda(x_{c,i} - \Delta x_i)$ and $\frac{1}{2}(d^2 - d)$ calculations for each $\lambda(x_{c,i} + \Delta x_i, x_{c,j} + \Delta x_j)$, where $i \neq j$.

For design optimization, the following derivatives of the buckling load are required:

$$\frac{\partial\lambda}{\partial x_i},\ \frac{\partial\lambda}{\partial\beta_j},\ \frac{\partial^2\lambda}{\partial x_i\,\partial\beta_j}\quad\text{and}\quad\frac{\partial^2\lambda}{\partial\beta_j\,\partial\beta_k}\quad\text{for } i=1,\ldots,d\quad\text{and}\quad j,k=1,\ldots,p. \tag{C.26}$$

The buckling loads are given by finite element calculation (see Appendix C.4) and thus the derivatives must be determined numerically. The simplest approximations of these derivatives for the current design variable vector $\boldsymbol{\beta}$ at the center point $\mathbf{x}_c$ are given by

$$\frac{\partial\lambda(\mathbf{x}_c,\boldsymbol{\beta})}{\partial x_i}\approx\frac{\lambda(x_{c,1},\ldots,x_{c,i}+\Delta x_i,\ldots,x_{c,d},\boldsymbol{\beta})-\lambda(\mathbf{x}_c,\boldsymbol{\beta})}{\Delta x_i} \tag{C.27}$$

$$\frac{\partial\lambda(\mathbf{x}_c,\boldsymbol{\beta})}{\partial\beta_j}\approx\frac{\lambda(\mathbf{x}_c,\beta_1,\ldots,\beta_j+\Delta\beta_j,\ldots,\beta_d)-\lambda(\mathbf{x}_c,\boldsymbol{\beta})}{\Delta\beta_j} \tag{C.28}$$

$$\frac{\partial^2\lambda(\mathbf{x}_c,\boldsymbol{\beta})}{\partial x_i\,\partial\beta_j}\approx\frac{\begin{array}{l}\lambda(x_{c,i}+\Delta x_i,\beta_j+\Delta\beta_j)\\ \quad-\lambda(x_{c,i}+\Delta x_i,\boldsymbol{\beta})-\lambda(\mathbf{x}_c,\beta_j+\Delta\beta_j)+\lambda(\mathbf{x}_c,\boldsymbol{\beta})\end{array}}{\Delta x_i\,\Delta\beta_j} \tag{C.29}$$

$$\frac{\partial^2\lambda(\mathbf{x}_c,\boldsymbol{\beta})}{\partial\beta_j^2}\approx\frac{\lambda(\mathbf{x}_c,\beta_j+\Delta\beta_j)-2\,\lambda(\mathbf{x}_c,\boldsymbol{\beta})+\lambda(\mathbf{x}_c,\beta_j-\Delta\beta_j)}{\Delta\beta_j^2} \tag{C.30}$$

$$\frac{\partial^2\lambda(\mathbf{x}_c,\boldsymbol{\beta})}{\partial\beta_j\,\partial\beta_k}\approx\frac{\begin{array}{l}\lambda(\beta_j+\Delta\beta_j,\beta_k+\Delta\beta_k)-\lambda(\beta_j+\Delta\beta_j)\\ \quad-\lambda(\beta_k+\Delta\beta_k)+\lambda(\boldsymbol{\beta})\end{array}}{\Delta\beta_j\,\Delta\beta_k}. \tag{C.31}$$

In each iteration the buckling loads $\lambda(\mathbf{x}_c,\boldsymbol{\beta})$, $\lambda(x_{c,i}+\Delta x_i,\boldsymbol{\beta})$, $\lambda(\mathbf{x}_c,\beta_j+\Delta\beta_j)$, $\lambda(\mathbf{x}_c,\beta_j-\Delta\beta_j)$, $\lambda(x_{c,i}+\Delta x_i,\beta_j+\Delta\beta_j)$ and $\lambda(\beta_j+\Delta\beta_j,\beta_k+\Delta\beta_k)$ have to be determined for $i=1,\ldots,d$ and $j,k=1,\ldots,p$. Thus, the number of buckling load calculations per iteration equals $1+d+dp+2p+\frac{1}{2}(p^2-p)$, if the Hessian is required. Otherwise, only $1+d+dp+p$ buckling load calculations are required in each iteration step.

The number of buckling load calculations in each iteration step increases quadratically with the number of design parameters. In the current example the use of the simple gradient method does not cause a drastic reduction of the required number of calculations, because the number of uncertainty parameters d exceeds the number of design parameters p significantly. If, for example, $p=10$ and $d=4$, the simple gradient method requires half as much buckling load calculation as the Newton method.

Table C.3. Number of buckling load calculations in each iteration step.

Optimization algorithm	Number of buckling load calculations per iteration step	for $[\beta_1, -\beta_1, \beta_2, -\beta_2]$	for $[\beta_1, \beta_2, \beta_3, \beta_4]$
Newton, conjugate gradient	$1 + d + dp + 2p + \frac{1}{2}(p^2 - p)$	39	70
Simple gradient	$1 + d + dp + p$	36	60

Table C.4. Experimentally determined buckling loads of the cylinders considered, from (2010).

	Z1			Z2						
Shell	5	7	8	0	1	2	3	4	5	6
Buckling load in kN	23.36	24.63	21.32	23.08	22.63	23.99	25.02	23.62	25.69	22.43

6. Experimentally Determined Buckling Loads

The buckling loads that have been observed within experiments performed by DLR (2010) are summarized in Table.

References

Everybody loves a buckling problem!
B. Budiansky and J.W. Hutchinson (1979)

GOOSE (Get Out Of Shells—Everybody!)" — attributed to M. Stein
N.F. Knight and J.H. Starnes, Jr. (1998)

How do we select in the mud the papers conveying innovative ideas?
P. Villaggio (2013)

There is a fair amount of bad, erroneous and downright mischievous material published in journals and books, so you must be on guard and develop your own critical faculties.
J.M.T. Thompson (2013)

Abramovich, H., Singer, J., and Yaffe, R. (1981). Imperfection characteristics of stiffened shells-group 1, TAE Report 406, Department of Aeronautical Engineering, Technion — Israel Institute of Technology, Haifa, Israel.

Abramovich, H., Yaffe, R., and Singer, J. (1987). Evaluation of stiffened shell characteristics from imperfection measurements. *Journal of Strain Analysis*, 22(1), 17–23.

Adali, S., Elishakoff, I., Richter, A., and Verijenko, V.E. (1994). Optimal design of symmetric angle-ply laminates for maximum buckling load with scatter in material properties. In J. Sobieski (Ed.), Paper AIAA-94-4365-CP, *A Collection of Technical Papers, 5th AIAA/USAF/NASA/ISSMO Symposium on Multidisciplinary Analysis and Optimization* (Part 2, pp. 1041–1045,.

Adali, S., Richter, A., and Verijenko, V.E. (1995). Minimum weight design of symmetric angle-ply laminates under multiple uncertain loads. *Structural Optimization*, 9, 89–95.

Adali, S., Richter, A., and Verijenko, V.E. (1997). Minimum weight design of symmetric angle-ply laminates with incomplete information on initial imperfections. *Journal of Applied Mechanics*, 64, 94–97.

Adams, E. and Kulisch, U. (Eds.) (1993). *Scientific Computing with Automatic Result Verification*. New York: Academic Press.

Aghajari, S., Abedi, K., and Showkati, H. (2006). Buckling and post buckling behavior of thin-walled cylindrical steel shells with varying thickness subjected to uniform external pressure. *Thin-Walled Structures*, 44, 904–909.

Alefeld, G. and Herzberger, J. (1983). *An Introduction to Interval Computation*. New York: Academic Press.

Alibrandi, U., Impollonia, N., and Ricciardi, G. (2010). Probabilistic eigenvalue buckling analysis solved through the ratio of polynomial response surface. *Computer Methods in Applied Mechanics and Engineering*, 199(9–12), 450–464.

Almroth, B.O. (1966). Influence of imperfections and edge constraint on the buckling of axially compressed cylinders, *NASA CR-432*.

Almroth, B.O. (1979). Design of composite material structures to buckling, *Report LMSC-D681425*, Lockheed Missiles & Space Co., Palo Alto, CA.

Almroth, B.O., Brogan, F.A., Miller, E., Zele, F., and Peterson, H.T. (1973). Collapse analysis for shells of general shape. II: User's manual for the STAGS — A computer code. Air Force Flight Dynamics Laboratory, Wright-Patterson AFB, *AFFDL-TR-71-8*.

Almroth, B.O. and Brush, D.O. (1963). Postbuckling behavior of pressure- or core-stabilized cylinders under axial compression. *AIAA Journal*, 1, 2338–2341.

Almroth, B.O., Burns, A.B., and Pittner, E.V. (1970). Design criteria for axially loaded cylindrical shells. *Journal of Spacecraft*, 7(6), 714–720.

Almroth, B.O., Holmes, A.M.C., and Brush, D.O. (1964). An experimental study of the buckling of cylinders under axial compression. *Experimental Mechanics*, 4, 263–270.

Almroth, B.O. and Rankin, C.C. (1983). Imperfection sensitivity of cylindrical shells. In W.F. Chen and A.D.M. Lewis (Eds.), *Recent Advances in Engineering Mechanics and Their Impact on Civil Engineering Practice* (Vol. 2, pp. 1071–1074). New York: ASCE Press.

Almroth, B.O., Stern, P., and Bushnell, D. (1981). Imperfection sensitivity of optimized structures, *AFWAL-TR-80-3128*, Flight Dynamics Laboratory, Wright Patterson Air Force Base, Ohio.

Amazigo, J.C. (1969). Buckling under axial compression of long cylindrical shells with random axisymmetric imperfections. *Quarterly of Applied Mathematics*, 26, 537–566.

Amazigo, J.C. (1970). Asymptotic analysis of the buckling of imperfect columns on nonlinear elastic foundations. *International Journal of Solids and Structures*, 6, 1341–1356.

Amazigo, J.C. (1971). Buckling of stochastically imperfect columns on nonlinear elastic foundations. *Quarterly of Applied Mathematics*, 28, 403–409.

Amazigo, J.C. (1974a). Asymptotic analysis of the buckling of externally pressurized cylinders with random imperfections. *Quarterly of Applied Mathematics*, 31, 429–442.

Amazigo, J.C. (1974b). Dynamic buckling of structures with random imperfections. In H.H.E. Leipholz (Ed.), *Stochastic Problems in Mechanics* (pp. 234–254). Waterloo, Canada: University of Waterloo Press.

Amazigo, J.C. (1976). Buckling of stochastically imperfect structures. In B. Budiansky (Ed.), *Buckling of Structures* (pp. 172–182). Berlin: Springer.

Amazigo, J.C. and Budiansky, B. (1972). Asymptotic formulas for the buckling stresses of axially compressed cylinders with localized or random axisymmetric imperfections. *Journal of Applied Mechanics*, 39, 179–184.

Amazigo, J.C. and Budiansky, B. (1973). Discussion on the paper "Buckling of long, axially compressed, thin cylindrical shell with random initial imperfections" by R.A. Van Slooten and T.T. Soong. *Journal of Applied Mechanics*, 40, 634–635.

Amazigo, J.C., Budiansky, B., and Carrier, G.F. (1970). Asymptotic analysis of the buckling of imperfect columns on nonlinear elastic foundations. *International Journal of Solids and Structures*, 6, 1341–1356.

Amazigo, J.C. and Frank, D. (1975). Dynamic buckling of externally pressurized imperfect cylindrical shells. *Journal of Applied Mechanics*, 42, 316–320.

Ambartsumian, S.A. and Belubekyan, M.V. (1994). On bending waves localized along the edge of a plate. *International Applied Mechanics,* 30, 135–140.

Amiro, I.Ya. and Etokov, V.I. (1976). Stability of imperfect cylindrical shells in axial compression. *International Applied Mechanics*, 12(1), 23–27.

Amiro, I.Ya, Grachev, D.A., Zarutskii, V.A., Pal'chevskii, A.S., and Sannikov, Yu.A. (1987). *Stability of Ribbed Shells of Revolution* (in Russian). Kiev: "Naukova" Dumka Publishers.

Amiro, I.Ya., Polyakov, P.S., and Palamarchuk, V.G. (1974). Stability of cylindrical shells of imperfect shape. *International Applied Mechanics*, 7(8), 838–842.

Anderson, P.W. (1958). Absence of diffusion in certain random lattices. *Physical Review*, 109, 1492–1505.

Andreev, L.V., Obodan, N.I., and Lebedev, A.G. (1988). *Stability of Shells during Nonaxisymmetric Deformation* (Russian). Moscow: "Nauka" Publishers.

Anisimov, V.Yu. (2006). Stability of transversely-isotropic shell of revolution under axial compression (in Russian). *Vestnik Leningradskogo Universiteta*, 1(1), 69–77.

Anonymous (1968). Buckling of Thin-Walled Circular Cylinders, NASA SP-8007.

Anonymous (2010). Space Engineering/Buckling of Structures, ECSS-E-HB-32-24A, European cooperation ECSS for space standardization, ESA-ESTEC, Noordwijk, The Netherlands.

Arbocz, J. (1968a). The effect of general imperfections on the buckling of cylindrical shells. Ph.D. Thesis, California Institute of Technology, Pasadena, CA.

Arbocz, J. (1968b). Buckling of Conical Shells under Axial Compression. NASA CR-1162.

Arbocz, J. (1981). Past, present, future of shell stability analysis. *Zeitschrift für Flugwissenschaften und Weltraumforschung*, 5(6), 335–348.

Arbocz, J. (1982). The imperfection data bank, a means to obtain realistic buckling loads. In E. Ramm (Ed.), *Buckling of Shells: A State-of-the-Art Colloquium.* Institut für Baustatik, Universität Stuttgart, Federal Republic of Germany.

Arbocz, J. (1983). Shell stability analysis: theory and practice. In J.M.T. Thompson and G.W. Hunt (Eds.), *Collapse: Buckling of Structures in Theory and Practice* (pp. 43–74). Cambridge, UK: Cambridge University Press.

Arbocz, J. (1987). Post-buckling behavior of structures. In J. Arbocz, M. Potier-Ferry, J. Singer, and V. Tvergaard (Eds.), *Buckling and Postbuckling* (pp. 84–143). Berlin: Springer.

Arbocz, J. (1989). Shell buckling research at Delft (1976–1988), Memorandum M-596, Faculty of Aerospace Engineering, Delft University of Technology.

Arbocz, J. (1991). Towards an improved design procedure for buckling critical structures. In J.F. Julien (Ed.), *Buckling of Shells and Structures, on Land, in the Sea, and in the Air* (pp. 270–276). London: Elsevier Applied Science.

Arbocz, J. (1998). Towards a probabilistic criterion for preliminary shell design. Proceedings 39th AIAA/ASME/ASCE/AHS/ASC Structures, Structural Dynamics, and Materials Conference and Exhibit and AIAA/ASME/AHS Adaptive Structures Forum.

Arbocz, J. (1998). Future directions and challenges in shell stability analysis. In N.F. Knight, Jr. and M.P. Nemeth (Eds.), *Stability Analysis of Plates and Shells: A Collection of Papers in Honor of Dr. Manuel Stein* (pp. 47–60). NASA/CP-1998-206280.

Arbocz, J. (2000). A comparison of probabilistic and lower bound methods for predicting the response of buckling sensitive structures. Proceedings, 41st AIAA/ASME/ASCE/AHS/ASC Structures, Structural Dynamics, and Materials Conference and Exhibit.

Arbocz, J. and Abramovich, H. (1979). The initial imperfection data bank at the Delft University of Technology — Part, 1. Report LR-290, Department of Aerospace Engineering, Delft University of Technology.

Arbocz, J. and Babcock, C.D., Jr. (1968). Experimental investigation of the effect of general imperfections on the buckling of cylindrical shells. NASA CR-1163.

Arbocz, J. and Babcock, C.D., Jr. (1974). A multi-mode analysis calculating buckling loads of imperfect cylindrical shells. GALCIT Report. SM-74-4. California Institute of Technology, Pasadena, CA, June.

Arbocz, J. and Babcock, C.D. (1976). Prediction of buckling Load based on experimentally measured initial imperfections. In B. Budiansky (Ed.), *Buckling of Structures* (pp. 291–311). Berlin: Springer.

Arbocz, J. and Babcock, C.D. (1978). Utilization of STAGS to determine knockdown factors from measured initial imperfections. Report LR-275, Department of Aerospace Engineering, Delft University of Technology, The Netherlands.

Arbocz, J. and Babcock, C.D., Jr. (1980). Computerized stability analysis using measured initial imperfections. Proceedings of the 12th Congress of the Council of the Aeronautical Sciences, ICAS-80-20.2 (pp. 688–701), München, Federal Republic of Germany.

Arbocz, J. and Hillburger, M.W. (2005). Toward a probabilistic preliminary design criterion for buckling critical composite shells. *AIAA Journal*, 43(8), 1823–1827.

Arbocz, J. and Hol, J.M.A.M. (1989). The role of experiments in improving the computational models for composite shells. In A.K. Noor, T. Belytschko and J.C. Simo (Eds.), *Analytical and Computational Models of Shells* (CED — Vol. 3, pp. 613–639). New York: ASME.

Arbocz, J. and Hol, J.M.A.M. (1990). Recent development in shell stability analysis. Report LR-633. Faculty of Aerospace Engineering, Delft University of Technology, Delft, The Netherlands.

Arbocz, J. and Hol, J.M.A.M. (1991). Collapse of axially compressed cylindrical shells with random imperfections. *AIAA Journal*, 29, 2247–2256.

Arbocz, J. and Hol, J.M.A.M. (1995). Collapse of axially compressed cylindrical shells with random imperfections. *Thin-Walled Structures*, 23, 131–158.

Arbocz, J. and Singer, J. (2000). Professor Bernard Budiansky's contributions to buckling and postbuckling of shell structures, Paper 2000-1322. 41st AIAA/ ASME/ ASCE/ AHS/ ASC Structures, Structural Dynamics and Materials Conference, Atlanta, GA.

Arbocz, J. and Stam, A.R. (2004). A probabilistic approach to design shell structures. In J.G. Teng and J.M. Rotter (Eds.), *Buckling of Thin Metal Shells* (pp. 455–489). London: Spon Press.

Arbocz, J., Starnes, J.H., Jr. and Nemeth M. (2000). A comparison of probabilistic and lower bound methods for predicting the response of buckling sensitive structures. Proceedings, 41st AIAA/ASME/AHS/ASC Structures, Structural Dynamics and Materials Conference and Exhibit. AIAA Press.

Arbocz, J. and Starnes, J.H., Jr. (2002). Future directions and challenges in shell stability analysis. *Thin-Walled Structures*, 40, 729–754.

Arbocz, J. and Starnes, J.H., Jr. (2002). On a high-fidelity hierarchical approach to buckling load calculations. In H.R. Drew and S. Pellegrino (Eds.), *New Approaches to Structural Mechanics, Shells and Biological Structures* (pp. 271–292). Dordrecht: Kluwer Academic Publishers.

Arbocz, J., Starnes, J.H., Jr., and Nemeth, M.P. (1998). Towards a probabilistic criterion for preliminary shell design, AIAA Paper 1998-2051, Proceedings of the 39th AIAA/ ASME/ASCE/AHS/ASC Structures, Structural Dynamics, and Materials Conference, Long Beach, CA, AIAA Press, Reston, VA.

Arbocz, J., Starnes, J., Jr., and Nemeth, M. (2001). On the accuracy of probabilistic buckling load predictions. Proceedings, 41st AIAA/ASME/ASCE/AHS/ASC Structures, Structural Dynamics, and Materials Conference and Exhibit, Paper AIAA-2000-1236. Reston, VA: AIAA Press.

Arbocz, J. and Williams, J.G. (1977). Imperfection surveys on a 10 ft diameter shell structure. *AIAA Journal*, 15, 949–956.

Argyris, J., Papadrakakis, M., and Stefanou, G. (2002). Stochastic finite element analysis of shells. *Computer Methods in Applied Mechanics and Engineering*, 191(41), 4781–4804.

Ariaratnam, S.T. and Xie, W.-C. (1995). Wave localization in randomly disordered nearly periodic long continuous beams. *Journal of Sound and Vibration,* 181(91), 7–22.

Arnold, L. (1982). Personal communication, Dec. 30, 1982.

Arnold, V.I. (1989). Bifurcatiions and singularities in mathematics and mechanics. In P. Germain, M. Piau, and D. Caillerie (Eds.), *Theoretical and Applied Mechanics* (pp. 1–25). Amsterdam: North-Holland.

Astapov, N.S. (1991). Postcritical behavior of a column on a nonlinear elastic foundation (in Russian). *Deformation and Collapse of Modern Materials and Structures*, 103, 15–19, Novosibirsk, 1991 (in Russian).

Astapov, N.S., Demeshkin, A.G., and Kornev, V.M. (1994). Instability of a column resting on an elastic foundation (in Russian). *Applied Mechanics and Technical Physics*, 35(5), 108–112.

Astapov, N.S. and Kornev, V.M. (1994). Postbuckling behavior of an ideal bar on an elastic foundation. *Journal of Applied Mechanics and Technical Physics*, 35(2), 286–296.

Augusti, G. (1964). Stabilità di strutture elastiche elementary in presenza di grandi spostamenti (in Italian). Publication 10.172, Faculty of Engineering, University of Naples, Italy.

Augusti, G. (1974). Probabilistic treatments of column buckling problems. In *Stochastic Problems in Mechanics* (pp. 255–274). Waterloo, Ontario, Canada: University of Waterloo Press.

Augusti, G. (1977). Il carattere aleatorio del collasso da instabilità (in Italian). UFIST Reprint No. 05, Department of Civil Engineering, University of Florence, Italy.

Augusti, G. and Baratta, A. (1972). Limit analysis of structures with stochastic strength variations. *Journal of Structural Mechanics*, 1, 43– 62.

Augusti, G. and Baratta, A. (1976). Reliability of slender columns, comparison of different approximations. In B. Budiansky (Ed.), *Buckling of Structures* (pp. 183–198). Berlin: Springer Verlag.

Augusti, G., Baratta, A., and Casciati, F. (1984). *Probabilistic Methods in Structural Engineering*. London: Chapman and Hall.

Augusti, G., Sepe, V., and Paolone A. (1988). An introduction to compound buckling and dynamic bifurcations. In J. Rondal (Ed.), *Compled Instabilities in Metal Structures: Theoritical and Design Aspects* (pp. 1–27). Vienna: Springer.

Babcock, C.D. (1967). The influence of the testing machine on the buckling of cyclindrical shells under compression, *International Journal of Solids and Structures*, 3(5), 809–810.

Babcock, C.D. (1974). Experiments in shell buckling. In Y.C. Fung and E.E. Sechler (Eds.), *Thin-Shell Structures: Theory, Experiment, and Design* (pp. 345–369). Englewood Cliffs, NJ: Prentice-Hall.

Babcock, C.D. (1983). Shell stability. *Journal of Applied Mechanics*, 50(4), 935–940.

Babcock, C.D. and Sechler, E.E. (1963). The effect of initial imperfection on the buckling stress of cylindrical shells. NASA TN D-2005.

Babenko, V.I. (1978). Asymptotic representation of solution of cylindrical shell equations in vicinity of branching point. *Differential Equations and some Methods of Functional Analysis* (in Russian; pp.13–29). Kiev: "Naukova Dumka" Publishers.

Babich, D.V. (1989). Joint effect of inhomogeneities of thickness and moduli of elasticity of nonlinear cylindrical shells. *Strength of Materials*, 12, 1634–1640.

Babich, D.V., Koshevoi, I.K., and Shpakova, S.G. (1985). Effect of nonuniformity of the material on the stress state and stability of shells. *Strength of Materials*, 11, 1603–1609.

Badino, M. (2005). The foundational role of ergodic theory. *Press Print* 292. Max-Plank-Institut für Wissenschaftgeschichte.

Baio, S., *In Praise of Science: Curiosity, Understanding, and Progress*. Cambridge, MA: MIT Press.

Baitsch, M. and Hartmann, D. (2006). Optimization of slender structures considering geometric imperfections. *Inverse Problems in Science and Engineering*, 14(6), 623–637.

Balabukh, L.I., Alfutov, N.A., and Usyukin, V.I. (1984). *Structural Mechanics of Spacrcraft* (in Russian). Moscow: "Vysshaya Shkola" Publishers.

Banichuk, N.V. and Barsuk, A.A. (2008). Localization of eigenforms and limit transition in problems of stability of rectangular plates. *PMM — Journal of Applied Mathematics and Mechanics,* 72(2), 302–307.

Bathe, K.-J., Chapelle, D., and Lee, P.-S. (2003). A shell problem "highly sensitive" to thickness changes. *International Journal for Numerical Methods in Engineering*, 57, 1039–1052.

Batista, R.C. (1979). Lower bound estimates for cylindrical shell buckling. Ph.D. Thesis, University College, London.

Batista, R.C. and Croll, J.G.A. (1979). A design approach for axially compressed unstiffened cylinders. *Stability Problems in Engineering Structures and Components*. London: Applied Science.

Bauer, S.M. (1978). About influence of localized axial initial imperfections of shells of revolution that are close to cylinders (in Russian). *Vestnik Leningradskogo Universiteta, Mathematika, Mekhanika, Astronomia*, 13, 131–134.

Bauer, S.M. (1984). About stability of shells of revolution with random axisymmetric imperfections, *Vestnik Leningradskogo Universiteta Mathematika, Mekhanika, Astronomia.*

Baul, A.B. (1984). Loss of stability of thin cylindrical shells during axial compression. *Stability and Vibrations of Mechanical Systems*, 7, 211–215.

Bayer, V. and Roos, D. (2008). Non-parametric structural reliability analysis using random fields and robustness evaluation. Retrieved May 10, 2012, http://citeseerx.ist.psu.edu/viewdoc/summary?doi=10.1.1.73.2333.

Bažant, Z.P. (2000a). Structural stability. *International Journal of Solids and Structures*, 37, 55–67.

Bažant, Z.P. (2000b). Stability of elastic, anelastic, and disintegrating structures: a conspectus of main results. *ZAMM-Zeitschrift fur angewandte Mathematik and Mechanik,* 80(11–12), 109–732.

Bažant, Z.P. (2009). Reminiscences and reflections of a mechanician by luck. Speech of acceptance of Timoshenko Medal, ASME International Mechanical Engineering Congress, Orlando, FL, November 17 (available on www.imechanica.org/7099).

Bažant, Z.P. and Cedolin, L. (1991). *Stability of Structures: Elastic, Inelastic, Fracture, and Damage Theories*. New York: Oxford University Press.

Beck, M. (1952). Die Knicklast des einseitig eingespannten tangential gedrückten Stabes (in German). *Zeitschrift Für angecsandte Mathematik and Physik*, 3, 225–228.

Becker, H., Gerard, G., and Winter, R. (1963). Experiments on axial compressive general instability of monolithic ring-stiffened cylinders. *AIAA Journal*, 1(7), 1614–1618.

Bellman, R. (1962). The roles of the mathematician in applied mechanics. *U.S. National Congress in Applied Mechanics* (pp. 195–204). New York: ASME Press.

Belubekyan, M.V. (2008). Problems of localized instability of plates. *Optimal Control, Stability and Robustness of Mechanical Systems* (pp. 95—99). Yerevan: Yerevan State University.

Belubekyan, M.V. and Chil-Akobyan, E.V. (2004). Problems of localized instability of plates with a free edge. *Proceedings, Armenian National Academy of Sciences, Mechanics,* 57(2), 34–49.

Benaroya, H. (Ed.) (1996). Special issue: Localization of vibration in structures. *Applied Mechanics Reviews*, 49(2), February.

Ben-Haim, Y. (1985). *The Assay of Spatially Random Material.* Dordrecht: Kluwer Academic.

Ben-Haim, Y. (1993). Convex models of uncertainty in radial pulse buckling of shells. *Journal of Applied Mechanics*, 60, 683–688.

Ben-Haim, Y. (1993). Failure of an axially compressed beam with uncertain initial deflection of bounded strain energy. *International Journal of Engineering Science*, 31, 989–1001.

Ben-Haim, Y. (1995). On convex models of uncertainty for small initial imperfections. *ZAMM-Zeitschrift für angewandte Mathematik und Mechanik*, 75, 901–908.

Ben-Haim, Y. (1996). *Robust Reliability in the Mechanical Sciences.* Berlin: Springer.

Ben-Haim, Y. (2001). New diretions in information, uncertainty and decision. *Mechanical Systems and Signal Processing*, 15, 453–455.

Ben-Haim, Y. (2006). *Info-Gap Decision Theory: Decisions under Severe Uncertainty*, 2nd ed. Amsterdam: Academic Press.

Ben-Haim, Y. (2011). Personal communication, March 16.

Ben-Haim, Y. (2012). Why risk analysis is difficult, and some thoughts on how to proceed. *Risk Analysis*, 32(10), 1638–1646.

Ben-Haim, Y. and Elishakoff, I. (1989). Dynamics and failure of thin bar with unknown-but-bounded imperfections, In D. Hui and N. Jones (Eds.), *Recent Advances in Impact Dynamics in Engineering Structures* (Vol. 105, pp. 89–96). New York: ASME Press.

Ben-Haim, Y. and Elishakoff, I. (1989). Non-probabilistic models of uncertainty in the buckling of shells with general imperfections: Theoretical derivation of the knockdown factor. *Journal of Applied Mechanics*, 56, 403–410.

Ben-Haim, Y. and Elishakoff, I. (1990). *Convex Models of Uncertainty in Applied Mechanics.* Amsterdam: Elsevier.

Bendat, J.S. (2005). A personal history of random data analysis. International Operational Model Analysis Conference, Copenhagen, April 26–27.

Bendich, N.N. and Kornev, V.M. (1971). On the eigenvalue density in elastic shell stability problems (in Russian). *Prikladnaya Matematika i Mekhanika (Journal of Applied Mathematics and Mechanics*), 35(2), 364–368.

Benvenuto, E. (1990). *An Introduction to the History of Structural Mechanics*, Vol. 1. Berlin: Springer.

Besseling, J.F., Ernst, L.J., de Koning, A.U., Riks, E., and van der Werff, K. (1979). Geometrical and physical nonlinearities, some development in the Netherlands. *Computer Methods in Applied Mechanics and Engineering*, 17/18, 131–157.

Biagi, M. and Del Medico, F. (2008). Reliability-based knockdown factors for composite cylindrical shells under axial compression. *Thin-Walled Structures,* 46(12), 1351–1358.

Bielewicz, E. and Górski, J. (2002). Shell with random geometric imperfections, simulation-based approach. *International Journal of Non-Linear Mechanics*, 37(4/5), 777–784.

Bielewicz, E., Górski, J., Schmidt, R., and Walukiewicz, H. (1994). Random fields in the limit analysis of elastic-plastic shell structures. *Computers & Structures*, 51, 267–275.

Bigi, D. and Riganti, R. (1987). Stochastic response of structures with small geometric imperfections. *Meccanica*, 22(1), 27–34.

Bigoni, D. and Noselli, G. (2011). Experimental evidence of flutter and divergence instabilities induced by dry friction. *International Journal of the Mechanics and Physics of Solids*, 59(10), 2208–2226.

Blachut, J. (2010). Buckling of axially compressed cylinders with imperfect length. *Computers and Structures*, 88(5–6), 365–374.

Bodner, S.R. and Rubin, M.B. (2005). Modeling the buckling of axially compressed elastic cylindrical shells. *AIAA Journal*, 43(1), 103–110.

Boganovich, A.E. and Yushanov, S.P. (1981). Analysis of the buckling of cylindrical shells with a random field of initial imperfections under axial dynamic compression. *Mechanics of Composite Materials*, 17(5), 552–560.

Bolotin, V.V. (1958). Statistical methods in the non-linear theory of elastic shells (in Russian). *Izvestiya Akademii Nauk SSSR, Otdelenie Tekhnicheskykh Nauk*, 3, 33–41, 1958 (English translation: *NASA TTF-85*, pp. 1–16, 1962).

Bolotin, V.V. (1963). *Non-Conservative Problems of the Theory of Elastic Stability.* New York: Pergamon Press.

Bolotin, V.V. (1965). The density of eigenvalues in vibration problems of elastic plates and shells. *Proceedings of Vibration Problems*, 4(6), 342–351.

Bolotin, V.V. (1966). Application of methods of the theory of probability in the theory of plates and shells. In S.M. Durgar'yan (Ed.), *Theory of Plates and Shells* (pp. 1–45). Jerusalem: Israel Program for Scientific Translation, Jerusalem.

Bolotin, V.V. (1967). Statistical aspects in the theory of structural stability. In G. Herrmann, (Ed.), *Dynamic Stability of Structures* (pp. 67–81). Oxford: Pergamon Press.

Bolotin, V.V. (1969). *Statistical Methods in Structural Mechanics*. San Francisco, CA: Holden-Day, Inc.

Bolotin, V.V. (1971). *Application of the Methods of the Theory of Probability and the Theory of Reliability to Analysis of Structures* (Section 1.14: Methods of solution to nonlinear stochastic boundary value problems) (pp. 101–105). Moscow: State Publishing House of Buildings; 2nd ed., 1981 (in Russian) (English translation: *FDT-MT-24-771-73*, Foreign Technology Division, Wright-Patterson AFB, Ohio, pp. 129–136, 1974.)

Bolotin, V.V. (1979). Reliability of structures. In J.F. Besseling and A.M.A. van der Heijden (Eds.), *Trends in Solid Mechanics* (pp. 79–91). Delft: Delft University Press.

Bolotin, V.V. and Grigoliuk, E.I. (1972). Stability of elastic and inelastic systems. *Mechanic in the USSR for 50 Years* (in Russian; Vol. 3, pp. 324–357). Moscow: "Nauka" Publishers.

Bolotin, V.V. and Makarov, B.P. (1968). Correlation theory of pre-critical deformations in thin elastic shells. *PMM — Prikladnaya Matematika i Mekhanika (Journal of Applied Mathematics and Mechanics)*, 32, 428–434.

Bolt, H.M. (1989). Secondary bifurcation and localization phenomena in nonlinear structural mechanics. Ph.D. Thesis. Imperial College, London.

Boltzmann, L. (1871). Einige allgemeine Sätze über Wärmegleichgewicht (in German). *Wissenschaftliche Abhandlungen*, 1, 259–287.

Born M. (1949). *Natural philosophy of cause and chance*, Clarendan Press, Oxford.

de Borst, R. (2002). Preface. *International Journal of Non-Linear Mechanics*, 37, 571–573.

de Borst, R. (2002). Warner Tjardus Koiter-Het instabiele hanteerbaar. *Delst Goud Leven en Werk van Achttien Markante Hoogleraren* (pp. 222–231). Delft: Beta Imaginations, Technical University .

de Borst, R. (2010). Computational methods in buckling and instability. In R. Blockley and W. Shyy (Eds.), *Encyclopedia of Aerospace Engineering*, Vol. 3, Structural Technology (pp. 1747–1754). New York: Wiley.

de Borst, R., Kyriakides S., and von Baten T. (2002). Preface, Professor Johann Arbocz anniversary issue. *International Journal of Non-Linear Mechanics*, 37(4–5), 571–1002.

Bourinet, J.-M. (2012). Private communication, January 5.

Bourinet, J.–M. (2012). Personal communication, January 20.

Bourinet, J.-M., Deheeger, F., and Lemaire, M. (2011). Assessing small failure probabilities by combined subset simulation and support vector machines. *Structural Safety*, 33(6), 343–353.

Bourinet, J.-M., Gayton, N., Lemaire, M., and Çombescure, A. (2000). Reliability analysis of stability of shells based on combined finite element and response surface methods. In M. Papadrakakis, A. Samartin and E. Onate (Eds.), *Computational Methods for Shell and Spatial Structures*. Athens, Greece: ISASR–NTUA.

Bradburry, R. (1952). The sound of thunder. *Collier's*, June 28 (see also *Golden Apples in the Sun*, Doubleday, New York, 1953 and *R Is for Rocket*, Simon & Schuster, New York, 1962).

Brar, G.S., Hari, Y., and Williams, D.K. (2008). Axially compressed cylindrical shell containing axisymmetric random imperfections: Fourier series techniques and ASME Section VIII Division 2 rules, Paper PVP 2008-61391, pp. 273–279, ASME Pressure Vessels and Piping Conference, Chicago, IL, July 27–31, 2008.

Brauns, Ya.A. and Rikards, R.B. (1971). Investigation of the initial imperfections and buckling modes of glass-reinforced plastic shells under hydrostatic pressure. *Mechanics of Composite Materials*, 7(6), 940–945.

Brendel, B. and Ramm, E. (1980). Linear and nonlinear stability analysis of cylindrical shells. *Computers & Structures*, 12, 549–558.

Broeck, J.A., van der (1947). Euler's classic paper "On the strength of columns". *Journal of Physics,* 12, 309–318.

Broggi, M. (2008). Buckling of cylindrical shells with random imperfections revisited, *Report,* Institute of Engineering Mechanics, University of Innsbruck, Innsbruck, Austria.

Broggi, M., Calvi, A., and Schuëller, G.I. (2011). Reliability assessment of axially compressed composite cylindrical shells with random imperfections. *International Journal of Structural Stability and Dynamics*, 11(2), 215–236.

Broggi, M. and Schuëller, G.I. (2011). Efficient modeling of imperfections for buckling analysis of composite cylindrical shells. *Engineering Structures,* 33(5), 1796–1806.

Brown, A.W.M. (2003). The ensemble statistics of the response of structural components with uncertain properties. Ph.D. Thesis, Fitzwilliam College, Cambridge University, Cambridge UK.

Bruhn, E.F. (1973). *Analysis and Design of Flight Vehicle Structures*, chapter 8. Indianapolis, IN: Jacobs.

Buchanan, J.Y. (1904). On a remarkable effect produced by the momentary relief of great pressure. *Proceedings of the Royal Society of London, A*, 459, 2097–2119.

Bucher, C. (2006). Applications of random field models in stochastic structural mechanics. *Advances in Engineering Structures, Mechanics and Construction* (pp. 471–484). Berlin: Springer.

Bucher, C. and Ebert, M. (2000). Load carrying behavior of prestressed bolted steel flanges considering random geometric imperfections. 8th ASCE Specialty Conference on Probabilistic Mechanics and Structural Reliability, PMC 2000-185.

Budiandsy, B. (1967). Dynamic buckling of elastic structures: criteria and estimates. In G. Hermann (Ed.), *Dynamic Stability of Structures* (pp. 83–106). New York: Pergamon Press.

Budiansky, B. (1974). Theory of buckling and post-buckling behavior of elastic structures. *Advances in Applied Mechanics* (Vol. 14, pp. 1–65). New York: Academic Press.

Budiansky, B. (1990). Reflections, *Applied Mechanics Division Newsletter*. ASME, Spring.

Budiansky, B. and Hutchinson, J.W. (1996a). A survey of some buckling problems. *AIAA Journal*, 4(9), 1505–1510 (Reprinted in *Journal of Spacecraft and Rockets*, 40(6), 918–923, 2003).

Budiansky, B. and Hutchinson, J.W. (1996b). Dynamic buckling of imperfection-sensitive structures. *Proceedings, XI International Congress on Applied Mechanics* (pp. 636–651). Berlin: Springer.

Budiansky, B. and Hutchinson, J.W. (1972). Buckling of circular cylindrical shells under axial compression. *Contributions to the Theory of Aircraft Structures* (pp. 239–259). Delft, The Netherlands: Delft University Press.

Budiansky, B. and Hutchinson, J.W. (1979). Buckling: progress and challenge. In J.F. Besseling and A.M.A. van der Heijden (Eds.), *Trends in Solid Mechanics* (pp. 93–116). Delft, The Netherlands: Delft University Press.

Bulgakov, B.V. (1940). Fehberanhaenfung ber Kreizelapparten (in German). *Ingenieur-Archiv*, 11, 461–469.

Bulgakov, B.V. (1946). On the accumulation of disturbances in linear systems with constant coefficients. *Doklady Akademii Nauk SSSR (Proceedings of the USSR Academy of Sciences)*, LI (5), 339–342.

Bulgakov, B.V. (1954). *Oscillations* (in Russian). Moscow: "Gostekhizdat" Publishers.

Bushnell, D. (1981). Buckling of shells — pitfall for designers. *AIAA Journal*, 19(9), 1183–1226.

Bushnell, D. (1985). *Computerized Buckling Analysis of Shells.* Dordrecht: Martinus Nijhoff Publishers.

Bushnell, D. (2012). Shell buckling. Retrieved on 12 December. http://shellbuckling.com/index.php.

Calladine, C.R. (1972). Structural consequences of small imperfections in elastic shells of revolution. *International Journal of Solids and Structures*, 8, 679–697.

Calladine, C.R. (1983). *Theory of Shell Structures*, Cambridge University Press, Cambridge, U.K.

Calladine, C.R. (1995). Understanding imperfection sensitivity in the buckling of thin-walled shells. *Thin-Walled Structures*, 23, 215–235.

Calladine, C.R. (2001). A shell buckling paradox resolved. In D. Durban, D. Givoli, and J. Simmonds (Eds.), *Advances in the Mechanics of Plates and Shells* (pp. 119–134). Dordrecht: Kluwer Academic.

Calladine, C.R. (2013). Personal communication, 6 November.

Calladine, C.R. and Barber, J.N. (1970). Simple experiments on self-weight buckling of open cylindrical shells. *Journal of Applied Mechanics*, 37, 1150–1151.

Calladine, C.R. and Robinson, J.M. (1980). A simplified approach to the buckling of thin elastic shells. In W.T. Koiter and G.K. Mikhailov (Eds.), *Theory of Shells* (pp. 173–196). Amsterdam: North-Holland Publishing.

van Campen, D.H. (1999). Warner Tjardus Koiter: 16 June 1914–2 September 1997. *Biographical Memoirs of Fellows of Royal Society of London*, 45, 269–273.

Casciati, F., Elishakoff, I., and Roberts, J.B. (Eds.) (1990). *Nonlinear Structural Systems under Random Conditions*. Amsterdam: Elsevier.

Cederbaum, G. and Arbocz, J. (1994). Reliability of axially compressed cylindrical shells. Report LR–767, Faculty of Aerospace Engineering, Delft University of Technology, The Netherlands, July.

Cederbaum, G. and Arbocz, J. (1996a). Reliability of shells via Koiter formulas. *Thin-Walled Structures*, 24, 173–187.

Cederbaum, G. and Arbocz, J. (1996b). On the reliability of imperfection-sensitive long isotropic cylindrical shells. *Structural Safety*, 18(1), 1–9.

Cederbaum, G. and Arbocz, J. (1977). Reliability of imperfection sensitive composite shells via the Koiter–Cohen criterion. *Reliability Engineering and System Safety*, 56, 257–263.

Challamel, N., Lanos, C., and Casandjian, C. (2006). Localisation in the buckling or in the vibration of a two-span weakened column. *Engineering Structures,* 28, 776–782.

Champneys, A.R. (2013). Personal communication, 11 October.

Champneys, A.R., Hunt, G.W., and Thompson, J.M.T. (Eds.) (1997). *Localization and Solitary Waves in Solid Mechanics*. Singapore: World Scientific, Singapore.

Champneys, A.R. and Toland, J.F. (1997). Bifurcation of a plethora of multi-modal homoclinic waves in solid mechanics. *Philosophical Transactions of the Royal Society of London*, A355, 2073–2213.

Charmpis, D. and Papadrakakis, M. (2005). Improving the computational efficiency in FEA of shells with uncertain properties. *Computer Methods in Applied Mechanics and Engineering*, 194(12–16), 1447–1478.

Chater, E. and Hutchinson, J.W. (1984). On the propagation of bulges and buckles. *Journal of Applied Mechanics*, 51, 269–277.

Chater, E., Hutchinson, J.W., and Neale, K.W. (1983). Buckle propagation on a beam on a nonlinear elastic foundation. In J.M.T. Thompson and G.W. Hunt (Eds.), *Collapse: The Buckling of Structures in Theory and Practice* (pp. 31–41). Cambridge: Cambridge University Press.

Chen, X.J., Zhou, K.M., and Aravena, J.L. (2002). Fast construction of robustness degradation function. *Proceedings of the 41st IEEE Conference on Decision and Control* (pp. 2242–2247). Las Vegas, NE.

Chernousko, F.L. (1981). Optimal guaranteed estimation of indeterminacies with the aid of ellipsoids (in Russian). *Tekhnicheskaya Kibernetika (Technical Cybernetics)*, 1–9.

Chernousko, F.L. (1994). *State Estimation of Dynamic Systems*. Boca Raton, FL: CRC Press.

Chernousko, F.L. (1999). What is ellipsoidal modeling and how to use it for control and state estimation? In I. Elishakoff (Ed.), *Whys and Hows in Uncertainty Modeling: Probability, Fuzziness, and Anti-Optimization* (pp. 127–188). Vienna: Springer.

Chilver, A.H. (1976). Design philosophy in structural stability. In B. Budiansky (Ed.), *Buckling of Structures* (pp. 331–345). Berlin: Springer.

Choi, C.K. and Noh, H.C. (1998). The stochastic analysis of the shape imperfection in cooling tower shells. *Structural Engineering Worldwide* (CD-ROM).

Choi, C.K. and Noh, H.C. (2000). Stochastic analysis of shape imperfection in RC cooling tower shells. *Journal of Structural Engineering*, 126(3), 417–423.

Chryssanthopoulos, M.K. (1998). Probabilistic buckling analysis of plates and shells. *Thin-Walled Structures*, 30(1–4), 135–157.

Chryssanthopoulos, M.K., Baker, M.Y., and Dowling, P.J. (1991). Statistical analysis of imperfections in stiffened cylinders. *Journal of Structural Engineering*, 117(7), 1979–1997.

Chryssanthopoulos, M.K. and Poggi, C. (1995a). Probabilistic imperfection sensitivity analysis of axially compressed composite cylinders. *Engineering Structures*, 17(6), 398–406.

Chryssanthopoulos, M.K. and Poggi, C. (1995b). Stochastic imperfections modeling in shell buckling studies. *Thin-Walled Structures*, 23, 179–200.

Cohen, A.G. (1966). Buckling of axially compressed cylindrical shells with ring stiffened edges. *AIAA Journal*, 4, 1859–1862.

Cohen, M.S. (2012). Bad advice, not young scientists, should hit the road. *Science*, 335, 794.

Coman, C. D. (2004a). Secondary bifurcations and localization in a three-dimensioanl buckling model. *ZAMM-Zeitschrift für angewandte Mathematic und Mekhanik*, 40, 758–761.

Coman, C.D. (2004b). Secondary bifurcations and localization in buckling pattern. *XXI ICTAM*, 15–21 August, Warsaw, Poland.

Coman, C.D. (2010). Localized elastic buckling: non-linearities versus inhomogeneities. *IMA Journal of Applied Mathematics*, 75, 461–474.

Coman, C.D. and Houghton, D.M. (2006). Localized wrinkling instabilities in radially stretched annular thin films. *Acta Mechanica*, 185, 179–200.

Combescure, A. and Gusik, G. (2001). Nonlinear buckling of cylinders under lateral pressure with nonaxisymmetric thickness imperfections using the COMI axisymmetric shell element. *International Journal of Solids and Structures*, 38, 6207–6226.

Coppa, A.P. (1966). Measurement of initial geometric imperfections of cylindrical shells. *AIAA Journal*, 4(1), 172–175.

Cornell, C.A. (1969). Probability based structural code. *ACI Journal*, 66, 974–985.

Coulumb, C.A. (1773). Essai sur une application des regles de maximis et minimis a quelques problemes de statique relatifs a l'architectture. *Mem. present Acad. Royal Sci.*

Cowell, R.G. (1986). Looping post-buckling paths of an axially loaded elastic cylindrical shell. *Dynamics and Stability of Systems*, 1(2), 115–123.

Cox, H.L. (1940). Stress analysis of thin metal construction. *Journal of the Royal Aeronautical Society*, 44 (351), 231–272.

Craig, K.J. and Roux, W.J. (2007). On the investigation of shell buckling due to random geometrical imperfections implemented using Karhunen–Loève expansions. *International Journal for Numerical Methods in Engineering*, 73(12), 1715–1726.

Croll, J.G.A. (1981). Explicit lower bounds for the buckling of the axially loaded cylinders. *International Journal of Mechanical Sciences*, 23(6), 331–343.

Croll, J.G.A. (1995). Towards a rationally based elastic-plastic shell buckling design methodology. *Thin-Walled Structures*, 23(1–4), 67–85.

Croll, J.G.A. (2006). Stability in shells. *Nonlinear Dynamics*, 43, 17–28.

Croll, J.G.A. and Gavrilenko, G.D. (1998). Substantiation of the method of reduced stiffness. *Strength of Materials*, 30(5), 481–496.

Croll, J.G.A. and Gavrilenko, G.D. (1999). Reduced stiffness method in theory of smooth shells and the classical analysis of stability. *Strength of Materials*, 31(2), 138–154.

Damil, N. and Potier-Ferry, M. (1992). Amplitude equations for cellular instabilities. *Dynamic and Stability of Systems*, 7(1), 1–34.

Dancy, R. and Jacobs, D. (1988). The initial imperfection data banks at the Delft University of Technology, Part II, Technical Report LR-559, Faculty of Aerospace Engineering, Delft University of Technology, Delft, The Netherlands.

Darevskii, V.M. (1963). Stability equations of cylindrical shells (in Russian). *Engineering Journal*, 3(4), 658–664.

Darevskii, V.M. (1980). Stability of nearly cylindrical shell. *Problems of Computation of Spatial Constructions* (in Russian; pp. 35–45). Moscow: Moscow Civil Engineering Institute.

Das, P.K., Thavalingam, A. and Bai, Y. (2003). Buckling and ultimate strength criteria of stiffened shells under combined loading for reliability analysis. *Thin-Walled Structures*, 41, 69–88.

Davidenko, D.F. (1951). On a new method of numerical solution of the systems of nonlinear equations (in Russian). *Doklady Akademii Nauk SSSR* (*Proceedings of the USSR Academy of Sciences)*, 88(4), 68–78.

Davis, M. (2013). *The Logician and the Engineer* (review of the book by P.J. Nahin). *Notices of the American Mathematical Society*, 60(9), 1170–1172.

Day, W.B. (1980). Buckling of a column with nonlinear restraints and random initial imperfections. *Journal of Applied Mechanics*, 47, 204–205.

Day, W., Karwowski, A.J., and Papanicolaou (1989). Buckling of randomly imperfect beams. *Acta Applicandae Mathematicae*, 17, 269–286.

De Groof, V., Oberguggenberger, M., Haller, H., Degenhardt, R., and Kling, A. (2012). Qualitative assessment of random field models in FE buckling analysis of composite cylinders. In J. Eberhardsteiner et al. (Eds.), *European Congress on Computational Methods in Applied Sciences and Engineering (ECCOMAS 2012)*, Vienna.

De Groof, V., Oberguggenberger, M., Haller, H., Degenhardt, R., and Kling, A. (2013). A case study of random field models applied to thin-walled composite cylinders in finite-element analysis. In G. Deodatis, B.R. Ellingwood, and D.M. Frangopol (Eds.), *Proceedings of the International Conference on Structural Safety, Reliability, Risk and Performance of Structures and Infrastructures*. Boca Raton, FL: CRC Press.

De Quincey, T. (1881). *The Works of Thomas De Quincey* (p. 535). Boston, MA: Houghton, Mifflin and Company.

Degenhardt, R., Bethge, A., and Kärger, L. (2007). Probabilistic aspects of buckling knock-down factors — test and analysis. Technical Report, ESA Contract 19709/06/NL/IA.

Degenhardt, R., Kling, A., Bethge, A., Orf, J., Kärger, L., Zimmerman, R., Rohwer, K., and Calvi, A. (2010). Investigations on imperfection sensitivity and deduction of improved knock-down factors for unstiffened CFRP cylindrical shells. *Composite Strucutres*, 92(8), 1939–1946.

Degenhardt, R., Rolfes, R., Zimmermann, R., and Rohwer, K. (2006). COCOMAT — Improved material exploitation of composite airframe structures by accurate simulation of postbuckling and collapse. *Composite Structures*, 73(2), 175–178.

Demi, M. and Wunderlich, W. (1997). Direct evaluation of the "worst" imperfection shape in shell buckling. *Computer Methods in Applied Mechanics and Engineers*, 149(1–4), 201–222.

Der Kiureghian, A. (2005). Non-ergodicity and PEER's framework formula. *Earthquake Engineering and Structural Dynamics*, 34(13), 1643–1652.

DIN 18800 (1990). *Stahlbauten — Stabilitätsfälle, Schalenbeulen* (in German). Berlin, Part 4.

Ding, X., Coleman, R. and Rotter, J.M. (1996). Technique for precise measurement of large-scale silos and tanks. *Journal of Survey Engineering*, 122, 14–25.

Donnell, L.H. (1933). Stability of thin-walled tubes under torsion. Report No. 479, NACA, Washington D.C.

Donnell, L.H. (1934). A new theory for the buckling of thin cylinders under axial compression and bending. *Transactions, ASME*, Ser. E., 56, 795–806.

Donnell, L.H. and Wan, C.C. (1950). Effects of imperfections on the elastic buckling of cylinders and columns under axial compression. *Transactions ASME*, 72(1), 73–83 (see Discussion, pp. 340–342).

Doup, M.R. (1997). Probabilistic analysis of the buckling of thin-walled shells using an imperfection database and two mode analysis. Memorandum M-808, Faculty of Aerospace Engineering, Delft University of Technology, Delft, The Netherlands, August.

Drenick, R.F. (1970). Model-free design of aseismic structures. *Journal of Engineering Mechanics Division*, 96, 483–493.

Drenick, R.F. (1977). On a class of non-robust problems in stochastic dynamics. In B.L. Clarkson (Ed.), *Stochastic Problems in Dynamics* (pp. 237–255). London: Pitman.

Dryden, H.L. (1963). Contributions of Theodore von Kármán to applied mechanics. *Applied Mechanics Reviews*, 16(8), 589–595.

Du, S., Ellingwood, B.R. and Cox, J.V. (2005). Solution methods and initialization techniques in SFE analysis of structural stability. *Probabilistic Engineering Mechanics*, 20(2), 179–187.

Dubourg, V. (2001). Adaptive surrogate models for reliability analysis and reliability-based design optimization. Ph.D. Thesis, Université Blaise Pascal, Clermont-Ferrand, France, 2011.

Dubourg, V., Bourinet, J.M., and Sudret, B. (2009a). FE-based reliability analysis of the buckling of shells with random shape, material and thickness imperfections. In H. Furuta, D. Frangopol, and M. Shinozuka (Eds.), Proceedings of the 10th International Conference on Structural Safety and Reliability (ICOSSAR 2009), Osaka, Japan, 2009a.

Dubourg, V., Bourinet, J.-M., and Sudret, B. (2009b). Analyse fiabiliste du flambage des coques avec prise en compte du caractère aléatoire et de la variabilité spatiale des défauts de forme et d'épaisseur, et des propriétés matériaux (in French). Proc. 19ème Congrès Français de Mécanique (CFM19), Marseille.

Dubourg, V., Noirfalise, C., Bourinet, J.-M., and Fogli, M. (2009). FE-based reliability of the buckling of shells with random shape, material and thickness imperfections. ICOSSAR, Osaka, Japan, 2009.

Dubourg, V., Sudret, B., and Deheeger, F. (2013). Metamodel-based importance sampling for structural reliability analysis. *Probabilistic Engineering Mechanics*, 33, 47–57.

Dym, C.L. and Hoff, N.J. (1968). Perturbation solutions of the buckling problems of axially compressed thin cylindrical shells of infinite or finite length. *Journal of Applied Mechanics*, 35, 754–762.

Earman, J. and Rédei, M. (1996). Why ergodic theory does not explain the success of equilibrium statistical mechanics. *British Journal of Philosophical Sciences*, 47, 63–78.

ECCS No.56 (1988). Buckling of steel shells: European recommendations, European Convention for Constructional Steelwork, Brussels, Belgium.

Eckmann, J.P. and Ruelle, D. (1985). Ergodic theory of chaos and strange attractors. *Reviews of Modern Physics*, 57(3), Part 1, 617–656.

Edlund, B.L.O. (2007). Buckling of metallic shells: buckling and postbuckling behaviour of isotropic shells, especially cylinder, *Structural Control and Health Monitoring*, 14, 693–713.

Edlund, B. and Leopoldson, U. (1973). Simulation of the load carrying capacity of cylindrical shells under axial compression (in Swedish). Report 573:7, Chalmers Technical University, Göteborg.

van Eekelen, A.J. (1994). Computational module for the buckling and postbuckling behavior of a cylindrical shell with a two-mode imperfection. Report-LR 773, Department of Aerospace Engineering, Delft University of Technology, Delft, The Netherlands, December.

Eggwertz, S. and Palmberg, B. (1985). Structural safety of axially loaded cylindrical shells. FFA TN 1985–50, The Aeronautical Research Institute of Sweden, Stockholm, Sweden.

Eglitis, E., Kalninš, K., and Ozolinš (2009). Experimental and numerical study on buckling of axially compressed composite cylinders. *Scientific Journal of Riga Technical University, Construction Science*, 10(10), 33–39.

El Bahoui, J., El Bakkali, L., Khamlichi, A., and Limam, A. (2009). Reliability assessment of compressed cylindrical shells with interacting localized defects. Paper S1704-P0346, 3rd International Conference on Integrity, Reliability and Failure, Porto, Portugal.

El Bahoui, J., Khamlichi, A., El Bakkali, L., and Liman, A. (2010). Reliability assessment of buckling strength for compressed cylindrical shells with interacting localized geometric imperfections. *American Journal of Engineering and Applied Sciences*, 3(4), 620–628.

El Naschie, M.S. (1975a). Local post buckling of compressed cylindrical shells. *Proceedings Institution of Civil Engineers, Part 2*, 59, 523–525.

El Naschie, M.S. (1975b). Localized diamond-shaped buckling patterns of axially compressed cylindrical shells. *AIAA Journal*, 13(4), 837–838.

El Nashie, M.S. (1976). Initial and post buckling of axially compressed orthotropic cylindrical shells. *AIAA Journal*, 14(10), 1502–1504.

El Naschie, M.S. (1977a). Reply by author to P. Kaoulla. *AIAA Journal*, 15(5), 757–758.

El Naschie, M.S. (1977b). Nonlinear isometric bifurcation and shell buckling. *Zeitschrift für angewandte Mathematik und Mechanik-ZAMM*, 57, 293–296.

El Naschie, M.S. (1989). On certain homoclinic soliton in elastic stability. *Journal of the Physical Society of Japan*, 58(12), 4310–4321.

El Naschie, M.S. (2000). A very brief history of localization, *Chaos, Solitons and Fractals*, 11(10), 1479–1480.

Elishakoff, I. (1977). On the role of cross-correlations in random vibrations of shells. *Journal of Sound and Vibration*, 50, 239–252.

Elishakoff, I. (1978a). Axial impact buckling of a column with random initial imperfections. *Journal of Applied Mechanics*, 45(2), 361–365.

Elishakoff, I. (1978b). Impact buckling of a thin bar via Monte Carlo method. *Journal of Applied Mechanics*, 45(3), 561–590.

Elishakoff, I. (1979a). Simulation of space-random fields for solution of stochastic boundary value problems. *Journal of Acoustical Society of America*, 61, 399–403.

Elishakoff, I. (1979b). Buckling of stochastically imperfect finite column on nonlinear elastic foundations: a reliability study. *Journal of Applied Mechanics*, 46(2), 411–416 (see also *TAE Report No. 328*, Department of Aeronautical Engineering, Technion — Israel Institute of Technology, Haifa, Israel, March 1978).

Elishakoff, I. (1980a). Hoff's problem in a probabilistic setting. *Journal of Applied Mechanics*, 47, 403–408.

Elishakoff, I. (1980b). Remarks on the static and dynamic imperfection-sensitivity of nonsymmetric structures. *Journal of Applied Mechanics*, 47(1), 111–115.

Elishakoff, I. (1982). Simulation of an initial imperfection data bank, part 1: isotropic shells with general imperfections. TAE Report 500, Department of Aeronautical Engineering, Technion-Israel Institure of Technology, Haifa, Israel, July.

Elishakoff, I. (1983a). A simple model explaining some recent random vibration results. *Proceedings of the 4th international Conference on Applications of Statistics and Probability in Soil and Structural Engineering* (pp. 493–507). Bologna, Italy: Pitagora Press.

Elishakoff, I. (1983a). *Probabilistic Theory of Structures*. New York: Wiley (2nd edition, Dover Publications, Mineola, NY, 1999).

Elishakoff, I. (1983b). How to introduce the imperfection sensitivity concept into design. In J.M.T. Thompson and G.B. Hunt (Eds.), *Collapse: Buckling of Structures in Theory and Practice* (pp. 345–357). Cambridge, UK: Cambridge University Press.

Elishakoff, I. (1985). Reliability approach to the initial imperfection sensitivity. *Acta Mechanica*, 55, 151–170.

Elishakoff, I. (1986). A model elucidating significance of cross-correlating in random vibration analysis. In I. Elishakoff and R.H. Lyon (Eds.), *Random Vibration: Status and Recent Developments* (pp 101–112). Amsterdam: Elsevier.

Elishakoff, I. (1988a). Stochastic simulation of an initial imperfection data bank for isotropic shells with general imperfections. In I. Elishakoff, J. Arbocz, C.D. Babcock, Jr., and A. Libai (Eds.), *Buckling of Structures — Theory and Experiment* (pp. 195–210). Amsterdam: Elsevier.

Elishakoff, I. (1988b). Random vibrations of multi-degree-of-freedom systems with associated effect of cross-correlations. In W. Schiehlen and W. Wedig (Eds.), *Analysis and Estimation of Stochastic Mechanical Systems*. Vienna: Springer.

Elishakoff, I. (1990). An idea of the uncertainty triangle (editorial). *Shock and Vibration Digest*, 22(10), 1.

Elishakoff, I. (1995a). Essay on uncertainties in elastic and viscoelastic structures: from A.M. Freudenthal's criticisms to modern convex modeling. *Computers & Structures*, 58, 871–896.

Elishakoff, I. (1995b). Convex modeling — a generalization of interval analysis for nonprobabilistic treatment of uncertainty. *International Journal of Reliable Computing*, Supplement.

Elishakoff, I. (1997). Buckling of structures with uncertain imperfections — personal perspective. Paper AIAA-97-1242. Proceedings, 38th AIAA/ASME/ASCE/SDM Conference, Kissimmee, FL.

Elishakoff, I. (1998). How to introduce the imperfection-sensitivity concept into design 2. In N.F. Knight, Jr. and M. Nemeth (Eds.), *Stability Analysis of Plates and Shells* (pp. 237–267). NASA CP-1998-206280.

Elishakoff, I. (1999). Are probabilistic and antioptimization methods interrelated? In I. Elishakoff (Ed.), *Whys and Hows in Uncertainty Modeling: Probability, Fuzziness, and Antioptimization* (pp. 285–318). Vienna: Springer.

Elishakoff, I. (2000). Uncertain buckling: its past, present and future. *International Journal of Solids and Structures*, 37 (46–47), 6869–6889.

Elishakoff, I. (2001). Elastic stability: from Euler to Koiter there was none like Koiter. *Meccanica*, 38, 375–380.

Elishakoff, I. (2003). Notes on the philosophy of the Monte Carlo method. *International Applied Mechanics*, 39(7), 3–14.

Elishakoff, I. (2004). *Safety Factors and Reliability: Friends or Foes?* Dordrecht: Kluwer Academic.

Elishakoff, I. (2005). Controversy associated with the so-called "follower forces": critical overview. *Applied Mechanics Reviews*, 58, 117–142.

Elishakoff, I. (2006). Spatial parametric resonance and other novel buckling problems inspired by James H. Starnes, Jr. In N.F. Knight, Jr., M.P. Nemeth and J.B. Malone (Eds.), *Collected Papers in Structural Mechanics Honoring Dr. James H. Starnes, Jr.* (pp. 149–156). NASA/TM–2006– 214276, February.

Elishakoff, I. (2012). Probabilistic resolution of the twentieth century conundrum in elastic stability. *Thin-Walled Structures*, 59, 35–37.

Elishakoff, I. (2013). A celebration of mechanics: from nano to macro: The J. Michael T. Thompson Festschrift issue. *Philosophical Transactions of the Royal Society A*, 371, number 1993, paper 20130121.

Elishakoff, I., Andriamasy, L., and Lemaire, M. (2010). Hybrid randomness of initial imperfection and axial loading in reliability of cylindrical shells. *Journal of Applied Mechanics*, 77930, Paper 031003.

Elishakoff, I. and Arbocz, J. (1980). Buckling of shells with general random imperfections. Paper presented at the 15th IUTAM Congress, Toronto Also: Memorandum M-401, Department of Aerospace Engineering, Delft University of Technology, The Netherlands, 1980a.

Elishakoff, I. and Arbocz, J. (1980a). Reliability of axially compressed cylindrical shell with random axisymmetric imperfections. Report LR-206, Department of Aerospace Engineering, Delft University of Technology, The Netherlands.

Elishakoff, I. and Arbocz, J. (1982a). Reliability of compressed cylindrical shells with random axisymmetric imperfection. *International Journal of Solids and Structures*, 18(7), 563–585.

Elishakoff, I. and Arbocz, J. (1982b). Monte Carlo method for the reliability of compressed structures with random imperfections. *Proceedings of the Danish Engineering Academy of Science* (pp. 389–424), Lyngby, Denmark.

Elishakoff, I. and Arbocz, J. (1982c). Stochastic buckling of shells with general imperfections. In F.H. Schroeder (Ed.), *Stability of the Mechanics of Continua* (pp. 306–317). Berlin: Springer.

Elishakoff, I. and Arbocz, J. (1985). Reliability of axially compressed cylindrical shells with general nonsymmetric imperfections. *Journal of Applied Mechanics*, 52, 122–128.

Elishakoff, I., Arbocz, J., and Starnes, J.H., Jr. (1992). Buckling of stiffened shells with random initial imperfections, thickness, and boundary conditions. Proceedings, 33rd AIAA/ASCE/AMS/ASC Structures, Structural Dynamics and Materials Conference (pp. 95–100), Dallas, TX.

Elishakoff, I., Arbocz, J., Babcock, C.D., Jr., and Libai, A. (Eds.) (1988). *Buckling of Structures — Theory and Experiment*. Amsterdam: Elsevier Science.

Elishakoff, I. and Archaud, E. (2013). Modified Monte-Carlo method for buckling analysis of nonlinear imperfect structures. *Archive of Applied Mechanics*, 83, 1327–1339.

Elishakoff, I. and Ben-Haim, Y. (1990a). Dynamics of thin cylindrical shell under impact with limited deterministic information on initial imperfections. *Journal of Structural Safety*, 8, 103–112.

Elishakoff, I. and Ben-Haim, Y. (1990b). Probabilistic and convex models of uncertainty in structural dynamics. *Proceedings of the 8th International Modal Conference* (pp. 1487–1492), Orlando, FL.

Elishakoff, I., Cai, G.Q., and Starnes, J.H., Jr. (1994a). Non-linear buckling of a column with initial imperfection via stochastic and non-stochastic convex models. *International Journal of Non-Linear Mechanics*, 29(1), 71–82.

Elishakoff, I., Cai, G.Q., and Starnes, J.H., Jr. (1994b). Probabilistic and convex models of uncertainty in buckling of structures. In G.I. Schuëller, M. Shinozuka and J.T.P. Yao (Eds.), *Structural Safety and Reliability* (pp. 761–766). Rotterdam, The Netherlands: A.A. Balkema.

Elishakoff, I. and Colombi, P. (1993). Combination of probabilistic and convex models of uncertainty when scarce knowledge is present on acoustic excitation parameters. *Computer Methods in Applied Mechanics and Engineering*, 104, 187–209.

Elishakoff, I., Haftka, R.T., and Fang, J.J. (1994). Structural design under uncertainty—optimization and anti-optimization. *Computers and Structures*, 53, 1401–1405.

Elishakoff, I. and Hasofer, A.M. (1985). On the accuracy of Hasofer–Lind reliability index. In I. Konishi, A.H.-S. Ang, and M. Shinozuka (Eds.), *Structural Safety and Reliability* (Vol. 1, pp. 1.229–1. 239), Kobe, Japan.

Elishakoff, I. and Impollonia, N. (2001). Does a partial elastic foundation increase the flutter velocity of a pipe converging fluid? *Journal of Applied Mechanics*, 68, 206–212.

Elishakoff, I., Kriegesmann, B., Rolfes, R., Hühne, C., and Kling, A. (2012). Optimization and anti-optimization of buckling load for composite cylindrical shell under uncertainties. *AIAA Journal*, 50(7), 1513–1524.

Elishakoff, I., Li, Y.W., and Starnes, J.H., Jr. (1994). A deterministic method to predict the effect of unknown-but-bounded elastic moduli on the buckling of composite structures. *Computer Methods in Applied Mechanics and Engineering*, 111, 155–167.

Elishakoff, I., Li, Y.W., and Starnes, J.H., Jr. (1995). Buckling mode localization in elastic plates due to misplacement in the stiffener location. *Choas, Solitons and Fractals*, 5, 1517–1531.

Elishakoff, I., Li, Y.W., and Starnes, J.H., Jr. (1996). Imperfection sensitivity due to the elastic moduli in the Roorda–Koiter frame. *Chaos, Solitons & Fractals*, 7(8), 1179–1186.

Elishakoff, I., Li, Y.W., and Starnes, J.H. Jr. (2001). *Non-Classical Problems in the Theory of Elastic Stability*. Cambridge, UK: Cambridge University Press, Cambridge.

Elishakoff, I., van Manen, S., Vermeulen, P.G., and Arbocz, J. (1987). First-order second-moment analysis of the buckling of shells with random imperfections. *AIAA Journal*, 25(8), 1113–1117.

Elishakoff, I., Marcus, S., and Starnes, J.H., Jr. (1996). On vibrational imperfection sensitivity of Augusti's model structure in the vicinity of a non-linear static state. *International Journal of Non-Linear Mechanics*, 31 (2), 229–236.

Elishakoff, I. and Ren, Y.J. (2003). *Finite Element Methods for Structures with Large Stochastic Variation*. New York: Oxford University Press.

Elishakoff, I. and Starnes, J.H., Jr. (1999). Safety factor and the nondeterministic approaches. AIAA Paper 1999-1614. *Proceedings of the 40th AIAA/ASME/ASCE/AHS/ASC Structures, Structural Dynamics, and Material Conference*, St. Louis, MI. Reston, VA: AIAA Press.

Elishakoff, I. and Ohsaki, M. (2010). *Optimization and Anti-Optimization of Structures under Uncertainty*. London: Imperial College Press.

Elishakoff, I., van Zanten, A.Th., and Crandall, S.H. (1979). Wide-band random asymmetric vibration of cylindrical shells. *Journal of Applied Mechanics*, 46, 417–423.

Ellinas, C.P. and Croll, J.G.A. (1983). Experimental and theoretical correlations for elastic buckling of axially compressed stringer stiffened cylinders. *Journal of Strain Analysis*, 18(1), 41–67.

Elseifi, M.A., Gürdal, Z., and Nikolaidis, E. (1999). Convex/probabilistic models of uncertainties in geometric probabilistic models of uncertainties on geometric imperfections of stiffened composite panels. *AIAA Journal,* 37, 468–474.

Ermolenko, V.M. (1977). About the critical load of axially-compressed orthotropic cylindrical shell (in Russian). *Dynamics of Continuous Medium* (Vol. 28, pp. 152–156). Novosibirsk: Institute of Hydrodynamics.

Esslinger, M. and Geier, B. (1972). Gerechte Nachbeullasten als untere Grenze der experimentallen axialen Beullasten von Kreiszylinddern (in German). *Der Stahlbau,* 41, 353–359.

Esslinger, M. and Geier, B. (1976). Calculated postbuckling loads as lower limits for the buckling loads of thin-walled circular cylinders. In B. Budiansky (Ed.), *Buckling of Structures* (pp. 274–290). Berlin: Springer.

Etokov, V.I. (1972). Post-critical strain of cylindrical shells taking into account geometric surface flexure. *International Applied Mechanics*, 8(2), 161–167.

Euler, L. (1744). *Methodus Inveniendi Lineas Curvas Maximive Porprietate Gaudentes* (in Latin) (Appendix: De Curvis Elasticis), Marcum Michaelem Bosquet, Lausanne.

Eulero, L. (1744). Methodus inveniendi linears curvas maximi minimive proprietate qaidentas, sive solution problematis isoperitrici lattessimo sensu accepti (in Latin). Addentamentum 1: de curvis elasticis, Laussane et Genva, Apid Marcum-Mihaelum, Bousquet et Socios, pp. 245–310.

Eurocode 3. (2002). *General Rules: Strength and Stability of Shell Structures*, Brussels, Belgium, Part 1.6.

Evans, A.G., Hutchinson, J.W., and Ashby, M.F. (1999). Multifunctionality of cellular metal systems. *Progress in Materials Science*, 43, 171–221.

Evensen, D.A. (1964). High-speed photographic observation of the buckling of thin cylinders. *Experimental Mechanics*, 4(4), 110–119.

Evkin, A.Yu. (1975). A new approach to asymptotic integration of the equations of shallow convex shell theory in the post-critical range. *International Applied Mechanics*, 11(11), 1155–1159.

Evkin, A.Yu., Krasovsky, V.L., and Manevich, L. (1978a). Buckling of axially compressed cylindrical shells under local quasi-statical loads (in Russian). *Mechanics of Solids*, 6, 96–100.

Evkin, A.Yu., Krasovsky, V.L., and Manevich, L.I. (1978b). Stability of longitudinally compressed cylindrical shells under quasi-static local disturbances. *Mechanics of Solids*, 13(6), 83–88.

Evkin, A.Yu., Krasovsky, V.L., and Manevich, L.I. (1978c). Stability of the axially compressed cylindrical shells during local quasistatic loading (in Russian). *Mechanics of Solids*, 6, 95–100.

Evkin, A.Yu., Prokopalo, E.F., and Shukurov, A.Kh. (1981). Investigation of post-critical equilibrium forms in axially compressed cylindrical shell (in Russian). *Structural Mechanics and Construction Analysis*, 6, 45–47.

de Faria, A.R. (2002). Buckling optimization of composite plates: uncertain loading combination. *International Journal for Numerical Methods in Engineering*, 53(3), 719–732.

de Faria, A.R. (2004). Compliance minimization of structures under uncertain loadings. *Latin American Journal of Solids and Structures*, 1, 363–378.

de Faria, A.R. (2007). Prebuckling enhancement of beams and plates under uncertain loading and arbitrary initial imperfections. *Journal of the Brazilian Society of Mechanical Sciences and Engineering*, 29(4), 388–395.

de Faria, A. R. and de Almeida, S.F.M. (2003a). Buckling optimization of plates with thickness subjected to nonuniform uncertain loads. *International Journal of Solids and Structures*, 40, 3955–3966.

de Faria, A.R. and de Almeida, S.F.M. (2003b). Buckling optimization of variable thickness plates subjected to nonuniform uncertain loads. *XXIV Iberian Latin-America Congress on Computational Methods in Engineering* (Auto Preto/MG, Brazil).

de Faria, A.R. and de Almeida, S.F.M. (2004). Buckling optimization of variable thickness composite plates subjected to nonuniform loads. *AIAA Journal*, 42 (2), 228–231.

de Faria, A.R. and de Almeida, S.F.M. (2006). The maximization of fundamental frequency of structures under arbitrary initial stress states. *International Journal for Numerical Methods in Engineering*, 65, 445–460.

de Faria, A.R. and Hansen, J.S. (2001a). Buckling optimization of composite axisymmetric cylindrical shells under uncertain loading conditions. *Journal of Applied Mechanics*, 68(4), 632–639.

de Faria, A.R. and Hansen, J.S. (2001b). On buckling optimization under uncertain loading combinations. *Structural and Multidisciplinary Optimization*, 21, 272–282.

Feinstein, G., Chen, Y.N., and Kempner, J. (1971). Buckling of clamped oval cylindrical shells under axial loads. *AIAA Journal*, 9(9), 1733–1738.

Feodosiev, V.I. (1969). *Desiat' Lekzii — Besed po Soprotivleniyu Materialov (Ten Lectures — Conversations on Strength of Materials)* (in Russian). Moscow: "Nauka" Publishers.

Fersht, R.S. (1968a). Buckling of cylindrical shells with random imperfections. Ph.D. Thesis, California Institute of Technology, Pasadena, CA.

Fersht, R.S. (1968b). Almost sure stability of long cylindrical shells with random imperfections. NASA CR-1161.

Fersht, R.S. (1974). Buckling of cylindrical shells with random imperfections. In Y.C. Fung and E.E. Sechler (Eds.), *Thin-Walled Structures: Theory, Experiment and Design* (pp. 325–341). Englewood Cliffs, NJ: Prentice Hall.

Feynman, R. (1969). What is science, *The Physics Teacher*, 7(6), 313–320.

Feynman R. (1981). Interview. *BBC's Horizon Program.*

Feynman, R. (1988). *What Do You Care What Other People Think?: Further Advenchies of a Curious Character.* New York: W.W. Norton and Company.

Fisher, R.A. (1921). On the mathematical foundations of theoretical statistics. *Philosophical Transactions of the Royal Society, Series A* 222, 309.

Flügge, W. (1932). Die stabilität der Kreiszylinderschalen (in German). *Ingenieur-Archiv*, 3, 463–506.

Fraser, W.B. (1965). Buckling of a structure with random imperfections. Ph.D. Thesis, Harvard University, Cambridge, MA.

Fraser, W.B. and Budiansky, B. (1969). The buckling of a column with random initial deflections. *Journal of Applied Mechanics*, 36, 233–240.

Freudenthal, A.M. (1956). Safety and probability of structural failure. *ASCE Proceedings.*

Fung, Y.C. and Sechler, E.E. (1960). Instability of thin elastic shells. In J.N. Goodier and N.J. Hoff (Eds.), *Structural Mechanics* (pp. 115–131). New York: Pergamon.

Galbraith, J. (1973). *Designing Complex Organization.* New York: Addison-Wesley.

GALCIT: The First 75 Years. Retrieved on 12 December 2010, http://www.galat.caltech.edu/history/inde.html.

Galilei, G. (1638). *Discorsi e Demostrazioni Matematiche, intorno à due nuove scienze Attenenti all Mecanica & Movimenti Locali.* Leyden: Elsevier (English translation: *Dialogues Concerning Two New Sciences*, Northwestern University Press, 1914; McGraw Hill, New York, 1963; *Two New Sciences*, The University of Wisconsin Press, 1974).

Galton, F. (1893). *Inquiries into Human Faculty and Its Development* (free Google book).

Gao, S.Q. (1991). A study of the random response for a shell of rotation having meridional random imperfections. *Journal of Beijing Institute of Technology*, 11(1), 25–32.

Garrett, D.J. (1992). Critical buckling load statistics of an uncertain column, probabilistic mechanics and structural and geotechnical reliability. *Proceedings of the Sixth Specialty Conference, ASCE* (pp. 563–566), Denver.

Gavrilenko, G.D. (1989). *Stability of Ribbed Cylindrical Shells during Nonhomogeneous Stress-Strain State* (in Russian). Kiev: "Naukova Dumka" Publishers.

Gavrilenko, G.D. (1999). *Stability of Ribbed Shells of Non-ideal Shape* (in Russian). Kiev: Ukraine's Institute of Mathematics Press.

Gavrilenko, G.D. (2003). Numerical and analytical approaches to the stability analysis of imperfect shells. *International Applied Mechanics*, 39(9), 1029–1045.

Gavrilenko, G.D. (2005). Lower bounds for the buckling of axially loaded stiffened shells. In W. Pietraszkiewicz and C. Szymczak (Eds.), *Shell Structures: Theory and Applications* (pp. 215–218). London: Taylor & Francis.

Gavrilenko, G.D. (2007). *Load Carrying Capacity of Imperfect Shells* (in Russian). Drepropetrovsk, Ukraine: "Barbiks" Publishing.

Gavrilenko, G.D. (2010). On a fast method for buckling load calculations of incomplete ribbed shells. In R. Kienzler, H. Altenbach, and I. Ott (Eds.), *Theories of Plates and Shells: Critical Review and New Applications*, (pp. 45–52). Berlin: Springer.

Gavrilenko, G.D. and Croll, J.G.A. (2001a). Validation of analytical lower bounds for the imperfection sensitive buckling of axially loaded rotationally symmetric shells. *Proceedings, 3rd International Conference on Thin-Walled Structures* (pp. 643–651), Crakow, Poland, June.

Gavrilenko, G.D. and Croll, J.G.A. (2001b). Development of numerical and analytical methods of carrying capacity calculation of shells. *Buckling Predictions of Imperfection Sensitive Shells* (p. 34), Kerkrade, The Netherlands, September.

Gavrilenko, G.D. and Krasovsky, V.L. (2004a). Calculation of load-carrying capacity of elastic shells with periodic dents (theory and experiment). *Strength of Materials*, 36(5), 511–517.

Gavrilenko, G.D. and Krasovsky, V.L. (2004b). Stability of circular cylindrical shells with a single local dent. *Strength of Materials*, 3, 52–64.

Gavrilenko, G.D. and Krasovsky, V.L. (2004c). Stability of circular cylindrical shells with a single local dent. *Strength of Materials*, 3, 260–268.

Gavrilenko, G.D. and Matsner, V.I. (2007). *Analytical Method of Determination of Upper and Lower Critical Loads for Elastic Stiffened Shells* (in Russian). Dnepropetrovsk, Ukraine: "Barbiks" Publishing.

Gavrilenko, G.D., Mastner, V.I., and Sytnik, A.S. (1999). Stability of ribbed cylindrical shells with a non-ideal shape. *International Applied Mechanics*, 35(12), 1222–1228.

Gavrilenko, G.D., Mastner, V.I. and Sytnik, A.S. (2000a). Minimum critical loads of ribbed shells with given initial defections. *International Applied Mechanics*, 36(11), 1482–1486.

Gavrilenko, G.D., Matsner, V.I., and Sitnik, A.S. (2000b). Minimum critical loads of ribbed shells with given initial imperfections. *International Applied Mechanics*, 36(11), 1482–1486.

Gavrilenko, G.D., Palchevskii, A.S., and Yakubovskii, Yu.E. (1985). Determination of critical loads for nonideal shell models (in Russian). *Problemy Prochnosti (Strength of Materials)*, I6, 68–72.

Gayton, N. and Lemaire, M. (2002). Probabilistic approach for the design of thin axisymmetric shells. *Integrating Structural Reliability Analysis and Advanced Structural Analysis*, ASRANet, Glasgow, Scotland, 2002.

Gayton, N., Lambelin, J.P., and Lemaire, M. (2002). Approche probabiliste du facteur de réduction de charge d'une coque mince (in French). *Mécanique et Industrie*, 3, 227–239.

Geizenblazen, R.E., Manevich, L.I., Mossakovskii, V.I., and Prokapalo, E.F. (1969). On effect of initial disturbances on stability of smooth cylindrical shells (in Russian). *Mechanics of Solids*, 4(6), 106–122.

Gerard, G. and Becker, H. (1957). *Handbook of Structural Stability*. Supplement to Part III — Buckling of curved plates and shells. NASA TN-3783, 1957.

Gleick, J. (2011). *The Information: A History, a Theory, a Flood.* New York: Vintage Books.

Godoy, L.A. (1996). *Thin-Walled Structures with Structural Imperfections: Analysis and Behavior.* Oxford: Pergamon Press.

Godoy, L.A. (2000). *Theory of Elastic Stability: Analysis and Sensitivity*. Philadelphia, PA: Taylor & Francis.

Godoy, L.A. (2002). On the path of progress in the theory of elastic stability. *Mechanics*, pp. XIII–XIV, September/October.

Godoy, L.A. (2006). Historical sense in the historians of the theory of elasticity. *Meccanica*, 41, 529–538.

Godoy, L.A. (2010). *Estabilidad de estructuras: una perspective histrica (Stability of Structures: A Hystoric Perspective)* (in Spanish). Barcelona: "CIMNE" Publishers.

Godoy, L.A. (2011). Structural stability concepts in medeaval and renaissance mechanics, *Latin American Journal of Solids and Structures*, Vol. 8, 83–105.

Godoy, L.A. and Flores, F.G. (1993). Thickness changes in pressurized shells. *International Journal of Pressure Vessels and Piping*, 55, 451–459.

Goldenveizer, A.L. (1983). Geometric theory of stability of shells. *Mechanics of Solids*, 18(1), 135–145.

Goltzer, Ya.M. (1986). On a measure of the closeness of neutral systems to internal resonance. *Journal of Applied Mathematics and Mechanics*, 50(6), 731–737.

Goncharenko, V.M. (1963). Determination of the probability of loss of stability by a shell. *AIAA Journal*, 1(7), 1614–1618.

Goree, W.S. and Nash, W.A. (1962). Elastic stability of circular cylindrical shells stabilized by a soft elastic core. *Proceedings of the Society for Experimental Stress Analysis*, 19, 142–149.

Górski, J. and Mikulski, T. (2006). Statistical description and numerical calculations of cylindrical vertical tanks with initial geometric imperfections. In W. Pietraszkiewicz and C. Szymczak (Eds.), *Shell Structures: Theory and Applications* (pp. 547–553). London: Taylor & Francis.

Górski, J. and Mikulski, T. (2008). Identification and simulation of initial geometrical imperfections of steel cylindrical tanks. *Journal of Theoretical and Applied Mechanics* (Warsaw), 46(2), 413–434.

Graham, L.L. and Siragy, E. (1999). Stochastic finite element analysis of structures with random imperfections. In P.D. Spanos (Ed.), *Computational Stochastic Mechanics* (pp. 515–521). Rotterdam, The Netherlands: Balkema.

Grant, A. (2011). The giant crusher that can destroy NASA rockets. *Discover*, July–August.

Gray, R.M. and Davisson, L.D. (1974). Source coding theorems without the ergodic assumption. *IEEE Transactions on Information Theory*, IT-20(4), 502–516.

Green, D.R. and Neslon, H.M. (1982). Compressive tests on large-scale, stringer stiffened tubes. In P.J. Dowling and J.E. Harding (Eds.), *Buckling of Shells in Offshore Structures* (pp. 25–43). London: Granada.

Grigoliuk, E.I. and Kabanov, V.V. (1969). *Stability of Circular Cylindrical Shells* (in Russian). Moscow: "VINITI" Publishers.

Grigoliuk, E.I. and Kabanov, V.V. (1978). *Stability of Shells* (in Russian). Moscow: "Nauka" Publishing House.

Grigoliuk, E.I. and Lopatnitsyn, Y.A. (2004). *Finite Deflections, Stability and Postbuckling Behaviour of Thin Shallow Shells* (in Russian). Moscow: MGTU "MAMI" Publishers.

Grigoliuk, E.E. and Shalashilin, V.I. (1991). *Problems of Nonlinear Deformation: The Continuation Method Applied to Nonlinear Problems in Solid Mechanics*. Dordrecht: Kluwer.

Grime, A.J. and Langley, R.S. (2008). Lifetime reliability based design of offshore vessel mooring, *Applied Ocean Research*, Vol. 30, 221–234.

Gristchak, V.Z. (1976). Asymptotic formula for the buckling stress of axially compressed circular cylindrical shells with more or less localized short-wave imperfections. Report WHTD-88, Department of Mechanical Engineering, Delft Univeristy of Technology, Delft, The Netherlands.

Gristchak, V.Z. (1980). Asymptotic formula for critical stress of axially compressed cylindrical shells with load imperfections. *Prochnost' i Dolgovechnost' Konstrukzii (Strength and Durability of Strctures)* (in Russian; pp. 113–120). Institute of Mechanics, Dnepropetrovsk Branch. Kiev: "Naukova Dumka" Publishers.

Gristchak, V.Z. and Golovan, O.A. (2003). Hybrid asymptotic method for the effect of local thickness defects and initial imperfections on the buckling of cylindrical shells. *Journal of Theoretical and Applied Mechanics*, 41(3), 509–520.

Grove, T. and Didriksen, T. (1976). Buckling experiments on four large axially stiffened and one ring stiffened cylindrical shells. Report 76-432, Det Norske Veritas.

Guduru, P.R. and Xia, Z. (2007). Shell buckling of imperfect multi-walled carbon nanotubes — experiments and analysis. *Experimental Mechanics*, 47, 153–161.

Guggenberger, W., Greiner, R., and Rotter, J.M. (2000). The behavior of locally supported cylindrical shells: unstiffened shells. *Journal of Constructional Steel Research*, 56(2), 175–197.

Gulyaev, V.I., Zarutskii, V.A., and Sivak, E.F. (1987). On the effect of method of load application on the stability of cylindrical shells under axial compression. *Strength of Materials and Theory of Structures* (in Russian; Vol. 50, pp. 11–14). "Budival'nik" Publishers.

Guran, A. and Lebedev, L.P. (2011). Basic concepts in the stability theory of thin-walled structures. In H. Altenbach and V.A. Eremeyev (Eds.), *Shell-Like Structures: Non-Classical Theories and Applications* (pp. 135–152). Berlin: Springer.

Gürdal, Z. (2012). Private communication, June 13.

Gurav, S.P., Kasyap, A., Sheplak, L., Gaffaseta, L., Haftka, R.T., Goosen, J.F.L., and van Keulen, F. (2004). Uncertainty-based design of a micro-piezoelectric composite

energy reclamation drive. Proceedings, 10th AIAA/ISSMO Multidisciplinary Analysis and Optimization Conference (pp. 3559–3570).
Gusic, G. (1999). Buckling of cylindrical shells under external pressure effect of thickness imperfection. INSA de Lyon, France.
Gusic, G., Combescure, A., and Jullien, J.F. (1998). Buckling of thin cylindrical shells under external pressure-influence of localized thickness variations. In I.A. Allison (Ed.), *Experimental Mechanics: Advances in Design, Testing and Analysis, 11th International Conference on Experimental Mechanics*. Rotterdam, The Netherlands: Balkema.
Gusic, G., Combescure, A., and Jullien, J.F. (2000). The influence of circumferential thickness variations on the buckling of cylindrical shells under lateral pressure. *Computers & Structures*, 74, 461–477.
Guz, A.N., Zarutskii, N.A., and Amiro, I.Ya., et al. (Eds.) (1984). *Experimental Investigations of Thin-Walled Structures* (in Russian). Kiev: "Naukova Dumka" Publishers.
Haftka, R.T. (2006). Reflections on Jim Starnes' technical contributions. *Collected Papers in Structural Mechanics Honoring Dr. James H. Starnes, Jr., NASA TM-2006-214276* (pp. 1–10), NASA Langley Research Center, 2006.
Hamming, R. (2013). You and your research. University of Virginia, 1986. Retrieved on 10 October 2013. http://www.cs.virginia.edu/~robins/YouAndYourResearch.html.
Hansen, J.S. (1977). General random imperfections in the buckling of axially loaded cylindrical shells. *AIAA Journal*, 15(5), 1250–1256.
Hansen, J.S. and Roorda, J. (1974a) On a probabilistic stability theory for imperfection sensitive structures. *International Journal of Solids and Structures*, 10, 341–359.
Hansen, J.S. and Roorda, J. (1974b). Reliability of imperfection sensitive structures. In S.T. Ariaratnam and H.H.E. Leipholz (Eds.), *Stochastic Problems in Mechanics* (Study No. 10, pp. 229–242). Ontario, Canada: University of Waterloo Press.
Hao, P., Wang, B., Li, G., Tian, K., Du, K., Wang, X.J., and Tang, X.H. (2013). Surrogate-based optimization of stiffened shells including load-carrying capacity and imperfection sensitivity. *Thin-Walled Structures,* 72, 164–174.
Härdle, W. and Simar, L. (2007). *Applied Multivariate Statistical Analysis*. Berlin: Springer Verlag.
Harris, L.A., Suer, H.S., Skene, W.T., and Benjamin, R.J. (1957). The stability of thin-walled unstiffened circular cylinders under axial compression including the effects of internal pressure. *Journal of Aeronautical Sciences*, 24, 587– 596.
Harte, M.A. (1997). Some tools for probabilistic analysis using an imperfection database. Master's Thesis, Faculty of Aerospace Engineering, Delft University of Technology, Delft, The Netherlands, January.
Hasofer, A.M. and Lind, N.C. (1974). Exact and invariant second–moment code format. *Journal of Engineering Mechanics*, 100, 111–121.
Hassanien, M.A.M. and Loskot, P. (2009). Non-ergodic error-rate analysis of finite length received sequences. *Proceedings, International Conference on Wireless Communications & Signal Processing*, Nanking, People's Republic of China.
Haynie, W.T. and Hilburger, M.W. (2010). Comparison of methods to predict lower buckling loads of cylinders under axial compression, 51st AIAA/ASME/ASCE/

AHS/ASC Structures, Structural Dynamics, and Materials Conference, AIAA Paper 2010-2532.

Haynie, W.T. and Hilburger, M.W. (2012). Validation of lower-bound estimates for compression-loaded cylindrical shells. AIAA Paper 2012-1689.

van der Heijden, A.M.A. (2009). *W.T. Koiter's Elastic Stability of Solids and Structures.* Cambridge, UK: Cambridge University Press.

Hemez, F.M. and Ben-Haim, T. (2004). Info-gap robustness for the correlation of tests, *Mechanical Systems and Signal Processing*, 18(6), 1443–1467.

Herakovich, C. (1997). Report of the chair. *Applied Mechanics Newsletter*, ASME, Summer.

Herrmann, G. and Bungay, R.W. (1964). Shell buckling and nonconservative forces. *AIAA Journal*, 2(6), 1165–1166.

Hickley, C. (2013). Marcel Reich-Ranici, German literary critic. *Fort Lauderdale's SunSentinel,* September 19.

Hilburger, M.W. (2011). Facebook, SBKF, shell buckling knockdown factor, https://www.facebook.com/shellbucklingknockdownfactor.

Hilburger, M.W. (2012). Developing the next generation shell buckling design factors and technologies. 53rd AIAA/ASME/ASCE/AHS/ASC Structures, Structural Dynamics and Materials Conference, Paper AIAA-2012-1686.

Hilburger, M.W., Nemeth, M.P., and Starnes, J.H., Jr. (2004). Shell buckling design criteria based on manufacturing imperfection signatures. NASA/TM-2004-212659.

Hilfer, R., Fractional dynamics, irreversibility and ergodicity breaking. *Chaos, Solitons & Fractals*, 5(8), 1475–1484.

Hirano, Y. (1983). Optimization of laminated composite shells for axial buckling. *Transactions of the Japan Society for Aeronautical and Space Sciences*, 26(73), 154–162.

Hlaváček, I., Chleboun, J., and Babuška, I. (2004). *Uncertain Input Data Problems and the Worst Scenario Method.* Amsterdam: Elsevier.

Hoff, N.J. (1951). The dynamics of the buckling of elastic columns. *Journal of Applied Mechanics*, 18, 68–74.

Hoff, N.J. (1954). Buckling and stability: Forty-First Wilbert Wright memorial lecture. *Journal of the Aeronautcal Society*, 58, 3–52.

Hoff, N.J. (1961). Buckling of thin shells. Proceedings of an Aerospace Symposium of Distinguished Lecturers in Honor of Dr. Theodore von Kármán on this 80th Anniversary, The Institute of Aerospace Sciences, New York.

Hoff, N.J. (1966). The perplexing behavior of thin circular cylindrical shells in axial compression. *Israel Journal of Technology*, 4(1), 1–28.

Hoff, N.J. (1967). Thin shells in aerospace structures: 4th von Kármán Lecture. *Astronautics & Aeronautics*, 2, 26–45.

Hoff, N.J. (1969). Some recent studies of the buckling of thin shells. *The Aeronautical Journal*, 73(708), 1057–1070.

Hoff, N.J., Madsen, W.A., and Mayers, J. (1966). Post-buckling equilibrium of axially compressed circular cylindrical shells. *AIAA Journal*, 4, 126–133.

Hoff, N.J., Nardo, S.V., and Erickson, B. (1952). The maximum load supported by an elastic column in a rapid compression test. *Proceedings of the First U.S. National Congress of Applied Mechanics* (pp. 419–423), NewYork.

Hoff, N.J., Nardo, S.V., and Erickson, B. (1955). An experimental investigation of the process of buckling of columns. *Proceedings of the Society for Experimental Stress Analysis*, 13(1), 201–208.

Horák, J., Lord, G.J., and Peletier, M.A. (2006). Cylinder buckling: the mountain pass as an organizing center. *SIAM Journal of Applied Mathematics*, 66(5), 1793–1823.

Horton, W.H. (1977). On the elastic stability of shells, NASA CR-145088.

Horton, W.H. and Craig, J.I. (1968). Experimental observations on the instability of circular cylindrical shells. *Israel Journal of Technology*, 6(1–2), 101–116.

Horton, W.H. and Durham, S.C. (1965). Imperfections, a main contributor to scatter in experimental values of buckling load. *International Journal of Solids and Structures*, 1, 59–72.

Huff, D. (1993). *How to Lie with Statistics*. New York: W.W. Norton & Company.

Hühne, C. and Rolfes, R. (2002). Stabilitätsanalyse perfekter und imperfekter Kreiszylinderschalen aus Faserverbundwerkstoff (in German). *14 deutschsprachiges ABAQUS Anwendertreffen*, Wiesbaden, Germany, pp. 1–10.

Hühne, C., Rolfes, R., Breitbach, and Teßmer, J. (2008). Robust design of composite cylindrical shells under axial compression–simulation and validation. *Thin-Walled Structures*, 46(7–9), 947–962.

Hui, D. (1988). Postbuckling behavior of infinite beams on elastic foundation using Koiter's improved theory. *International Journal of Non-Linear Mechanics*, 23(2), 113–123.

Hunt, G.W. (1983). Elastic stability in structural mechanics and applied mathematics. In J.M.T. Thompson and G.W. Hunt G.W. (Eds.), *Collapse: The Buckling of Structures in Theory and Practice* (pp. 125–147). Cambridge: Cambridge University Press.

Hunt, G.W. (2002). Review: Non-Classical Problems in Theory of Elastic Stability, *Chemistry & Industry*, Issue 3, 20–21, Feb. 4.

Hunt, G.W. (2006). Buckling in space and time. *Nonlinear Dynamics*, 43, 29–46.

Hunt, G.W. (2011). Reflections and symmetries in space and time. *IMA Journal of Applied Mathematics*, 76, 2–26.

Hunt, G.W. (2012a). Personal communication, June 23.

Hunt, G.W. (2012b). Personal communication, August 31.

Hunt, G.W. (2012c). Personal communication, September 3.

Hunt, G.W. (2012d). Personal communication, 22 September.

Hunt, G.W. (2013). Personal communication, 15 October.

Hunt, G.W. and Blackmore, A. (1996). Principles of localized buckling for a strut on an elastoplastic foundation. *Journal of Applied Mechanics*, 63(1), 234–239.

Hunt, G.W., Bolt, H.M., and Thompson, J.M.T. (1989). Structural localisation phenomena and the dynamical phase-space analogy. *Proceedings of the Royal Society of London, A*, 4(25), 245–267.

Hunt, G.W., Lawther, R., and Costa, P.E. (1997). Finite element modeling of spatially chaotic structures. *International Journal for Numerical Methods in Engineering*, 40, 2237–2256.

Hunt, G.W., Lord, G.J., and Champneys, A.R. (1999). Homoclinic and heteroclinic orbits underlying the post-byckling of axially-compressed cylindrical shells. *Computer Methods in Applied Mechanics and Engineering*, 170, 239–251.

Hunt, G.W., Lord, G.J., and Peletier, M.A. (2003). Cylindrical shell buckling: a characterization of localization and periodicity. *Discrete and Continuous Dynamical Systems*, Ser.B., 3, 505–518.

Hunt, G.W. and Neto, E.L. (1991). Localized buckling in long axially loaded cylindrical shells. *Journal of Mechanics and Physics of Solids*, 39, 881–894.

Hunt, G.W. and Neto, E.L. (1993). Maxwell critical loads for axially loaded cylindrical shells. *Journal of Applied Mechanics*, 60, 702–706.

Hunt, G.W., Peletier, M.A., Champneys, A.R., Woods, P.D., Wadee, M.A., Budd, C.J., and Lord, G.J. (2000). Cellular buckling of long structures. *Nonlinear Dynamics*, 21, 3–29.

Hunt, G. and Virgin, L. (2013). Michael Thompson: some personal recollections. *Philosophical Transactions of the Royal Society A*, 371, number 1993, paper 20120449.

Hunt, G.W. and Wadee, M.K. (1991). Comparative Lagrangian formulations for localized buckling. *Proceedings of the Royal Society of London*, A434, 485–502.

Hunt, G.W., Wadee, M.K., and Shiacolas, N. (1993). Localized elasticae for the strut on the linear foundation, *Journal of Applied Mechanics*, 63(1), 1033–1038.

Hunt, G.W., Williams, K.A.J., and Cowell, R.G. (1986). Hidden symmetry concepts in the elastic buckling of axially-loaded cylinders. *International Journal of Solids and Structures*, 22(12), 1501–1515.

Hussain, K.S. (1996). Probabilistic analysis of the stability of imperfection sensitive arch and shell structures, Ph.D. Thesis, Rice University, Houston, TX.

Hutchinson, J.W., Private communication to R.C. Tennyson and D.B. Muggeridge, Harvard University, 1969 (see Muggeridge D.B., 1969, Ref. 27, and Tennyson R.C. and Muggeridge D.B., 1969, Ref. 11).

Hutchinson, J.W. (2010). Knockdown factors for buckling of cylindrical and spherical shells subject to reduced biaxial membrane stress. *International Journal of Solids and Structures*, 47, 1443–1448.

Hutchinson, J.W. (2011). Personal communication, January 19.

Hutchinson, J.W. (2012). Personal communication, June 5.

Hutchinson, J.W. (2013). The role of nonlinear substrate elasticity in the wrinkling of thin films. *Philosophical Transactions of the Royal Society A: Mathematical, Physical & Engineering Sciences*, 371, number 1993, paper 20120422.

Hutchinson, J.W. and He, M.Y. (2000). Buckling of cylindrical sandwich shells with metal foam cores. *International Journal of Solids and Structures*, 37, 6777–6794.

Hutchinson, J.W. and Koiter, W.T. (1970). Postbuckling theory. *Applied Mechanics Reviews*, 23(12), 1353–1366.

Hutchinson, J.W., Tennyson, R.C., and Muggeridge, D.B. (1971). Effect of local axisymmetric imperfection on the buckling behavior of a circular cylindrical shell under axial compression. *AIAA Journal*, 9(1), 48–52.

Ikeda, K. and Murota, K. (1991). Random initial imperfections of structures. *International Journal of Solids and Structures*, 28(8), 1003–1021.

Ikeda, K. and Murota, K. (2002). *Imperfect Bifurcation in Structures and Materials: Engineering Use of Group-Theoretic Bifurcation Theory*. Berlin: Springer.

Ikeda, K., Murota, K., and Elishakoff, I. (1996). Reliability of structures subject to normally distributed initial imperfections. *Computers & Structures*, 59, 463–469.

Il'in, V.A., Ovcharov, N.P., and Sukallo, A.A. (1975). Statistical prognosis of critical loads in stability analysis of imperfect cylindrical shells (in Russian). *Mechanics of Solids*, 5, 191–194.

Imbert, K. (1971). The effect of imperfections on the buckling of cylindrical shells. Aeronautical Engineer Thesis, California Institute of Technology, Pasadena, CA, June.

Ingerle, K. (2013). Stability of massless non-conservative systems, *Journal of Sound and Vibration*, 332(19), 4529–4540.

Ivanova, J. and Pastrone, F. (1988). Buckling of cylindrical plastic shells under axial compression. *Bolletino U.M.I.*, 2-B(7), 529–540.

Ivanova, J. and Pastrone, F. (2002). *Geometric Methods for Stabiliy of Nonlinear Elastic Thin Shells*. Dordrecht: Kluwer Academic.

Ivanova, J. and Trendafilova, I. (1992). A stochastic approach to the problem of stability of a spherical shell with initial imperfections. *Probabilistic Engineering Mechanics*, 7(4), 227–223.

Jabareen, M. and Sheinman, I. (2009). Dynamic buckling of a beam on a nonlinear elastic foundation under step loading. *Journal of Mechanics of Materials and Structures*, 4(7), 1365–1373.

Jamal, M., Lahlou, L., Midani, Zahrouni, H., Limam, A., Damil, N., and Potier-Ferry, M. (2003). A semi-analytical buckling analysis of imperfect cylindrical shells under axial compression. *International Journal of Solids and Structures*, 40, 1311–1327.

James, R.D. (1979). Co-existent phases in the one-dimensional static theory of elastic bars. *Archive for Rational Mechanics and Analysis*, 72, 99–140.

Jia, X. and Mang, H.A. (2010). Conversion of imperfection-sensitive elastic structures into imperfection-insensitive ones by adding tensile members. Proceedings of the International Association for Shell and Spatial Structures (IASS) Symposium, Spatial Structures — Permanent and Temporary, November 8–12.

Johns, K.C. (1976). Some statistical aspects of coupled buckling structures. In B. Budiansky (Ed.), *Buckling of Structures* (pp. 172–182). Berlin: Springer Verlag.

Jones, R.M. (2006). *Buckling of Bars, Plates, and Shells*. Blacksbourg, VA: Bull Ridge Publishing.

Ju, G.T. and Kyriakides, S. (1992). Bifurcation and localization instabilities in cylindrical shells under bending — 2. Predictions. *International Journal of Solids and Structures*, 29(9), 1143–1171.

Judt, T. (2012). *Thinking the Twentieth Century* (p. 396). London: William Heinemann.

Kafka, F. (1967). *Briefe an Felice* (in German). E. Heller and J. Born (Eds.), Frankfurt am Main, Federal Republic of Germany.

Kala, Z. (2007). Stability problems of steel structures in the presence of stochastic and fuzzy uncertainty. *Thin-Walled Structures*, 45 (10–11), 861–865.

Kalman, R.E. (1995a). Randomness and probability. *Mathematica Japonica*, 41, 41–58.

Kalman, R.E. (1995b). Addendum to "Randomness and probability." *Mathematica Japonica*, 41(2), 463.

Kalman, R.E. (1996). Probability in the real world as a system attribute. *CWI Quarterly*, 913, 181–204.

Kan, S.N. (1963). Towards the question of stability of circular cylindrical shells in compression (in Russian). *Structural Mechanics and Analysis of Constructions*, 6, 31–34.

Kan, S.N. (1964). Load carrying capacity of circular cylindrical shells under compression (in Russian). Theory of Shells and Plates, Proceedings of the Fourth All-Union Conference (pp. 489–494), Armenian Academy of Sciences Press, Erevan, Armenia.

Kan, S.N. (1996a). Load-carrying capacity of circular cylindrical shells at compression. In S.M. Durgar'yan (Ed.), *Theory of Shells and Plates* (pp. 444–449). Jerusalem: Israel Program for Scientific Translations.

Kan, S.N. (1996b). Stability of isotropic and orthotropic shells with initial imperfections. *Theory of Shells and Plates, Proceedings of the Sixth All-Union Conference* (in Russian; pp. 919–926). Moscow: "Nauka" Publishers.

Kanemitsu, S. and Nojima, N.-M. (1939). Axial compression tests of thin circular cylinder. M.Sc. Thesis, California Institute of Technology, Pasadena, CA.

Kaoulla, P. (1977). Comment on "Localized diamond-shaped buckling patterns of axially compressed cylindrical shells." *AIAA Journal*, 15(5), 757.

Kaplan, Yu.I., Kan, N.S., and Shternis, A.Z. (1974). Stability of orthotropic cylindrical shells with variable thickness. *International Applied Mechanics*, 10(5), 129–132.

Karam, G.N. and Gibson, L.J. (1995a). Elastic buckling of cylindrical shells with elastic cores, I. Analysis. *International Journal of Solids and Structures*, 32, 1259–1263.

Karam, G.N. and Gibson, L.J. (1995b). Elastic buckling of cylindrical shells with elastic cores, II. Experiments. *International Journal of Solids and Structures*, 32, 1285–1306.

Karaulov, D.S. (1991). Critical load as a function Poisson's ratio in axially compression of a cylindrical shell. *Leningrad University Mechanics Bulletin*, 4, 10–13.

von Kármán, T. (1910). Untersuchungen über Knickfestigkeit (in German), *Mitteilungen über Forschungsarbeiten, Verein Deutscher Ingenieure*, 81.

von Kármán, T., Dunn, L.G., and Tsien, H.S. (1940). The influence of curvature on the buckling characteristics of structures. *Journal of Aeronautical Sciences*, 7, 276–281.

von Kármán, T. and Tsien, H.S. (1939). The buckling of spherical shells by external pressure. *Journal of Aeronautical Sciences*, 7, 43–50.

von Kármán, T. and Tsien, H.S. (1941). Buckling of thin cylindrical shells under axial compression. *Journal of Aeronautical Sciences*, 8, 302–335.

Khamlichi, M., Bezzazi, A., and Limam, A. (2004). Buckling of elastic cylindrical shells considering the effect of localized axisymmetric imperfections. *Thin-Walled Structures*, 42, 1035–1047.

Khelil, A. (2002). Buckling of steel shells subjected to non-uniform axial and pressure loadings. *Thin-Walled Structures*, 40, 955–970.

Kil'chevskii, N.A. (1942). On the axisymmetric deformations and elastic stability of a circular tube under action of longitudinal compressive forces (in Russian). *Applied Mechanics and Mathematics*, 6(6), 497–508.

Kil'chevskii, N.A. and Nikulinskaya, S.N. (1965). About axisymmetric form of stability loss of circular cylindrical shells (in Russian). *Applied Mechanics and Mathematics*, 1(11), 1–6.

Kim, S.-E. and Kim, C.-S. (2002). Buckling strength of the cylindrical shell and tank subjected to axially compressive loads. *Thin-Walled Structures*, 40, 329–353.

Kirkpatrick, S.W. and Holmes, B.S. (1989). Effect of initial imperfections on dynamic buckling of shells. *Journal of Engineering Mechanics*, 115(5), 1075–1093.

Kirste, L. (1954). Abwickbare verformung duennwandiger Kreiszylinder (in German). *Osterreichische Ingenieur-Archiv*, 8(2–3), 149–151.

Kleijnen, J.P.C. (2009). Kriging metamodeling in simulation: a review. *European Journal of Operational Research*, 192, 707–716.

Klompé, A.W.H. (1988). UNIVIMP: A universal instrument for the survey of initial imperfections of thin-walled shells. Report LR-570, Faculty of Aerospace Engineering, Delft University of Technology, Delft, The Netherlands, December.

Knight, N.F., Jr. and Starnes, J.H., Jr. (1998). Developments in cylindrical shell stability analysis. In N.F. Knight, Jr. and M.P. Nemeth (Eds.), *Stability Analysis of Plates and Shells: A Collection of Papers in Honor of Dr. Manuel Stein* (pp. 31–46). NASA/CP-1998-206280.

Koiter, W.T. (1945). Over de stabiliteit van het elastisch evenwicht (On the stability of elastic equilibrium) (in Dutch), Thesis, Technische Hooge School, Delft, The Netherlands, H.J. Paris, Amsterdam, 1945 (in Dutch) (English translation, (a) NASA TFF-10,833, 1967; Technical Report No. AFFDL-TR-70-25, 1970) (English translation is available online: http://imechanica.org/node/1400).

Koiter, W.T. (1963). Elastic stability and postbuckling behavior. In R.E. Langer (Ed.), *Proceedings of the Symposium on Nonlinear Problems* (pp. 257–275). Madison: University of Wisconsin Press.

Koiter, W.T. (1963). The effect of axisymmetric imperfections on the buckling of cylindrical shells under axial compression. *Koninklijke Nederlandse Academie van Wetenschappen (Proceedings of the Royal Netherlands Academy of Sciences)*, B66, 265–279.

Koiter, W.T. (1996a). Postbuckling analysis of a simple two-bar frame. *Recent Progress in Applied Mechanics* (pp. 337–354). Stockholm: Ahmquist and Wicksel.

Koiter, W.T. (1996b). Purpose and achievements of research in elastic stability. *Recent Advances in Engineering Science* (pp. 197–218). Raleigh, NC: Caroline State University.

Koiter, W.T. (1974). The influence of more or less localized short-wave imperfections on the buckling of circular cylindrical shells under axial compression (in first approximation). Report 534, Department of Technical Mechanics, Faculty of Mechanical Engineering, Delft University of Technology, May.

Koiter, W.T. (1976a). Current trends in the theory of buckling. In B. Budiansky (Ed.), *Buckling of Structures* (pp. 1–16). Berlin: Springer, Berlin.

Koiter, W.T. (1976b). General theory of mode interaction in stiffened plate and shell structures. Report WTHD 91, Delft University of Technology, the Netherlands.

Koiter, W.T. (1978). The influence of more or less localized imperfections on the buckling of circular cylindrical shells under axial compression. *Kompleksnyi Analiz I Ego Prilozheniya (Complex Analysis and Its Applications), Collection of Papers Dedicated to 70th Birtch Anniversary of I.N. Vekua* (pp. 242–244). Moscow: Nauka Publishing House.

Koiter, W.T. (1979). Forty years of retrospect, the bitter and the sweet. In J.F. Besseling and A.M.A. van der Heijden (Eds.), *Trends in Solid Mechanics* (pp. 237–246). Delft, The Netherlands: Delft University Press, Sijthoff & Noordhoff International Publishers.

Koiter, W.T. (1985). Elastic stability, 28th Ludwig Prandtl Memorial Lecture. *Zeitschrift für Flugwissenschaften und Weltraumforschung*, 9, 205–210.

Koiter, W.T. (1996). Unrealistic follower forces. *Journal of Sound and Vibration*, 194, 636–638.

Koiter, W.T., Elishakoff, I., Li, Y.W., and Starnes, J.H., Jr. (1994a). Buckling of an axially compressed cylindrical shell of variable thickness. *International Journal of Solids and Structures*, 31, 797–805.

Koiter, W.T., Elishakoff, I., Li, Y.W., and Starnes, J.H., Jr. (1994b). The combined effect of the thickness variation and axisymmetric initial imperfection on the buckling of the isotropic cylindrical shell under axial compression. Proceedings, 35th AIAA/ASME/ASCE/AHS/ASC Structural Dynamics and Materials Conference (pp. 277–289), Hilton Head, SC.

Kolanek, K. and Jendo, S. (2008). Random field models of geometrically imperfect structures with "clamped" boundary conditions. *Probabilistic Engineering, Mechanics,* 23(2–3), 219–226.

Kolesnikov, M. and Schmidt, R. (2011). On the numerical analysis of the stress–strain state of a cylindrical shell under non-axial compression. In W. Szczesniak (Ed.), *Theoretical Foundations of Civil Engineering — Polish-Ukrainian Transactions* (Vol. 19, pp. 113–122), Warsaw.

Koryo, M. (1969). Proposition of pseudocylindrical concave polyhedral shells. University of Tokyo, Institute of Space and Aeronautical Sciences, Report 442, Vol. 34(9), 141–163.

Kornev, V.M. (1969). On the instability modes of elastic shells under intensive loading (in Russian). *Mekhanika Tverdogo Tela (Mechanics of Solids)*, 2, 129–135.

Kornev, V.M. (1972). On approximations in stability and vibrations problems of thin elastic shells upon condensation of eigenvalues. *Mekhanika Tverdogo Tela (Mechanics of Solids)*, 2, 119–129.

Kornev, V.M. (1975). On the solution of shell stability problems taking into account the eigenvalue density. *Teoriya Plastin i Obolochek* (*Theory of Plates and Shells)* (in Russian; pp. 132– 134). Leningrad: "Sudostroenie" Publishers.

Kornev, V.M. (1976). Specifics of the instability problem of thin-walled shells. *Dynamics of Continuous Medium* (in Russian), Vol. 25, 61–74. Novosibirsk: Institute of Hydrodynamics.

Kornev, V.M. (1980). Optimization providing structural instability in connection with the density of eigenvalues. In A. Bosznay (Ed.), *Bracketing of Eigenfraquences of Continuous Structures* (pp. 261–271). Budapest: Académiai Kiado.

Kornev, V.M. and Ermolenko, V.M. (1980). Sensibility of shells to buckling disturbances in connection with parameters of critical loading spectrum. *International Journal of Engineering Science*, 18(2), 379–388.

Kotz, S. and Nadarajah, S. (1999). *Extreme Value Distributions Theory and Applications*. London: Imperial College Press.

Kounadis, A.N. (1993). Nonlinear dynamic buckling of discrete structural systems under impact loarding. *International Journal of Solids and Structures*, 30(21), 2895–2909.

Kounadis, A.N., Mallis, J., and Raftoyiannis, J. (1991). Dynamic buckling estimates for discrete systems under step loading. *ZAMM — Zeitschrift für angewandte Mathematic and und Mechanik*, 71 (10), 391–402.

Kounadis, A.N., Mallis, J., and Sbarounis, A. (2006). Postbuckling analysis of columns resting on elastic foundations. *Archive of Applied Mechanics*, 75, 395–404.

Kounadis, A.N., Raftoyiannis, J., and Mallis, J. (1989). Dynamic buckling of an arch model under impact loading. *Journal of Sound and Vibration*, 134(2), 193–202.

Koz, V.M. and Lipovskii, D.E. (1966). Experimental investigation of stability of cylindrical shells with initial perturbation taken into account. Proceedings, Sixth All-Union Conference on Theory of Shells and Plates. Moscow: Baku, "Nauka" Publishing House.

Krasovsky, V.L. (1990). Behavior and stability of compressed thin-walled cylinders with local imperfections (in Russian). Proceedings, XV All-Union Conference on Theory of Shells and Plates (pp. 303–308).

Krasovsky, V.L. (1994). Nonlinear effects in the behavior of cylindrical shells under nonuniform axial compression. Experimental results. The 25th Israel Conference on Mechanical Engineering (pp. 623–625), Haifa, Israel.

Krasovsky, V.L. (2001). Dangerous loads and stability safety factor for axially compressed thin-walled cylinders (in Russian). *Perspective Problems of Engineering Science* (pp. 177–188). Gaudeamus.

Krasovsky, V.L. (2002). Quality of thin-walled cylinders and starting mechanism of their buckling under axial loading (in Russian). *Theoretical Foundations of Civil Engineering*, 2, 696–715.

Krasovsky, V.L. (2012a). Experimental investigation of bucking of compressed cylindrical shells (quality of shells and mechanisms of buckling). In Z. Kolakowski and K. Kowal-Michalska (Eds.), *Statics, Dynamics and Stability of Structural Elements and Systems* (pp. 447–476). Lodz, Poland: Lodz University of Technology, Lodz, Poland.

Krasovsky, V.L., Kolesnikov, M., and Schmidt, R. (2008). The effect of static resonance in cylindrical shells under non-homogeneous loading (experimental and numerical research). *Theoretical Foundations of Civil Engineering*, 16, 189–200.

Krasovsky, V.L., Marchenko, V., and Schmidt, R. (2011). Deformation and buckling of axially compressed cylindrical shells with local loads in numerical simulation and experiments. *Thin-Walled Structures*, 19, 576–580.

Kriegesmann, B. (2012b). Probabilistic design of thin-walled fiber composite structures. Ph.D. Dissertation, Institute of Structural Analysis, Leibniz Universität, Hannover, Federal Republic of Germany, May.

Kriegesmann, B., Hilburger, M., and Rolfes, R. (2012). The effects of geometric and loading imperfections on the response and lower-bound buckling load of a compression-loaded cylindrical shell. AIAA Paper 2012-1864.

Kriegesmann, B., Jansen, E.L., and Rolfes, R. (2011). Semi-analytical probabilistic analysis of axially compressed stiffened composite panels. In A.J.M. Ferreira (Ed.), 16th International Conference on Composite Structures, Porto, Portugal.

Kriegesmann, B., Jansen, E.L., and Rolfes, R. (2012). Semi-analytical probabilistic analysis of axially compressed stiffened composite shells. *Composite Structures*, 94(2), 654–663.

Kriegesmann, B., Rolfes, R., Hühne, C., and Kling, A. (2010). Probabilistic second-order third-moment approach for design of axially compressed composite shells. Proceedings, 51st AIAA/ASME/ASCE/AHS/ASC Structures, Structural, Dynamics, and Materials Conference, AIAA 2010-2534, Orlando, FL.

Kriegesmann, B., Rolfes, R., Hühne, C., and Kling, A. (2011). Fast probabilistic design procedure for axially compressed composite shells. *Composite Structures*, 93(12), 3140–3149.

Kriegesmann, B., Rolfes, R., Hühne, C., Teßmer, J., and Arbocz, J. (2010). Probabilistic design of axially compressed composite cylinders with geometric and loading imperfections. *International Journal of Structural Stability and Dynamics*, 10(4), 623–644.

Kriegesmann, B., Rolfes, R., Jansen, E.L., Elishakoff, I., Hühne, C., and Kling, A. (2011). Design optimisation of composite cylindrical shells under uncertainty. Composites 2011.

Kukudzhanov, S.N. (2009). On the stability of nearly cylindrical long shells of revolution. *Mechanics of Solids*, 44(4), 543–551.

Kunzevich, V.M. and Lychak, M. (1992). *Guaranteed Estimates: Adaptation and Robustness in Control Systems*. Berlin: Springer.

Kurer, K.-E. (2008). *The History of the Theory of Structures: From Arch Analysis to Computational Mechanics* (pp. 741–742). Berlin: Ernst & Sohn Verlag.

Kuznetsov, V.K. and Lipovtsev, Yu.V. (1970). Influence of load imperfactions on stability of cylindrical shells under axial compression (in Russian). *Mechanics of Solids*, 1, 134–136.

Kyriakides, S. and Ju, G.T. (1992). Bifurcation and localization instabilities in cylindrical shells under bending — 1. Experiments. *International Journal of Solids and Structures*, 29(9), 1117–1142.

Lagaros, N.D. and Papadopoulos, V. (2006). Optimum design of shell structures with random geometric, material and thickness imperfections. *International Journal of Solids and Structures*, 43(22–23), 6948–6964.

Lagrange, R. and Averbuch, D. (2012). Solution methods for the growth of a repeating imperfection in the line of a strut on a nonlinear foundation. *International Journal of Mechanical Sciences*, 6, 48–58.

Lancaster, E.R., Calladine, C.R., and Palmer, S.C. (2000). Paradoxical buckling behavior of a thin cylindrical shell under axial compression. *International Journal of Mechanical Sciences*, 42, 842–865.

Laura, P.A.A., Rossit, C.A., and Bambill, D.V. (2002). Book review, "*Nonclassical Problems in the theory of Elastic Stability*." *Ocean Engineering*, 30, 151–152.

Lebowitz, J.L. and Penrose, O. (1973). Modern ergodic theory. *Physics Today*, 2, 1–7.

Lechner, B. and Pircher, M. (2005). Analysis of imperfection measurements of structural members. *Thin-Walled Structures*, 43(3), 351–374.

Lee, M.C.W., Mikulik, Z., Kelly, D.W., Thomson, R.S., and Degenhardt, R. (2010). Robust design — a concept for imperfection insensitive composite structures. *Composite Structures*, 92, 1496–1477.

Lee, S.H. and Waas, A.M. (1996). Initial post-buckling behavior of a finite beam on an elastic foundation. *International Journal of Non-Linear Mechanics*, 31, 313–328.

Leibovich, Y. (1992). Philosophical essay. *Makhshavot (Thoughts)* (in Hebrew).

Leiman, S. (1993). Dwarfs on the shoulders of giants, available at the website www.jewishlink.com (accessed on 13 Jan. 2014).

Leizerakh, V.M. (1969). Statistical analysis of random imperfections in cylindrical shells with the aid of computers (in Russian). *Proceedings of the Scientific-Technical Conference* (pp. 45–55). Moscow Power Engineering Institute, Section of Energetic Machines, Subsection of Dynamics and Strength of Machines.

Leizerakh, V.M. and Makarov, B.P. (1973). Correlation statistical analysis of deformation of compressed shells with initial imperfections (in Russian). *Proceedings of the Eighth All-Union Conference on Plates and Shells* (pp. 320–324). Moscow: "Nauka" Publishers.

Leizerakh, V.M. and Sekletov, S.V. (1977). Experimental investigation of the stability of fiberglass cylindrical shells subject to prolonged application of twisting moments. *Mechanics of Composite Materials*, 18(6), 692–696.

Li, C.-X. and Li, J.-H. (2011). Simulation of non-Gaussian random field for boundary in-plane imperfection. *Advenced Science Letters,* 4(3), 943–947.

Li, L.U. (2010). Stability analysis of Nantong Sports and exhibition center based on measured geometrical imperfection. *Advanced Materials Research*, 163–167, 426–432.

Li, Y.W., Elishakoff, I., and Starnes, J.H., Jr. (1995a). Buckling mode localization in multi-span periodic structure with a disorder in a single span. *Chaos, Solitons and Fractals*, 5, 955–969.

Li, Y.W., Elishakoff, I., and Starnes, J.H., Jr. (1995b). Effect of the thickness variation and initial imperfection on buckling of composite cylindrical shells. *International Journal of Solids and Structures*, 56, 65–74.

Li, Y.W., Elishakoff, I., Starnes, J.H., Jr., and Bushnell, D. (1997). Effect of the thickness variation and initial imperfection on buckling of composite cylindrical shells: asymptotic analysis and numerical results by BOSOR4 and PANDA2. *International Journal of Solids and Structures*, 34, 3755–3767.

Li, Y.W., Elishakoff, I., Starnes, J.H., Jr., and Shinozuka, M. (1995). Nonlinear buckling of a structure with random imperfection and random axial compression by a conditional simulation technique. *Computers & Structures*, 56(1), 59–64.

Li, Y.W., Elishakoff, I., Starnes, J.H., Jr., and Shinozuka, M. (1996). Prediction of natural frequency and buckling load variability due to uncertainty in material properties. *Fields Institute Communications*, 9, 139–154.

Lilly, W.E. (1908). The design of struts. *Engineering*, 85, 37–40.

Limam, A., Jalal, E.B., Adelatif, K., and Larbi, E.B. (2011). Effect of multiple localized geometric imperfections on stability of thin axisymmetric cylindrical shells under axial compression. *International Journal of Solids and Structures*, 48, 1034–1043.

Lindberg, H.E. (1991). Dynamic response and buckling failure measures for structures with bounded and random imperfections. *Journal of Applied Mechanics*, 58, 1092–1095.

Lindberg, H.E. (1992a). An evaluation for convex modeling for multimode dynamic buckling. *Journal of Applied Mechanics*, 59, 923–936.

Lindberg, H.E. (1992b). Convex models for uncertain imperfection control in multimode dynamic buckling. *Journal of Applied Mechanics*, 59, 937–945.

Lindley, D.V. (1987). The probability approach to the treatment of uncertainty in artificial intelligence and expert systems. *Statistical Science*, 2, 17–29.

Lipovtsev, Yu.V. (1968). Stability of viscoelastic and elastic shells in the presence of local stress. *Mechanics of Solids*, 3(5), 198–204.

Lipvaskii, D.E., Altukher, G.M., Koz, V.M, Nazarov, V.A., Todchuk, V.A., and Shun, V.M. (1977). Statistical estimate of the effect of random disturbances of the stability of stiffened shells based on experimental data (in Russian). *Raschet Prostranstvennykh Konstuktsii (Analysis of Space Structures)* (Vol. 17, pp. 32–44). Moscow: "Stroiizdat" Publishers.

Lizin, V.T. and Pyatkin, V.A. (1985). *Design of Thin-Walled Structures* (in Russian). Moscow: "Mashinostroenie" Publishers.

Lockhart, D. and Amazigo, J.C. (1975). Dynamic buckling of externally pressurized imperfect cylindrical shells. *Journal of Applied Mechanics,* 42, 316–320.

Lopanitsyn, Ye.A. and Frolov, A.B. (2008). Stability of round cylindrical shells under uniform compression. *Izbrannye Problemy Prochnosti Sovremennogo Mashinostroeniya, Posviashennyi 85-Letiu Chlena-Korrespondenta RAN E.I. Grigoliuka (Some Problems of Durability of the Modern Machinery Manufacturing, Collection of Scientific Papers Dedicated to 85th Anniversary of Corresponding Member of RAS E.I. Grigoliuk)* (in Russian; pp. 176–189). Moscow: Nauka" Publishing House.

Lopanitsyn, Ye.A. and Matveyev, Ye.A. (2001a). The possibility of a theoretical confirmation of the experimental values of the external critical pressure of thin-walled cylindrical shells. *Journal of Applied Mathematics and Mechanics*, 75, 580–588.

Lopanitsyn, Ye.A. and Matveyev, Ye.A. (2001b). Effect of methods for fixing and loading of cylindrical shells on their stability and supercritical behavior. *Journal of Machinery, Manufacture and Reliability,* 40(3), 117–126.

Lopanitsyn, Ye.A. and Matveyev, Ye.A. (2012). A modification of the method of continuous prolongation for finding the bifurcation solutions of steady-state self-adjoint boundary value problems. *Journal of Applied Mathematics and Mechanics*, 76, 715–721.

Lord, G.J., Champneys, A.R., and Hunt, G.W. (1997). Computation of localized post buckling in long axialy compressed cylindrical shells. *Philosophical Transactions of the Royal Society of London, A*, 355(1732), 2137–2150.

Lorenz, R. (1908). Achsensymmetrische verzerrungen in dünnwandigen hohlzylinder (in German). *Zeitschrift Vereinigung Deutscher Ingrieure*, 52, 1766–1793.

Lorenz, R. (1911). Die nicht Achsensymmetrische Knickung Dünnwandiger Hohlzylinder (in German). *Physikalische Zeitschrift Leipzig*, 13, 241–260.

Lukoshevichius, R.S., Rikards, R.B., and Teters, G.A. (1977). Probabilistic analysis of stability and mass minimization of composite cylindrical shells with random initial imperfections (in Russian). *Mekhanika Polimerov (Polymer Mechanics)*, 1, 80–89.

Lukoshevichius, R.S., Rikards, R.B., and Teters, G.A. (1997). Probability analysis of the stability and minimization of the mass of cylindrical shells made of composite material having random initial imperfections. *Mekhanica Polimerov (Polymer Mechanics)*, 13(1), 76–85.

Lundquist, E.E. (1933). Strength tests of thin-walled duralumin cylinders in compression. NACA TR-473.

Lundquist, E.E. (1935). Strength tests of thin-walled duralumin cylinders in combined transverse shear and bending. NAIA Technical Note, No. 523.

Luongo, A. (1992). Mode localization by structural imperfections in one-dimensional continuous systems. *Journal of Sound and Vibration*, 155(2), 249–271.

Luongo, A. (1993). On the amplitude modulation and localization phenomena in interactive buckling problems. *International Journal of Solids and Strcutures*, 27, 1943–1954.

Luongo, A. (2001). Mode localization in dynamics and buckling of linear imperfect continuous structures. *Nonlinear Dynamics*, 25, 133–156.

Mack, R.P. (1971). *Planning and Uncertainty: Decision Making in Business and Government Administration*. New York: Wiley 1971.

Maiboroda, A.L. (1993). Buckling of convex shells under nonaxisymetric loading. In R. Vaillancourt and A.L. Smirnov (Eds.), *Asymptotis Methods in Mechanics* (pp. 217–226). Providence, RI: American Mathematical Society.

Makarov, B.P. (1962a). Application of statistical method for analysis of nonlinear shell stability problems. *Theory of Plates and Shells: Proceeding of the 2nd All-Union Conference* (in Russian; pp. 363–367). Kiev: Ukrainian SSR Academy of Sciences Press.

Makarov, B.P. (1962b). Application of statistical method for analysis of experimental data on shell stability (in Russian). *Izvestiya Akademic Nauk SSSR, Otdelenie Tekhnicheskikh Nauk, Mekhanika I Mashinostoenie (Proceeding of the USSR Academy of Sciences, Section of Technical Sciences, Mechanics and Mechanical Engineering)*, 1, 157–158.

Makarov, B.P. (1963). Analysis of nonlinear shell stability problems with the aid of statistical methods (in Russian). *Inzhenernyi Zhurnal (Engineering Journal)*, 3(1), 100–106.

Makarov, B.P. (1969). Statistical analysis of stability of imperfect cylindrical shells (in Russian). Proceeding of the 7th All-Union Conference on the Theory of Plates and Shells (pp. 387–391). Moscow: "Nauka" Publishing House.

Makarov, B.P. (1970). Statistical analysis of nonideal cylindrical shells (in Russian). *Izvestiya Akademii Nauk SSSR, Mekhanika Tverdogo Tela*, 1, 97–104 (English translation: *Mechanics of Solids*, 5(1), 86–91, 1970).

Makarov, B.P. (1971). Statistical analysis of deformation of imperfect cylindrical shells (in Russian). *Raschety na Prochnost* (*Strength Analysis*), 15, 240–256.

Makarov, B.P. (1978). Statistical analysis of stability of imperfect cylindrical shells. Foreign Technology Division, Wright-Potterson AFB, Ohio, No. ADA066679.

Makarov, B.P. (1983). *Nelineinye Zadachi Statisticheskoi Dinamiki Mashin i Priborov (Nonlinear Problems of Statistical Dynamics of Machine and Devices)* (in Russian). Moscow: "Mashinastroenie" Publishers.

Makarov, B.P. and Leizerakh, V.M. (1969). Towards the question of cylindrical shell stability with random initial imperfections (in Russian). *Proceedings of the Scientific-Technical Conference* (pp. 35–44), Moscow Power Engineering Institute, Section of Energetic Machines, Subsection of Dynamics and Strength of Machines.

Makarov, B.P. and Leizerakh, V.M. (1970). About stability of nonideal cylindrical shell under axial compression (in Russian). In V.V. Bolotin (Ed.), *Proceedings of the Moscow Power Engineering Institute, Dynamics of Stength of Machines*, 74, 24–29.

Makarov, B.P. and Leizerakh, V.M. and Sudakova, N.I. (1968). Statistical investigation of initial imperfections in cylindrical shells. In V.V. Bolotin and A. Čyras (Eds.), *Reliability Problems in Structural Mechanics*, Proceedings of the Second All-Union Conference on Reliability Problems in Structural Mechanics (pp. 175–182), Vilnius, Lithuania.

Mallock, A. (1908). Note on the instability of tubes subjected to end pressure and on the folds in a flexible material. *Proceedings, Royal Society,* 81(A-549), 388–393.

Mamai, V.I. (2011). Load-carrying capacity of thin-walled shells with local imperfections. *Mechanics of Solids*, 46(2), 179–183.

Mandal, P. and Calladine, C.R. (2000). Buckling of thin cylindrical shells under axial compression. *International Journal of Solids and Structures*, 37, 4509–4525.

Manevich, L.I., Milzin, A.M., Mossakovskii, V.I., Prokopalo, E.F., Smelyi, G.N., and Sotnikov, D.I. (1968). Experimental investigation of stability of cylindrical shells of various scales (in Russian). *Mekhanika Tverdogo Tela (Mechanics of Solids)*, 5.

Manevich, L.M., Mossakovskii, V.I., and Prokopalo, E.E. (1975). Experimental investigation of transcritical behavior of shells. *Mechanics of Solids*, 10(1), 145–150.

Manevich, L.I. and Prokopalo, E.F. (1973). About statistical properties of load carrying capactiy of cylindrical shells (in Russian). In V.V. Bolotin and A. Čyras (Eds.), *Reliability Problems in Structural Mechanics*, Proceedings of the Second All-Union Conference on Reliability Problems in Structural Mechanics (pp. 182–187), Vilnius, Lithuania (also Document AD076 0946, Foreign Technology Division, Wright-Patterson AFB, Ohio, 16 May 1973).

Mang, H.A. (2012). Comparison of loss of stability with oncology. Personal communication, February 16.

Mang, H.A., Schranz, C., and Mackenzie-Helnwein, P. (2006). Conversion from imperfection-sensitive into imperfection-insensitive elastic structures I: Theory. *Computer Methods in Applied Mechanics and Engineering,* 195, 1422–1457.

Mangelsdorf, C.P. (1972). Koiter's buckling mode for short cylindrical shells. *AIAA Journal,* 10(1), 13–14.

Mann-Nachbar, P. and Nachbar, W. (1965). The preferred mode shape in the linear buckling of circular cylindrical shells under axial compression. *Journal of Applied Mechanics,* xx, 793–802.

Marek, P. (2000). Personal communication, 19 May.

Marek, P., Guštar, M. and Anagnos, Th. (1995). *Simulation-Based Reliability Assessment for Structural Engineers*. Boca Raton: CRC Press.

Maspoli, M., Hilburger, M., Bogge, M., Haynie, W., and Kriegesmann, B. (2012). Validation of lower-bound estimates for compression-loaded cylindrical shells. 53rd AIAA/ASME/ASCE/AHS/ASC Structures, Structural Dynamics and Materials Conference, Paper AIAA 2012-1689.

Massey, F.J., Jr. (1951). The Kolmogorov-Smirnov test for goodness of fit. *Journal of American Statistical Association,* 46, 68–78.

Masur, E.F. (1973). Buckling of shells — general introduction and review. *ASME National Structural Engineering Meeting*, April 9–13, Preprint 2000, San Francisco.

Maxwell, J.C. (1875). On the dynamical evidence of the molecular constitution of bodies. *Nature*, 11, 357–359.

Meimand, V.Z., Graham-Brady, L., and Schaffer, B.W. (2013). Imperfection sensitivity and reliability using simple bar-spring models for stability. *International Journal of Structural Stability and Dynamics*, 13(3), paper 1250075.

Michel, G., Gusic, G., and Jullien, J.F. (2000). Buckling of thin cylindrical shells under lateral pressure: influence of localized thickness variations. In D. Camotim, D. Dubina and J. Rondal (Eds.), *Proceedings of the Third International Conference on Coupled Instabilities in Metal Structure* (pp. 427–434). London: Imperial Collge Press.

Mieder, W. (1986). *The Prentice-Hall Encyclopedia of World Proverbs: a Treasury of Wit and Wisdom through the Ages* (p. 318). Englewood Cliffs, NJ: Prenctice Hall.

Mikhasev, G.I. (1984). Local loss of stability of shells of zero curvarture with variable thickness and elastic modulus. *Vestnik Leningradskogo Universiteta Mekhanika*, 2, 39–43.

Mikhasev, G.I. and Tovstik, P.E. (2009). *Localized Vibrations and Waves in Thin Shells: Asymptotic Methods* (in Russian). Moscow: "Fizmatlit" Publishers.

Mil'tsyn, A.M. (1992). The influence of technological imperfections on the stability of thin shells (multivariate approach). *Mechanics of Solids*, 6, 181–188.

Mil'tsyn, A.M. (1993). Nonlinear interaction of technological imperfections and their influence on the stability of thin shells (multivariate approach). *Mechanics of Solids*, 1, 178–184.

Miller, R.K. and Hedgepeth, J.M. (1979). The buckling of lattice columns with stochastic imperfections. *International Journal of Solids and Structures,* 15, 73–84.

Moiseev, N.D. (1949). *Essays on Development of Theory of Stability* (in Russian; p. 84). Moscow: "GITTL" Publishers.

Moore, R. (1979). *Methods and Applications of Interval Analysis.* Philadelphia, PA: SIAM.

Morring, F., Jr. (2011). NASA recalculates to save weight on launchers, *Aviation Week.*

Morris, N.F. (1996). Shell stability: the long road from theory of practice. *Engineering Structures*, 18(10), 801–806.

Mossakovskii, V.I. (1970). The effect of inhomogeneity of the stress state and of initial imperfections on stability of a cylindrical shell (in Russian). Proceedings, Seventh All Union Conference on Theory of Shells and Plates (pp. 831–839). Moscow: "Nauka" Publishing House.

Mossakovskii, V.I., Manevich, L.I., and Evkin, A.Yu. (1975). Investigation of postbuckling equilibrium forms of a compressed cylindrical shell. *International Applied Mechanics*, 11(11), 1155–1159.

Mossakovskii, V.I., Manevich, L.I., and Prokopalo, E.F. (1972). Investigation of post-critical behavior of cylindrical shells (in Russian). *Proceedings, USSR Academy of Sciences*, 206(2), 297–298, 1972.

Mossakovskii, V.I., Mil'tsyn, A.M., and Olevskii, V.I. (1991). Deformation and stability of technologically imperfect cylindrical shells in a nonuniform stress state. *Strength of Materials*, 12, 1745–1750.

Mossakovskii, V.I., Mil'tsyn, A.M., Selivanov, Yu.M., and Olevskii, V.I. (1994). Automating the analysis of results of a holographic experiment. *Strength of Materials*, 26(5), 385–391.

Most, T., Bucher, C., and Schorling, Y. (2004). Dynamic stability analysis of non-linear structures with geometrical imperfections under random loading. *Journal of Sound and Vibration*, 276 (1–2), 381–400.

Muggeridge, D.B. (1969). The effect of initial axisymmetric shape imperfections on the buckling behavior of circular cylindrical shells under axial compression. UTIAS Report No. 148, Institute for Aerospace Studies, University of Toronto, Canada, December.

Murray, D.W., Local buckling, strain localization, wrinkling and postbuckling response of line pipe, *Engineering Structures*, Vol. 19(5), 360–371, 1997.

Mushtari, H.M. (1951). On the elastic equilibrium of a thin shell with initial imperfections of a middle surface form (in Russian). *Applied Mechanics and Mathematics*, 15(6), 743–750.

Musschenbroeck, P. van (1729). Physicae experiementales et geometricae, de magnete, tuborum capillarium uitreorumque speculorum attractione, magnitudine térrea, cohaerentia corporum firmorum. (in Latin), Dissertationes, Samuel Luchtmans, Leyden.

Nahim, P.J. (1998). *An Imaginary Tale.* Princeton, NJ: Princeton University Press.

Nahin, P.J. (2008). *Digital Dice: Computational Solutions to Practical Probability Problems* (p. 2). Princeton, NJ: Princeton University Press.

Náprstek, J. (1999). Strongly non-linear stochastic response of a system with random initial imperfections. *Probabilistic Engineering Mechanics*, 14(1–2), 141–148.

Natke, H.G. and Ben-Haim, Y. (Eds.) (1997). *Uncertainty: Models and Measures*. Berlin: Akademie Verlag.

Naughton, J. (2011). James Gleick: "Information poses as many challenges as opportunities", *The Guardian*, 9 April.

Nayfeh, A.H. and Hawwa, M.A. (1994). Use of mode localization in passive control of structural buckling. *AIAA Journal*, 32(10), 2131–2133.

Needleman, A. and Tvergaard, V. (1982). Aspects of plastic post-buckling behavior. In H.G. Hopkins and M.J. Sewell (Eds.), *Mechanics of Solids. The Rodney Hill 60th Anniversary Volume* (pp. 453–498). New York: Pergamon Press.

Nemeth, M.P. and Starnes, J.H., Jr. (1998). The NASA monographs on shell stability design recommendations: a review and suggested improvements, NASA/TP-1998-206290, January.

Netto, T., Kyriakides, S., and Ouyang, X. (1999). On the initiation and propagation of buckles on a beam on a nonlinear foundation. *Journal of Applied Mechanics*, 66, 418–426.

Newton, I. (1960). *The Correspondence of Isaac Newton: Vol. II, 1676–1687*, H.W. Turnbuld (Ed.) (pp. 182–183). Cambridge, UK: Cambridge University Press.

Nguyen, H.L.T. and Dang, T.M.T. (2006). Buckling of initially imperfect rectangular thin plate with variable thickness. *Vietnam Journal of Mechanics*, 28(2), 103–110.

Nguyen, H.L.T., Elishakoff, I., and Nguyen, V.T. (2009). Buckling under the external pressure of cylindrical shells with variable thickness. *International Journal of Solids and Structures*, 46, 4163–4168.

Nguyen, H.L.T, Elishakoff, I., Nguyen, V.T., and Nguyen, Q.M. (2012). Buckling of cylindrical shells with initial imperfection and variable thickness under external pressure. *Vietnam Congress in Applied Mechanics*, Hanoi.

Nguyen, T.H.L. and Thach, S.S.H. (2006a). Stability of cylindrical panel with variable thickness. *Vietnam Journal of Mechanics,* 28(1), 56–65.

Nguyen, H.L.T. and Thach, S.S.H. (2006b). Influence of thickness variation and initial geometric imperfection on the buckling of cylindrical panel. Proceedings of the 8th Vietnamese Conference on Mechanics of Solids (pp. 491–499). Thai Nguyen.

Niedenfuhr, F.W. (1963). Scatter of observed buckling loads of pressurized shells. *AIAA Journal*, 1, 1923–1925.

Niedenfuhr, F.W. (1964). Reply by another to G. Herrmann and R.W. Bungay. *AIAA Journal*, 2(16), 1166.

Nieuwendijk, H.L. van der (1997). Preliminary study of a new design criterion for shells under axial compression. Memorandum M-850, Faculty of Aerospace Engineering, Delft University of Technology, Delft, The Netherlands.

Noh, H.C. (2006). Effect of multiple uncertain material properties on the response variability of in plane and plate structures. *Computer Methods in Applied Mechanics and Engineering*, 195 (19–22), 2697–2718.

Noirfalise, C., Bourinet, J.-M., Fogli, M., and Cochelin, B. (2008). Reliability analysis of the buckling of imperfect shells based on the Kahrunen–Loève expansion. 8th World Congress on Computational Mechanics, WCCM8, Venice, Italy.

Nordgren, R.P. and Conte, J.P. (1999). On one-dimensional random fields with fixed end values. *Probabilistic Engineering Mechanics*, 14, 301–310.

North, D.C. (2006). On Kenneth Binmore's *Natural Justice, Analyse & Kritik*, 28, 102–103.

Novikov, V.V. (1980). Internal resonances and stability of elastic shells. *Journal of Applied Mechanics and Technical Physics*, 21(4), 574–576.

Novikov, V.V. (1988). On the instability of elastic shells as the manifestation of internal resonance. *Journal of Applied Mathematics and Mechanics*, 52(6), 797–802.

Nowak, A.S. and Collins, R.R. (2000). *Reliability of Structures* (p. 104). Boston, MA: McGraw-Hill.

Obraztsov, I.F., Nerubaylo, B.V., and Andrianov, I.V. (1991). *Asymptotic Methods in Structural Mechanics of Thin-Walled Structures*. Moscow: "Mashinostroyenie" Publishers.

Obrecht, H., Rosenthal, B., Fuchs, P., Lange, S., and Marusczyk, C. (2006). Buckling, postbuckling and imperfection-sensitivity: old questions and some new answers. *Computational Mechanics*, 37, 498–506.

Oden, J.T. and Bathe, K.J. (1978). A commentary on computational mechanics. *Applied Mechanics Reviews*, 31(8), 1053–1058.

Ohira, H. (1961). Local buckling theory of axially compressed cylinders. Proceedings of the 11th Japan National Congress for Applied Mechanics (pp. 37–40), Tokyo.

Ohira, H. (1965). Local buckling theory for an axially compressed circular cylinder of finite length. Proceedings of the 14th Japan National Congress for Applied Mechanics (pp. 58–62), Tokyo.

Ohsaki, M. and Ikeda, K. (2007). *Stability and Optimization of Structures: Generalized Sensitivity Analysis*. Berlin: Springer Verlag.

Oldfather, W.A., Ellis, C.A., and Brown, D.M. (1933). Leonhard Euler's elastic curves. *Isis*, 20(1), 72–160.

Oloffson, P. (2007). *Probabilities: The Little Numbers That Rule Our Lives* (p. 212). New York: Wiley-Interscience.

Onkar, A.K., Upadhyay, C.S., and Yadav, D. (2006). Generalized buckling analysis of laminated plates with random material properties using stochastic finite elements. *International Journal of Mechanical Sciences*, 48(7), 780–798.

Orf, J. (2008). Einfluss von Imperfektionen auf das Beulverhalten unversteifter CFK-Zylinder. Bauhaus-Universität Weimar, Institut für Strukturmechanik (in German).

Öry, H. and Reimedes, H.G. (1987). Stresses in — and stability of thin-walled shells under non-ideal load distribution. ECCS Colloquium on Stability of Plate and Shell Structures (pp. 555–559), Ghent University, Belgium.

Oyesanga, M.O. (1993). End shortening relation and stability of columns on nonlinear foundation. *Journal of Applied Mechanics*, 60(4), 1050–1052.

Palassopoulos, G.V. (1980). A probabilistic approach to the buckling of thin cylindrical shells with general random imperfections: solution of the corresponding deterministic problem. In W.T. Koiter and G.K. Mikhailov (Eds.), *Theory of Shells* (pp. 417–443). Amsterdam: North-Holland Publishing Company.

Palassopoulos, G.V. (1991). Reliability-based design of imperfection-sensitive structures. *Journal of Engineering Mechanics*, 117(6), 1220–1240.

Palassopoulos, G.V. (1992). Response variability of structures subjected to bifurcation buckling. *Journal of Engineering Mechanics*, 118(6), 1164–1183.

Palassopoulos, G.V. (1993). New approach to buckling of imperfection-sensitive structures. *Journal of Engineering Mechanics*, 119, 850–869.

Palassopoulos, G.V. (1997). Buckling analysis and design of imperfection-sensitive structures. In A. Haldar, A. Guran and B.M. Ayyub (Eds.), *Uncertainty Modeling in Finite Element, Fatigue and Stability of Systems* (pp. 311–356). Singapore: World Scientific.

Palmer, A. (2002). Pogorelov's theory of creases, and point loads on thin cylindrical shells. In H.R. Drew and S. Pellegrino (Eds.), *New Approaches to Structural Mechanics, Shells and Biological Structures* (pp. 341–354). Dordrecht: Kluwer Academic.

Pantelides, C.P. (1996). Buckling and post-buckling of stiffened elements with uncertainty. *Thin-Walled Structures,* 26, 1–17.

de Paor, C., Cronin, K., Gleeson, J., and Kelliher, D. (2012). Statistical characterization and modeling of random geometric imperfections in cylindrical shells. *Thin-Walled Structures*, 58, 9–17.

de Paor, C., de Kelliher, D., Cronin, K., and Wright, W.M.D. (2010). The computation of accurate buckling pressures of imperfect thin-walled cylinders. Proceedings of the APME Pressure Vessels & Piping Division, K-PVP Conference, PVP 2010, 18–29 July, Bellevue, WA, 2010. http://rennes.ucc.ie/~bill/repository/2010_PVP_2580.pdf.

Papadopoulos, V., Charmpis, D.C., and Papadrakakis, M. (2009). A computationally efficient method for the buckling analysis of shells with stochastic imperfections. *International Journal of Solids and Structures*, 43(5), 687–700.

Papadopoulos, V. and Iglesis, P. (2007). The effect of non-uniformity of axial loading on the buckling behavior of shells with random imperfections. *International Journal of Solids and Structures*, 44(18–19), 6299–6317.

Papadopoulos, V. and Papadrakakis, M. (2004). Finite-element analysis of cylindrical panels with random initial imperfections. *Journal of Engineering Mechanics*, 130(8), 867–876.

Papadopoulos, V. and Papadrakakis, M. (2005). The effect of material and thickness variability on the buckling load of shells with random initial imperfections. *Computer Methods in Applied Mechanics and Engineering*, 194(12–16), 1405–1426.

Papadopoulos, V., Stefanou, G., and Papadrakakis, M. (2009). Buckling analysis of imperfect shells with stochastic non-gaussian material and thickness properties. *International Journal of Solids and Structures*, 46(14–15), 2800–2808.

Papadrakakis, M. (2012). Private communication, 12 October.

Patelli, E. and Schuëller, G.I. (2011). Soft-computing approach to solve ill-posed inverse problems: Applications to random materials and imperfect cylindrical shells. *Inverse Problems in Science and Engineering*, 19(1), 87–102.

Perry, S.H. (1966). Statistical variation of buckling strength. Ph.D. Dissertation, University of London, London.

Pfeifer, S. (2012). IAEA admits complacency. *Financial Times*, 12 March.

Phadke, M.S. (1989). *Quality Engineering using Robust Design.* Englewood Cliffs, NJ: Prentice Hall.

Pierre, C. (1988). Mode localization and eigenvalue loci veering phenomena in disordered structures. *Journal of Sound and Vibration*, 126(3), 485–502.

Pierre, C. and Plaut, R.H. (1989). Curve veering and mode localization in a buckling problem. *Journal of Applied Mathematics and Physics (ZAMP)*, 40, 758–761.

Pignataro, M. (1988). W.T. Koiter (1914–1997), *Meccanica*, 33(6), 605–606.

Pignataro, M., Rizzi, N., and Luongo, A. (1991). *Stability, Bifurcation and Postcritical Behaviour of Elastic Structures*. Amsterdam: Elsevier.

Platonov, V.V. (2011). Stability and vibrations of cylindrical shells under axial compression in non-classical setting (in Russian). *Vestnik, St. Petersburg State Univerity*, 1(1), 132–137.

Plouffe, B. (2007). Friedrich Dürrenmatt's *Der Auftrag*: an exeption within an Oeuvre of cultural pessimism and worst possible outcome. *Seminar: A Journal of Germanic Studies*, 43(1), 19–36.

Pogorelov, A.V. (1960). About post-critical deformations of compresed cylindrical shells (in Russian). *Proceedings USSR Academy of Sciences*, 134(1), 62–63.

Pogorelov, A.V. (1964). *Post-Buckling Behaviour of Cylindrical Shells.* NASA Technical Translation TTF-90, Washington, D.C.

Pogorelov, A.B. (1966). *Geometric Theory of Stability of Shells* (in Russian). Moscow: "Nauka" Publishers (English translation available at Foreign Technology Division, FDT-ID(RS)T-2219-78), Retrieved on 10 September 2013, http://www.dtic.mil/cgi-bin/GetTRDoc?Location= U2&doc=GetTRDoc.pdf&AD=ADA075193.

Pogorelov, A.V. (1967). *Geometric Methods in Nonlinear Theory of Elastic Shells*. Moscow: "Nauka" Publishers.

Pogorelov, A.B. (1971). Geometric methods in nonlinear theory of shells (in Russian). *Prochnost' i Plastichnost' (Strength and Plasticity)* (pp. 90–96). Moscow: "Nauka" Publishers.

Pogorelov, A.V. (1988). *Bending of Surfaces and Stability of Shells*. Providence, RI: American Mathematical Society.

Pogorelov, A.V. and Babenko, V. (1992). Geometric method in the stability theory of thin shells. *International Applied Mechanics*, 28(10), 3–22.

Potier-Feery, M. (1983). Amplitude modulation, phase modulation and localization of buckling patterns. In J.M.T. Thompson and G.W. Hunt (Eds.), *Collapse: The Buckling of Structures in Theory and Practice* (pp. 149–159). Cambridge: Cambridge University Press.

Price, S. (2011). *Worst-Case scenario? Governance, Meditation and the Security Regime* (p. 4). London: Zed Books.

Pugsley, A. and Macauley, M. (1960). Large-scale crumpling of thin cylindrical column. *Quarterly Journal of Mechanics and Applied Mathematics*, 13(1), 1–9.

Prigarin, S.M. (2001). *Spectral Models of Random Fields in Monte Carlo Methods*. Utrecht, The Netherlands: VSP.

Qiria, V.S. (1951). Motion of bodies in resisting media (in Russian). *Proceedings, Tbilisi State University,* 44, 1–20.

Qiu, Z.P. (2005). Convex models and interval analysis method to predict the effect of uncertain-but-bounded parameters on the buckling of composite structures. *Computer Methods in Applied Mechanics and Engineering*, 194 (18–20), 2175–2189.

Qiu, Z.P., Elishakoff, I., and Starnes, J.H., Jr. (1996). The bound set of possible eigenvalues of structures with uncertain but non-random parameters. *Chaos, Solitons & Fractals*, 7(11), 1845–1857.

Qiu, Z.P., Ma, L.H., and Wang, X.J. (2006). Ellipsoidal-bound convex model for non-linear buckling of a column with uncertain initial imperfection. *International Journal of Non-Linear Mechanics*, 48(8), 919–925.

Qiu, Z.P. and Wang, X.J. (2005). Two non-probabilistic set-theoretical modes for dynamic response and buckling failure measures of bars with unknown-but-bounded initial imperfections. *International Journal of Solids and Structures*, 42(3–4), 1039–1054.

Rabotnov, Yu.N. (1946). Local shell buckling (in Russian). *Proceedings, USSR Academy of Sciences*, 52(2), 111–112.

Raftoyiannis, J. and Kounadis, A.N. (1998). Dynamic buckling of limit-point systems under step loading. *Dynamics and Stability of Systems*, 3(3–4).

Rais-Rohani, M. (2001). Reliability analysis and reliability-based design optimization of circular composite cylinders under axial compression. Report 20020066511, NASA Langley Research Center.

Ramm, E. (Ed.) (1982). *Buckling of Shells*. New York: Springer.

Ramm, E. and Wall, W.A. (2002). Shells in advanced computational environment. In H.A. Mang, F.G. Ramerstorfer, and J. Eberhardsteiner (Eds.), Fifth World Congress on Computational Mechanics (pp. 1–22), Vienna, Austria.

Ramm, E. and Wall, W.A. (2004). Shell structures — a sensitive interrelation between physics and numerics. *International Journal for Numerical Methods in Engineering*, 60(1), 381–427.

Raviprakash, A.V., Adhithya Plato Sidharth, A., Prabu, B., and Alagumurthi, N. (2010). Structural reliability of thin plates with random geometrical imperfections subjected to uniform axial compression. *Jordan Journal of Mechanical and Industrial Engineering*, 4(2), 270–279.

Raviprakash, A.V., Prabu, B., and Alagumruthi, N. (2010). Mean value first order second moment analysis of buckling of axially loaded thin plates with random geometrical imperfections, *International Journal of Engineering, Science and Technology*, 2(4), 150–152.

Rawson, H. (1997). *If It Ain't Broke: The Unwritten Laws of Life* (p. 68). London: Penguin.

Reissner, E. (1969). A note on imperfection sensitivity of thin plates on a nonlinear elastic foundation. In H.E. Leipholz (Ed.), *Instability of Continuous Systems* (pp. 15–18). Berlin: Springer.

Reissner, E. (1970). On postbuckling behavior and imperfection sensitivity of thin elastic plates on a nonlinear elastic foundation. *Studies in Applied Mathematics*, XLIX(1), 45–57.

Reissner, E. (1971). A note on imperfection sensitivity of thin plates on a nonlinear elastic foundation. In H.H.E. Leipholz (Ed.), *Instability of Continuous Systems*. Berlin: Springer.

Reitinger, R. and Ramm, E. (2008). Buckling and imperfection sensitivity in the optimization of shell structures. *Thin-Walled Structures*, 23(1), 159–177.

Rice, J.R. (2008). *Bernard Budiansky 1925–1999. A Biographical Memoir*. Washington, DC: National Academy of Sciences.

Ricks, E. (2004). *Buckling, Encyclopedia of Computational Mechanics*.

Rikards, R.B. (1976). Optimization models including statistical parameters for shells mode of composite materials. *Mechanics of Composite Materials*, 12(6), 916–924.

Riks, E. (1984). Bifurcation and stability, a numerical approach. Presented at the ASME Winter Annual Meeting Symposium on Innovative Methods for Nonlinear Problems, New Orleans.

Robsertson, A. (1929). The strength of tubular struts. ARC Report and Memorandum No. 1185.

Rolfes, R., Hühne, C., Kling, A., Temmen, H., Geier, B., Klein, H., Teßmer, J., and Zimmermann, R. (2004). Advances in computational stability analysis of thin-walled aerospace structures regarding postbuckling. Robust Design and Dynamic Loading, *Thin-Walled Structures: Advances in Research, Design and Manufacturing Technology*, 3, 2004.

Roop, L. (2011). NASA in Huntsville to test a new rocket skin in "world's largest can crusher." *The Huntsville Times*, 21 March.

Roorda, J. (1965a). The buckling behavior of imperfect structural systems. *Journal of Mechanics of Physics of Solids*, 13, 267–280.

Roorda, J. (1965b). Stability of structures with small imperfections. *Journal of Engineering Mechanics*, 91, 87–106.

Roorda, J. (1969). Some statistical aspects of the buckling of imperfection-sensitivity in elastic post-buckling. *Journal of Mechanics and Physics of Solids*, 17, 111–123.

Roorda, J. (1971). Buckling of shells: an old idea with a new twist. *Journal of Engineering Mechanics*, 98, 531–538.

Roorda, J. (1975). The random nature of column failure. *Journal of Structural Mechanics*, 3(3), 1975.

Roorda, J. (1980). *Buckling of Elastic Structures*. Solid Mechanic Division, University of Waterloo, Ontario, Canada.

Roorda, J. and Hansen, J.S. (1971). Random buckling behavior in axially loaded cylindrical shells with axisymmetric imperfections. *Journal of Spacecraft and Rockets*, 9(2), 88–91.

Rota, G.-C. (1997). Ten lessons I wish I had been taught. *Notices of the American Mathematical Society*, 44(1), 22–25.

Rotter, J.M. (1998). Shell structures: the new European standard and current research needs. *Thin-Walled Structures*, 31, 3–23.

Roux, W., Stander, N., Günther, F., and Müllerschön, H. (2006). Stochastic analysis of highly non-linear structures. *International Journal for Numerical Methods in Engineering*, 65(8), 1221–1242.

Rumsfeld, D.H. (2002). Donald Rumsfeld press conference as US Secretary of Defense. http://www.defense.gov/transcripts/transcript.aspx?transcripted=2636.

Rzhanitsin, A.R. (1947). *Structural Design with Consideration of Plastic Properties of Materials* (in Russian). Moscow: "Stroivoenmorizdat" Publishers.

Sachenkov, A.V. (1962). About an approach to solution of nonlinear stability problems of thin shells. *Nonlinear Theory of Plates and Shells* (in Russian; pp. 3–41). Kazan: Kazan University Press.

Salencon, J. (1900). An introduction to the yield theory and its application to soil mechanics. *European Journal of Mechanics A*, 9(5), 477–500.

Salerno, G., Colletta, G., and Casciaro, R. (1996). Attractive post critical paths and stochastic imperfection sensitivity and analysis of elastic strectires within an FEM contest. ECCOMAS Conference on Numerical Methods in Engineering No. 2 (pp. 190–196). Chichester: Wiley.

Samuelson, L.A. and Eggwertz, S. (1992). *Shell Stability Handbook*. Essex, England: Elsevier Science.

Sanders, J.L., Jr. (1970). Random deflection of a string on an elastic foundation. *SIAM Journal on Applied Mathematics*, 22(3), 406–418.

Schenk, C.A. and Schuëller, G.I. (2002). Buckling analysis of cylindrical shells with cutouts including random geometric imperfections. In H.A. Mang, F.G. Ramemerstorfer and J. Eberhandsteiner (Eds.), *Fifth World Congres on Computational Mechanics*, Vienna, Austria, July.

Schenk, C.A. and Schuëller, G.I. (2002). Buckling analysis of cylindrical shells with random boundary and geometric imperfections. Proceeding of the 8th International Conference on Structural Safety and Reliability (ICO ASSAR, '01; CD-ROM), Newport Beach, CA, June 17–21, 2001. Lisse, The Netherlands: Swets & Zeitlinger.

Schenk, C.A. and Schuëller, G.I. (2003). Buckling analysis of cylindrical shells with random geometric imperfections. *International Journal of Non-Linear Mechanics*, 38, 1119–1132.

Schenk, C.A. and Schuëller, G.I. (2005). *Uncertainty Assessment of Large Finite Element Systems*. Berlin: Springer Verlag.

Schenk, C.A. and Schuëller, G.I. (2007). Buckling analysis of cylindrical shells with cutouts including random boundary and geometric imperfections. *Computer Methods in Applied Mechanics and Engineering*, 196(35–36), 3424–3434.

Schenk, C.A., Schuëller, G.I., and Arbocz, J. (2000). On the analysis of cylindrical shells with random imperfections. ASS-IACM 2000, Fourth International Colloquium on Computation of Shell & Spatial Structures, Chania-Crete, Greece, 5–7 June.

Schenk, C.A., Schuëller, G.I., and Arbocz, J. (2001). Buckling analysis of cylindrical shells with random imperfections. In G.I. Schuëller and P.D. Spanos (Eds.), Proceedings of the International Conference on Monte Carlo Simulation (pp. 137–141). Lisse, The Netherlands: Swets & Zeitlinger BV.

Scheurkogel, A. and Elishakoff, I. (1985). On ergodicity assumption in an applied mechanics problem. *Journal of Applied Mechanics*, 52, 133–136.

Scheurkogel, A., Elishakoff, I., and Kalker, J. (1981). On the error that can be induced by an ergodicity assumption. *Journal of Applied Mechanics*, 48, 654–656.

Schiffner, K. (1965). Untersuchung des Stabilitaetsverhaltens duennwandiger nichtlineare axialsymmetrischer Belastung mittels einer nichtlinearen Schallentheorie (in German), Jahrbuch 1965 des WGLR (pp. 448–453).

Schillinger, D. (2004). Stochastic FEM based stability analysis of I-sections with random inperfections, Diploma Thesis, Departement of Civil Engineering, University of Stuttgart, Federal Republic of Germany.

Schillinger, D., Papadopoulos, V., Papadrkakis, M., and Bishoff, M. (2010). Buckling analysis of imperfect I-section stochastic shell elements. *Computational Mechanics*, 46(3), 495–510.

Schmidt, H. (2000). Stability of steel shell structures: General report. *Journal of Constructional Steel Research*, 55, 159–181.

Schorling, Y. (1997). Beitrag zur Stabilitätsuntersuchung von Strukturen mit räumlich korrelierten geometrischen Imperfektionen (in German). Ph.D. Thesis, Bauhaus University Weimar, Federal Republic of Germany.

Schorling, Y. and Bucher, C. (1999). Stochastic stability for structures with geometric imperfections. In B.F. Spencer, Jr. and E.A. Johnson (Eds.), *Stochastic Structural Dynamics* (pp. 343–348). Rotterdam: Balkema.

Schorling, Y., Bucher, C., and Purkert, G. (1998). Stochastic analysis for randomly imperfect structures. In R. Melchers and M. Stewart (Eds.), *Application of Statistics and Probability* (Vol. 2, pp. 1027–1032). Rotterdam: Balkema.

Schranz, C., Krenn, B., and Mang, H.A. (2006). Conversion from imperfection-sensitive into imperfection-insensitive elastic structures II: Numerical investigation. *Computer Methods in Applied Mechanics and Engineering*, 195, 1458–1479.

Schweppe, F.C. (1973). *Uncertain Dynamic Systems*. Englewood Cliffs, NJ: Prentice-Hall.

Sebek, R.W.L. (1981). Imperfections surveys and data reduction of ARIANE interstages I/II and II/III, Ir. Thesis, Faculty of Aerospace Engineering, Delft University of Technology, Delft, The Netherlands.

Sechler, E.E. (1974). The historical development of shell research and design. In Y.C. Fung and E.E. Sechler (Eds.), *Thin Shell Structures, Theory, Experiment and Design* (pp. 3–25). Englewood Cliffs, NJ: Prentice-Hall.

Seide, P. (1962). The stability under axial compression and lateral pressure of circular cylindrical shells with a soft elastic core. *Journal of Aerospace Sciences*, 29, 851–862.

Seide, P. (1974). A reexamination of Koiter's theory of initial post-buckling behavior and imperfection-sensitivity of structures. In Y.C. Fung and E.E. Sechler (Eds.), *Thin Shell Structures: Theory, Experiment, Design* (pp. 59–80). Englewood Cliffs, NJ: Prentice Hall.

Seide, P., Weingarten, V.I., and Morgan, E.J. (1960). Final report on development of design criteria for elastic stability of thin shell structures. STL/TR-60-0000-19425, Space Technology Laboratories, Inc., Los Angeles, CA.

Semenyuk, N.P. (1987a). Stability of cylindrical shells of composite materials with imperfections, *(Prikladnaya Mekhanika) International Applied Mechanics*, Vol. 23(1), 37–43 (in Russian).

Semenyuk, N.P. (1987b). Load-carrying capacity of cylindrical shells of composite materials under compressive loads. *Strength of Materials*, 3, 409–415.

Semenyuk, V.P. and Zhukova, N.B. (2006). Initial postbuckling behavior of cylindrical composite shells under axisymmetric deformation, *Prikladnaya Mekhanika (International Applied Mechanics)*, Vol. 42(4), 461–471.

Sendlebeck, R.L., Carlson, R.L., and Hoff, N.J. (1967). An experimental study of the effect of length on the buckle pattern of axially compressed circular cylindrical shells. Report SUDAAR No. 318, Department of Aeronautics and Astronautics, Stanford University, Stanford, CA, June.

Sheinman, I. and Adan, M. (1991). Imperfection sensitivity of a beam on a nonlinear elastic foundation. *International Journal of Mechanical Sciences*, 33 (9), 753–760.

Sheinman, I., Adan, M., and Altus, E. (1993). On the role of the displacement function in nonlinear analysis of beams on an elastic foundation. *Thin-Walled Structures*, 15, 109–125.

Sheinman, I. and Firer, M. (1994). Buckling analysis of laminated cylindrical shells with arbitrary noncircular cross section. *AIAA Journal*, 32(3), 648–654.

Sheinman, I. and Goldfeld, Y. (2001). Buckling of laminated cylindrical shells in terms of different shell theories and formulations. *AIAA Journal*, 39(9), 1773–1781.

Sheinman, I. and Goldfeld, Y. (2003). Imperfection sensitivity of laminated cylindrical shells according to different shell theories. *Journal of Engineering Mechanics*, 129(9), 1408–1053.

Sheinman, I. and Goldfeld, Y. (2004). Shell theory accuracy with regard to initial postbuckling behavior of cylindrical shell. *AIAA Journal*, 42(2), 429–432.

Sheinman, I. and Simitses, G.J. (1983). Buckling and postbuckling of imperfect cylindrical shells under axial compression. *Computers & Structures*, 17(2), 277–285.

Sheuer, E.M. and Stoller, D.S. (1962). On the generation of the random vectors. *Technometrics*, 4, 278–281.

Shinozuka, M. (1970). Maximum structural response to seismic excitations. *Journal of Engineering Mechanics Division*, 96, 729–738.

Shinozuka, M. and Deodatis, G. (1991). Simulation of stochastic processes by spectral decomposition. *Applied Mechanics Reviews*, 44(4), 191–204.

Shinozuka, M. and Yamazaki, I. (1988). Stochastic finite element analysis: an introduction. In S.T. Ariacatnam, G.I. Schuëller and I. Elishakoff (Eds.), *Stochastic Structural Dynamics* (pp. 241–292), London: Elsevier.

Shirshov, V.P., Local shell buckling. In *Proceedings, 2nd All-Union Conference on Shells and Plates Theory*, pp.314–317, Academy of Ukraine Publishers, Kiev, 1962 (in Russian).

Shkutin, L.I. (1967). The effect of precritical strains on the stability of a longitudinally compressed cylindrical shell. *International Applied Mechanics*, 3(1), 74–75.

Silver, N. (2012). *The Signal and the Noise: Why So Many Predictions Fail — but Some Don't*, Penguin Press, New York.

Simitses, G.J. (1986). Buckling and postbuckling of imperfect cylindrical shells: a review. *Applied Mechanics Reviews*, 39(1), 1517–1524.

Simitses, G.J. (1990). *Dynamic Stability of Suddenly Loaded Structures*. Berlin: Springer.

Simitses, G.J., Shaw, D., and Sheinman, I. (1985). Stability of cylindrical shells by various nonlinear shell theories. *ZAMM — Zeitschrift für angewandte Mathematik and Mechanik*, 65, 159–166.

Simitses, G.J., Sheinman, I., and Shaw, D. (1985). The accuracy of Donnell's equations for axially-loaded, imperfect orthotropic cylinders. *Computers & Structures*, 20(6), 939–945.

Simmonds, J.G. (1998). Some comments on the status of shell theory at the end of the 20th century: Complaints and correctives. In N.F. Knight, Jr. and M.P. Nemeth (Eds.), *Stability Analysis of Plates and Shells: A Collection of Papers in Honor of Dr. Manuel Stein* (pp. 9–18). NASA/CP-1998-206280.

Simmonds, J.G. (2000). Lyell Sanders' contributions to shell theory. 41st AIAA/ASME/ASCE/AHS/ASC Structures, Structural Dynamics, and Materials Conference, Atlanta, GA.

Simmonds, J.G. (2011). Personal communication, 19 December.

Singer, J. (1982). Buckling experiments in shells — a review of recent developments. *Solid Mechanics Archives*, 7, 213–313.

Singer, J. (1997). Experimental studies in shell buckling. Paper AIAA-1997-1075. Proceedings of the 38th AIAA/ASME/ASCE/AHS/ASC Structures, Structural Dynamics, and Materials Conference and Exhibit, Kissimmee, FL.

Singer, J. (1999). On the importance of shell buckling experiments. *Applied Mechanics Reviews*, 52(6), R17–R25.

Singer, J. and Abramovich, H. (1995). The development of shell imperfection measurement techniques. *Thin-Walled Structures,* 23(1–4), 379–398.

Singer, J., Abramovich, H., and Weller, T. (2002). The prerequisites for an advanced design methodology in shells prone to buckling. In H.R. Drew and S. Pellegrino (Eds.), *New Approaches to Structural Mechanics* (pp. 393–411). Dordrecht, the Netherlands: Kluwer Academic Publishers.

Singer, J., Abramovich, H., and Yaffe, R. (1978). Initial imperfection measurements of integrally stringer-stiffened shells, TAE Report 330, Technion — Israel Institute of Technology, Haifa, Israel.

Singer, J., Abramovich, H., and Yaffe, R. (1979). Initial imperfection measurements of integrally stringer-stiffened shells. *Israel Journal of Technology*, 17, 324–338.

Singer, J., Arbocz, J., and Babcock, C.D., Jr. (1971). Buckling of imperfect stiffened cylindrical shells under axial compression. *AIAA Journal*, 9(1), 68–75.

Singer, J., Arbocz, J., and Weller, T. (1997). *Buckling Experiments: Experimental Methods in Buckling of Thin-Walled Structures*, Vol. 2. New York: Wiley.

Sivak, V.F. (2000). Experimental investigation of the nonlinear resonance properties of cylindrical shells. *International Applied Mechanics*, 36(2), 247–250.

Sliz, R. and Chang, M.-Y. (2011). Reliable and accurate prediction of the experimental buckling of thin-walled cylindrical shell under an axial load. *Thin-Walled Structures*, 49(3), 409–421.

van Slooten, R.A. and Soong, T.T. (1972). Buckling of long, axially compressed thin cylindrical shell with random initial imperfections. *Journal of Applied Mechanics*, 39(4), 1066–1071.

van Slooten, R.A. and Soong, T.T. (1973). Authors' closure. *Journal of Applied Mechanics*, 40, 635.

Smith, T.E. and Herrmann, G. (1972). Stability of a beam on an elastic foundation subjected to a follower force. *Journal of Applied Mechanics*, 39, 628–629.

Sniedovich, M. (2007). The art and science of modeling decision-making under severe uncertainty. *Decision Making in Manufacturing and Services*, 1(2), 111–136.

Sniedovich, M. (2008). Wald's maximin model: a treasure in disguise! *The Journal of Risk Finance,* 9(3), 287–291.

Sniedovich, M. (2010). A bird's view of info-gap decision theory. *The Journal of Risk Finance*, 11(3), 269–283.

Sniedovich, M. (2012a). Black swans, new Nostradamuses, voodoo decision theories, and the science of decision making in the face of severe uncertainty. *International Transactions in Operational Research*, 19(1–2), 253–281.

Sniedovich, M. (2012b). Fooled by local robustness. *Risk Analysis*, 32(10), 1630–1637.

Sobey, A.J. (1964). The buckling strength of a circular cylinder loaded in axial compression. Reports and Memoranda No. 3366, 1–22, Aeronautical Research Council.

Softpedia (2011). What is ergodicity? Retrieved on 29 August 2011, http://news.softpedia.com/news/what-is-ergodicity-15686.shtml.

Sosa, E.M., Godoy, L.A., and Croll, J.G.A. (2006). Computation of lower-bound, elastic buckling loads using general-purpose finite element codes. *Computers & Structures*, 84, 1934–1945.

Southwell, R. (1913). On the collapse of tubes by external pressure, Parts I, II, III. *Philosophical Magazine*, Ser. 6, Vol. 25, No. 149, pp. 687–697; Vol. 26, No. 153, pp. 502–510 (also Vol. 26, No. 153, pp. 502–510; Vol. 29, No. 169, pp. 67–76, 1915).

Southwell, R.V. (1914). On the general theory of elastic stability. *Philosophical Transactions of the Royal Society of London, Series A*, 213, 187–202.

Spanos, P.D. and Zeldin, B.A. (1998). Monte Carlo treatment of random fields: A broad perspective, *Applied Mechanics Reviews*, 51(3), 219–237.

Spencer, H.H. (1978). Are measurements of geometric imperfection of plates and shells useful?, *Experimental Mechanics*, 30, 107–111.

Srubshchik, L. (1981). *Buckling and Post-Buckling Behavior of Shells* (in Russian). Rostov State University Publishing House.

Stam, A.R. (1996). Stability of imperfect cylindrical shells with random properties, Proceedings, AIAA-96-1462-CP. A Collection of Technical Papers, 37th AIAA/ASME/ASCE/AHS/ASC Structures, Structural Dynamics, and Materials Conference (pp. 1307–1314), Salt Lake City, UT, 1996.

Stam, A.R. (1998). A multi-mode random imperfection model in shell stability analysis. *Stability Analysis of Plates and Shells: A Collection of Papers in Honor of Dr. Manuel Stein* (pp. 187–196). NASA/CP-1998-206280.

Stam, A.R. (1999). Probabilistic sensitivities in nonlinear stability analysis, Paper AIAA-1999-1608, Proceedings of the 40th AIAA/ASME/ASCE/AHS/ASC Structures, Structural dynamics, and Materials Conference and Exhibit, St. Louis, MO, 1999.

Stam, A.R. and Arbocz, J. (1997). Staiblity analysis of anisotropic cylindrical shells with random layer properties. In O.D. Ditlevsen and J.C. Mitteau (Eds.), Euromech 372, Reliability in Nonlinear Structural Mechanics (pp. 77–81), University of Blaise Pascal and IFMA-French Institute for Advanced Mechanics, Clermont-Ferrand, France, 21–24 October.

Starnes, J.H., Jr. and Hilburger, M. (2002). Using high-fidelity analysis methods and experimental results to account for the effects of imperfections on the buckling response of composite shell structures. RTO AVT Symposium on "Reduction of Military Vehicle Acquisition Time and Cost through Advanced Modelling and Virtual Simulation," Paris, France, April.

Steele, C.R. (1989). Asymptotic analysis and computation for shells. In A.K. Noor, T. Belytschko, and J.C. Simo (Eds.), *Analytical and Computational Models of Shells* ASME CED, Vol. 3.

Stefanou, G. and Papadrakakis, M. (2004). Stochastic finite element analysis of shells with combined random material and geometric properties. *Computer Methods in Applied Mechanics and Engineering*, 193, 139–160.

Stefanou, G., Papadopoulos, V., and Papadrakakis, M. (2009). Buckling load variability of cylindrical shells with stochastic imperfections. *International Journal of Solids and Structures*, 46, 2800–2808.

Stefanou, G., Papadopoulos, V., and Papadrakakis, M. (2011). Buckling load variability of cylindrical shells with stochastic imperfections. *International Journal of Reliability and Safety*, 5(2), 191–208.

Steinböck, A., Hoefinger, G., Jia, X., and Mang, H.A. (2009). Three pending questions in elastic stability. *Journal of the International Association for Shell and Spatial Structures*, 50(1), 51–64.

Sternberg, E. (1985). Ruminations of a reclusive elastician. ASME Winter Annual Meeting, Miami Beach, FL, 1985, imechanica.org/node/177.

Stroud, W.J., Dale Davis, D., Maring, L.D., Krishnamurthy, T., and Elishakoff, I. (1992). Reliability of stiffened structural panels: two examples. In T.A. Cruse (Ed.), *Reliability Techniques* (pp. 199–216). New York: ASME Press, New York.

Stroud, W.J., Krishnamurthy, T., Sykes, N.P., and Elishakoff, I. (1993). Effect of bow-type initial imperfection on the reliability of minimum-weight stiffened structural panels. NASA TP-3263.

Stull, C.J., Nichols, J.M., and Earls, C.J. (2011). Stochastic inverse identification of geometric imperfections in shell structures. *Computer Methods in Applied Mechanics and Engineering*, 200(25–28), 2256–2267.

Sugiyama, Y., Langthjem, M.A., and Ryu, B.Y. (1999). Realistic follower forces. *Journal of Sound and Vibration*, 225(4), 779–782.

Szczepanski, J. and Wajnryb, E. (1995). Do ergodic or chaotic properties of the reflection law imply ergodicity or chaotic behavior of a particle's motion. *Chaos, Solitons & Fractals*, 5(1), 77–89.

Takano, A. (2012a). Statistical knockdown factors of buckling anisotropic cylinders under axial compression. *Journal of Applied Mechanics*, 79, paper 051004.

Takano, A. (2012b). Personal communication, 12 June.

Takewaki, I. and Ben-Haim, T. (2005). Info-gap robust design with load and model uncertainities, *Journal of Sound and Vibration*, 288(3), 551–570.

Teng, J.G. (1996). Buckling of thin shells: recent advances and trends. *Applied Mechanics Reviews*, 49(4), 263–274.

Teng, J.G., Lin, X., Rotter, J.M., and Ding, X.L. (2005). Analysis of geometric imperfections in full-scale welded steel silos. *Engineering Structures*, 27, 938–950.

Tennyson, R.C. (1963). A note on the classical buckling load of circular cylindrical shells under axial compression. *AIAA Journal*, 1(2), 475–476.

Tennyson, R.C. (1964a). An experimental investigation of the buckling of circular cylindrical shells in axial compression using the photoelastic technique. UTIAS Report No. 102, Institute for Aerospace Studies, University of Toronto, Canada, November.

Tennyson, R.C. (1964b). Buckling of circular cylindrical shells in axial compression. *AIAA Journal*, 2(7), 1351–1353.

Tennyson, R.C. (1967). Photoelastic circular cylinders in axial compression. ASTM STP419.

Tennyson, R.C. (1969). Buckling modes of circular cylindrical shells under axial compression. *AIAA Journal*, 7, 1481–1487.

Tennyson, R.C. (1980). Interaction of cylindrical shell buckling experiments with theory. In W.T. Koiter and G.K. Mikhailov (Eds.), *Theory of Shells* (pp. 65–116).

Tennyson, R.C. (1995). Composites in space: challenges and opportunities, *Proceedings, ICCM*, 1–35.

Tennyson, R.C., Booton, M., and Caswell, R.D. (1971). Buckling of imperfect elliptical cylindrical shells under axial compression. *AIAA Journal*, 9(9), 250–255.

Tennyson, R.C. and Muggeridge, D.B. (1969). Buckling of axisymmetric imperfect circular cylindrical shells under axial compression. *AIAA Journal*, 7 (11), 2127–2131.

Tennyson, R.C. and Muggeridge, D.B. (1973). Buckling of laminated anisotropic imperfect circular cylinder, under axial compression, *Journal of Spacecraft and Rockets*, Vol. 10(2), 143–148.

Tennyson, R.C., Muggeridge, D.B., and Caswell, R.D. (1971a). Buckling of a cylindrical shells having axisymmetric imperfection distributions. *Journal of Spacecraft and Rockets*, 9(5), 924–930.

Tennyson, R.C., Muggeridge, D.B., and Caswell, R.D. (1971b). New design criteria for predicting buckling of cylindrical shells under axial compression. *Journal of Spacecraft and Rockets*, 8(10), 1062–1067.

Thompson, J.M.T. (1967). Towards a general statistical theory of imperfection — sensitivity in elastic post-buckling. *Journal of the Mechanics and Physics of Solids*, 15, 413–417.

Thompson, J.M.T. (1982). *Instabilities and Catastrophes in Science and Engineering.* Chichester, UK: Wiley, Chichester.

Thompson, J.M.T. (2013a). Advise to a young researcher: with reminiscences of a life in science. *Philosophical Transactions of the Royal Society A: Mathematical ,Physical & Engineering Sciences*, 371, number 1993, paper 20120425.

Thompson, J.M.T. (2013b). Personal communication, 29 November.

Thompson, J.M.T. and van der Heijden, G.H.M. (2014). Quantified 'shock-sensitivity' above the Maxwell load, *International Journal of Bifurcation and Chaos*, available on http://arxiv.org/abs/1311.7390 (retrieved on 18 December, 2013).

Thompson, J.M.T. and Hunt, G.W. (1967). *A General Theory of Elastic Stability*. London: Wiley.

Thompson, J.M.T. and Hunt, G.W. (Eds.) (1983). *Collapse: The Buckling of Structures in Theory and Practice*. Cambridge, UK: Cambridge University Press, Cambridge.

Thompson, J.M.T. and Hunt, G.W. (1984). *Elastic Instability Phenomena*. Chichester: Wiley.

Thompson, J.M.T. and Virgin, L.N. (1988). Spatial chaos and localization phenomena in nonlinear elasticity. *Physics Letters A*, 126, 491–496.

Timoshenko Lectures. (2010). Retieved on 20 December 2010, http://imechanica.org/node/177.

Timoshenko, S.P. (1910). Einige Stabilitätsprobleme der Elastizitätstheorie (in German). *Zeitschrift für Mathematik und Physik*, 58, 337–357.

Timoshenko, S.P. (1968). *As I Remember, The Autobiography of Stephen P. Timoshenko*. Princeton, NJ: Van Nostrand.

Tootkaboni, M., Graham-Brady, L., and Schafer, B.W. (2009). Geometrically non-linear behavior of structural systems with random material property: an asymptotic spectral stochastic approach. *Computer Methods in Applied Mechanics and Engineering*, 198(37–40), 3171–3185.

Tovstik, P.E. (1982). Towards the local stability loss of shells (in Russian). *Vestnik Leningradskogo Universiteta Mathematika Mechanika, Astronomia*, 3, 73–85.

Tovstik, P.E. (1983). Some problems of the stability of cylindrical and conical shells. *PMM-Journal of Applied Mathematics and Mechanics*, 47(5), 815–822.

Tovstik, P.E. (1984). Local loss of stability by cylindrical shells under axial compression. *Leningrad University Mechanics Bulletin*, 1, 46–54.

Tovstik, P.E. (1991). On the forms of local buckling of thin elastic shells. *Transactions of the Canadian Society for Mechanical Engineering*, 15(3), 199–211.

Tovstik, P.E. (1995). *Stability of Thin Shells: Asymptotic Methods* (in Russian). Moscow: "Nauka" Publishers.

Tovstik, P.E. (1995). Axisymmetric deformation of thin shells of revolution under axial compression. *Vestnik Leningradskogo Universiteta*, 1, 95–102.

Tovstik, P.E. (1997). Stability of shells of revolution with a dent in middle surface (in Russian). Proceedings, 18th Conference on Shells and Plates, Vol. 2, pp. 120–127.

Tovstik, P.E. (2005). Local buckling of plates and shallow shells on an elastic foundation. *Mechanics of Solids*, 40(1), 120–131.

Tovstik, P.E. (2009). Stability of a transversally isotropic cylindrical shell under axial compression. *Mechanics of Solids*, 44(4), 552–564.

Tovstik, P.E. and Smirnov, A.L. (2001). *Asymptotic Methods in the Buckling Theory of Elastic Shells*. Singapore: World Scientific.

Trefethen, L.N. (1988). Maxims about numerical mathematics, science, and life. *SIAM Review*, 8 January.

Trendafilova, I. and Ivanova, J. (1995). Loss of stability of thin, elastic, strongly convex shells of revolution with initial imperfections, subjected to uniform pressure: a probabilistic approach. *Thin-Walled Structures*, 23(1–4), 201–214.

Trinckher, V.K. (1965). New method for determination of postcritical equilibrium of cylindrical shell during axial compression (in Russian). *Vestnuk Mosovskogo Universiteta*, 1, 76–82.

Troshchenko, V.T., Lepikhin, P.P., Khamaza, L.A., and Babich, Yu.N. (2009). Computerized data bank "strength of materials". *Strength of Materials*, 3, 235–242.

Truesdell, C. (1960). *The Rational Mechanics of Flexible or Elastic Bodies 1638–1788, Introduction to Leonhardi Euleri Opera Omnia*, Vol. X.

Truesdell, C.A. (1968). *Essays on the History of Mechanics*, Springer, Berlin.

Tsien, H.S. (1942a). Buckling of a column with non-linear lateral supports. *Journal of Aeronautical Science*, 9(4), 119–124.

Tsien, H.S. (1942b). Theory of buckling of thin shells. *Journal of Aeronautical Science*, 373.

Tsien, H.S. (1947). Lower buckling load in the nonlinear buckling theory of the thin shells. *Quarterly of Applied Mathematics*, 5, 236–237.

Turcic, F. (1990). Resistance of axially compressed cylindrical shells determined for the measured geometric imperfections. In M. Iványi and B. Veröci (Eds.), *Stability of Steel Structures, International Colloquium, East European Session* (pp. III-209–III-215), Budapest, Hungary.

Tvergaard, V. (1976). Buckling behavior of plate and shell structures. In W.T. Koiter (Ed.), Proceedings of the 14th International Congress of Theoretical and Applied Mechanics (pp. 233–247). Amsterdam: North-Holland.

Tvergaard, V. (1976). Buckling behavior of plate and shell structures. In W.T. Koiter (Ed.), *Theoretical and Applied Mechanics* (pp. 233–247). Amsterdam: North Holland.

Tvergaard, V. and Needleman, A. (1983). On the development of localized buckling patterns. In J.M.T. Thompson and G.W. Hunt (Eds.), *Collapse: The Buckling of Structures in Theory and Practice* (pp. 1–17). Cambridge: Cambridge University Press.

Tvergaard, V. and Needleman, A. (1980). On the localization of buckling patterns. *Journal of Applied Mecahnics*, 47, 613–619.

Tvergaard, V. and Needleman, A. (2000). Buckling localization in a cylindrical panel under axial compression. *International Journal of Solids and Structures*, 37, 6825–6842.

Ulam, S. (1976). *Adventures of a Mathematician* (pp. 196–198). New York: Charles Scribner's Sons.

Vanin, G.A. and Semeniyk, N.P. (1987). *Stability of Shells Made of Composite Materials with Imperfections* (in Russian). Kiev: "Naukova Dumka" Publishers.

Vanin, G.A., Semeniyk, N.P., and Emelyanov, R.F. (1978). *Stability of Shells of Reinforced Materials* (in Russian). Moscow: "Nauka" Publishers.

Verderaime, V. (1994). Illustrated structural application of universal first order reliability methods. NASA Technical Paper 3501.

Verduyn, W.D. and Elishakoff, I. (1982). A testing machine for statistical analysis of small imperfect shells. Report LR-357, Faculty of Aerospace Engineering, Delft University of Technology, The Netherlands; also in A.A. Betzer (Ed.), *Proceedings of the Seventh International Conference on Experimental Stress Analysis* (pp. 545–557), Ayalon Offset Ltd., Israel, 1982.

Vermeulen, P.G. (1982). Some considerations about the simulation technique of initial imperfections via the Monte Carlo method. Memorandum M-448, Department of Aerospace Engineering, Delft University of Technology, Delft, The Netherlands, December 1982.

Vermeulen, P.G. (1983). Second progress report on the project "Stability of shells with stochastic initial imperfections," Memorandum M-471, Department of Aerospace Engineering, Delft University of Technology, Delft, The Netherlands, June.

Vermeulen, P.G. (1984). Initial implementation of the LEVEL-II approach to the stochastic buckling of shells. Memorandum M-503, Department of Aerospace Engineering, Delft University of Technology, Delft, The Netherlands, January.

Vermeulen, P.G., Elishakoff, I., and van Geer, J. (1984). The statistical analysis of initial imperfection measurements (shell profiles), Report LR-442, Department of Aerospace Engineering, Delft University of Technology, Delft, The Netherlands, September.

Videc, B.P. (1974). Nonlinear stochastic boundary value problems for ordinary differential equations, Ph.D. Thesis, Harvard University, Cambridge, MA.

Videc, B.P. and Sanders, J.L. (1976). Application of Khas'minskii limit theorem to the buckling problem of column with random initial deflection. *Quarterly of Applied Mathematics*, 33, 422–428.

Villaggio, P. (1997). *Mathematical Models for Elastic Structures*. Cambridge, UK: Cambridge University Press.

Villaggio, P. (2001). Distortions in the history of mechanics. *Meccanica*, 36, 589–592.

Villaggio, P. (2011). Sixty years of solid mechanics. *Meccanica*, 46, 1171–1189.

Villaggio, P. (2013). Crisis in mechanics literature?, *Meccanica*, 48, 765–767.

Vladimirov, S., Konoch, V., and Mossakovskii, V. (1969). Experimental investigation of local stability of a cylindrical shell under axial compression. *Mechanics of Solids*, 4(4), 158–161.

Vladimirov, B.A., Konoch, V.I., Mossakovskii, V.I. and Smelyi (1969). Experimental investigation of local loss of stability of a cylindrical shell under axial compression. *Mechanics of Solids*, 4(4), 161–165.

Voblykh, V.A. (1965). Effect of initial deflections on the magnitude of critical load in circular cylindrical shells (in Russian). *Prikladnaya Mekhanika* (*Applied Mechanics*), 1(3), 17–26.

Volmir, A.S. (1956). *Gibkie Plastinki i Obolochki (Flexible Plates und Shells)* (p. 399). Moscow: "GITTL" Publishers (also *Biegsame Platten and Schalen*, VEB Verlag Für Bauwesen, Berlin, 1962, in German).

Volmir, A.S. (1964). *Stability of Elastic Systems*. Moscow: "Fizmatgiz" Publishers (English translation, Wright-Patterson Air Force Base, FDM-MT-1964, 1964).

Vorovich, I.I. (1959). Statistical method in the theory of shell stability (in Russian). *PMM-Prikladnag Matematika I Mehamika (Applied Mathematics and Mechanics)*, 23, 885–892.

Vorovich, I.I. (1996). Epilogue to the book by A.V. Pogorelov, *Geometric Theory of Shell Stability* (pp. 291–296) (in Russian: English translation available at Foreign Technology Division, FDT-ID(RS)T-2219-78. Retrieved on 10 September 2013, http://www.dtic.mil / cgi-bin / GetTRDoc?Location=U2&doc=GetTRDoc.pdf&AD=ADA075193).

Vorovich, I.I. (1966). Certain problems of the use of statistical methods in the theory of stability of plates and shells. In S.M. Durgar'yan (Ed.), *Theory of Shells and Plates* (pp. 46–75). Jerusalem: Israel Program for Scientific Translations.

Vorovich, I.I. and Minakova, N.I. (1967). Stability of non-shallow spherical arch (in Russian). *Mechanics of Solids*, 1.

Vorovich, I.I. and Minakova, N.I. (1973). Stability problems and numerical methods in theory of shells (in Russian). *Itogi Nauki, Mekhanika*, 7, 5–86.

Vorovich, I.I. and Shepeleva, V.G. (1969). Investigation of nonlinear stability of shallow shells of double curvature in high approximations (in Russian). *Mechanics of Solids*, 3, 70–73.

Vorovich, I.I. and Zipalova, V.F. (1965). Toward solution of boundary value problems of theory of elasticity with the aid of transfer to the Cauchy Problem (in Russian). *PMM-Journal of Applied Mathematics and Mechanics*, 29(5).

Vorovich, I.I. and Zipalova, V.F. (1966). Nonlinear deformation analysis of a spherical shell in high approximations (in Russian). *Mechanics of Solids*, 2, 150–153.

Vrancken, D. (1988). Probabilistic buckling analysis of cylindrical shells. Memorandum M-860, Faculty of Aerospace Engineering, Delft University of Technology, The Netherlands, November.

de Vries, J. (2001). Imperfection database. *Third European Conference on Launcher Technology* (pp. 323–332), Strasbourg, France.

de Vries, J., Research on the Yoshimura buckling pattern of small cylindrical thin walled shells. In K. Fletcher (Ed.), *European Conference on Spacecraft Structures, Materials & Mechanical Testing*, Noordwijk, The Netherlands (CD-ROM).

de Vries, J. (2006). Local buckling load of small cylindrical shells: an energy approach. In W. Pietraszkiewicz and C. Szymczak (Eds.), *Shell Structures: Theory and Applications* (pp. 207–210). London: Taylor & Francis.

de Vries, J. (2009). The initial imperfection data bank and its applications, Ph.D. Thesis, Department of Aerospace Engineering, Delft University of Technology, Delft, The Netherlands.

Wadee, M.A. (2000). Effects of periodic and localized imperfections on struts on nonlinear foundations and compression sandwich panels. *International Journal of Solids and Structures*, 37, 1191–1209.

Wadee, M.K. (1999). The elastic strut on an elastic foundation: a model localized buckling problem. In A.R. Champneys, G.W. Hunt and J.M.T. Thompson (Eds.), *Localization and Solitary Waves in Solid Mechanics* (pp. 31–53). Singapore: World Scientific.

Wadee, M.K. and Bassom, A.P. (2000a). Characterization of limiting homoclinic behavior in a one-dimensional elastic buckling model. *Journal of the Mechanics and Physics of Solids*, 48, 2297–2313.

Wadee, M.K. and Bassom, A.P. (2000b). Restabilization in structures susceptible to localized buckling: an approximate method for the extended post-buckling regime. *Journal of Engineering Mathematics*, 38, 77–90.

Wadee, M.K. and Bassom, A.P. (2011). Unfolding of homoclinic and heteroclinic behavior in a multiply-symmetric strut buckling problem. *Quarterly Journal of Mechanics and Applied Mathematics*, 65(1), 141–160.

Wadee, M.K., Coman, C.D., and Bassom, A.P. (2003). Solitary wave interaction phenomena on a strut buckling model incorporating restablisation, *Physica D*, 163, 26–48.

Wadee, M.K., Coman, C.D., and Bassom, A.P. (2004). Numerical stability criteria for localized post-buckling solutions in a strut-on-foundation model. *Journal of Applied Mechanics*, 71, 334–341.

Wadee, M.K., Higuchi, Y., and Hunt, G.W. (2000). Galerkin approximations to static and dynamic localization problems. *International Journal of Solids and Structures*, 37, 3015–3029.

Wadee, M.K., Hunt, G.W. and Whiting, A.I.M. (1997). Asymptotic and Rayleigh–Ritz routes to localized buckling solutions in an elastic stability problem. *Proceedings of the Royal Society of London*, A453, 2085–2107.

Wald, A. (1945). Statistical decision functions which minimize the maximum risk. *The Annals of Mathematics*, 46(2), 265–280.

Wald, A. (1950). *Statistical Decision Functions*. New York: Wiley.

Walker, A.C., Andronicou, A., and Sridharan, S. (1982). Experimental inverstigation of the buckling of stiffened shells using small scale models. In J.E. Harding and P. Dowling, (Eds.), *Buckling of Shells in Offshore Structures* (pp. 25–43). London: Granada.

Walukiewicz, H., Bielewicz, Z., and Górski, J. (1997). Simulation of nonhomogeneous random fields for structural applications. *Computers & Structures*, 64(1–4), 491–490.

Wang, B., Hao, P., Du, K.F., and Li, G. (2011). Knockdown factor based on imperfection sensitivity analysis for stiffened shells. *International Journal of Aerospace and Lightweight Strcutures,* 1(2), 315–333.

Wang, B., Hao, P., Li, G., Fang, Y., Wang, X.J., and Zhang, X. (2013). Determination of realistic worst imperfection for cylindrical shells using surrogate model. *Structural and Multidisciplinary Optimization*, 48, 777–794.

Wang, H. (2009). Buckling design for fibre reinforced laminated cylindrical shells, Ph.D. Thesis, Department of Civil Environmental, and Geomatic Enginnering, University College London, London.

Wang, H. and Croll, J.G.A. (2008). Optimisation of shell buckling using lower buckling capacities. *Thin-Walled Structures*, 46(7–9), 1011–1020.

Wang, H. and Croll, J.G.A. (2013). Lower-bound analysis of fiber-reinforced polymeric laminated cylindrical shells. *AIAA Journal*, 51(1), 218–225.

Wang, X.J., Elishakoff, I., Qui, Z.P., and Ma, L.H. (2009). Comparison of probabilistic and two nonprobabilistic methods for uncertain imperfection sensitivity of a column on a nonlinear mixed quadratic–cubic foundation. *Journal of Applied Mechanics*, 76(1), Paper 011007.

Wang, X.J., Wang, L. and Elishakoff, I. (2011). Probability and convexity are not antagonistic. *Acta Mechanica*, 219, 45–64.

Watson, J.G. and Reiss, E.L. (1982). A statistical theory for imperfect bifurcation. *SIAM Journal of Applied Mathematics*, 42(1), 135–148.

Weingarten, V.I., Morgan, E.J., and Seide, P. (1965). Elastic stability of thin-walled cylindrical and conical shells under axial compression. *AIAA Journal*, 3(5), 500–505.

Weingarten, V.I., Morgan, E.J., and Seide, P. (1965). Elastic stability of thin-walled cylindrical and conical shells under combined internal pressure and axial compression. *AIAA Journal*, 3(6), 1118–1125.

Wentzel, E.C. (1980). *Operations Research: Problems, Principles, Methodology* (p. 28). Moscow: "Mir" Publishers.

Whiting, A.I.M. (1997). A Galerkin procedure for localized buckling of a strut on a nonlinear elastic foundation. *International Journal of Solids and Structures*, 34, 727–739.

Wilson, W.M. and Newmark, N.M. (1993). The strength of thin cylindrical shells as columns, Bulletin No. 255, Engineering Experimental Station, University of Illinois.

Wilson, W.M. and Newmark, N.M. (1933). The strength of thin cylindrical shells as columns. Engineering Experimental Station Bulletin N°. 255, Urbana, IL, 28 February.

Xie, W.C. (1995). Buckling mode localization in randomly disordered multispan continuous beam. *AIAA Journal*, 33, 1142–1149.

Xie, W.C. (1997). Buckling mode localization of nonhomogeneous beams on elastic foundation. *Chaos, Solitons and Fractals*, 8, 411–431.

Xie, W.C. (1988). Buckling mode localization in rib-stiffened plates with randomly misplaces stiffness. *Computers and Structures*, 67, 175–189.

Xie, W.C. (Ed.) (2000). Special issue. *Chaos, solitons and Fractals*, 11(10).

Xie, W.C. and Elishakoff, I. (2000). Buckling mode localization in rib-stiffened at plates with misplaced stiffeness — Kantorovich approach. *Chaos, Solitons and Fractals,* 11(10), 1559–1574.

Xu, Y.H. and Bai, G.L. (2013). Random buckling bearing capacity of super-large cooling towers considering stochastic material properties and wind loads. *Probabilistic Engineering Mechanics*, 33, 18–25.

Yakushev, V.L. (1988). *Nonlinear Deformations and Stability of Thin Shells* (in Russian). Moscow: "Nauka" Publishers.

Yakushev, V.L. (2010). Stability of thin-walled structures with initial imperfections taken into account (in Russian). *Structural Mechanics and Analysis of Constructions*, 1, 43–46.

Yamada, S. and Croll, J.G.A. (1999). Contributions to understanding the behavior of axially compressed cylinders, *Journal of Applied Mechanics*, 66, 299–309.

Yamaki, N. (1984). *Elastic Stability of Circular Cylindrical Shells.* Amsterdam: North-Holland, Amsterdam.

Yao, J.C. (1962). Buckling of axially compressed long cylindrical shell with elastic core. *Journal of Applied Mechanics*, 29, 329–334.

Yasin, E.M. (1967). Stability of cylindrical shell in stress state of axisymmetric edge effect (in Russian). *Aviation Technique*, 1, 57–60.

Yiang, C., Li, W.X., Han, X., Liu, L.X., and Le, P.H. (2011). Structural reliability analysis based on random distributions with interval parameters. *Computers & Structures*, 89(23–24), 2292–2302.

Yithak, E., Elishakoff, I., and Baruch, M. (1988). Dynamic imperfection sensitivity of a simple spring model structure. *Journal of Mechanics of Structures and Machines*, 16, 187–199.

Young, T. (1807). *A Course of Lectures on Natural Philosophy and the Mechanical Arts*. London, 1807.

Zarutskii, V.A. and Sivak, V.F. (1991). Method of analyzing experimental data obtained in tests of cylindrical shells in axial compression. *International Applied Mechanics*, 30(10), 45–50.

Zarutskii, V.A. and Sivak, V.F. (1993). Method of analyzing experimental data from tests of the stability of cylindrical shells in axial compression. *International Applied Mechanics*, 30(10), 777–781.

Zarutskii, V.A. and Sivak, V.F. (1997). Empirical formula for the stability design of shells. *International Applied Mechanics*, 33(7), 532–536.

Zarutskii, V.A. and Sivak, V.F. (1995). Method of predicting the critical stresses in a longitudinally compressed cylindrical shells. *International Applied Mechanics*, 31(4), 261–264.

Zhao, H., Xu, Y.Z., and Bai, G.L. (2011). The influence of stochastic imperfection field on the bearing capacity of hyperbolic cooling towers. *Advanced Materials Research*, 374–377, 2297–2300.

Zhiteckij, L.W. (1996). Adaptive control of systems subjected to bounded disturbances. In M. Milanese, J. Norton, H. Pret-Lahanier, and E. Walter (Eds.), *Bounding Approaches to System Identification* (pp. 383–407). New York: Plenum Press.

Zhu, E., Mandal, P., and Calladine, C.R. (2002). Buckling of thin cylindrical shells: an attempt to resolve a paradox. *International Journal of Mechanical Sciences*, 44, 1583–1601.

Zhu, L.P., Elishakoff, I., and Starnes, J.H., Jr (1996). Derivation of multi-dimensional ellipsoidal convex model for experimental data. *Mathematical Computing and Modeling*, 24(2), 103–114.

Zingales, M. and Elishakoff, I (2000). A note on localization of the bending response in presence of axial load. *International Journal of Solids and Structures*, 37, 6739–6753.

Zipalova, V.F. and Nenastieva, V.M. (1966). High approximations investigation of stability of the spherical arch with various boundary conditions (in Russian). *Mechanics of Solids*.

Zipalova, V.F. and Nenastieva, V.M. (1971). About application of the Galerkin's method in temperature-force problems in nonlinear stability of shells (in Russian). *Mechanics of Solids*, 1, 100–103.

Zimmermann, R. (1992). Optimierung axialgedrückter CFK-Zylinderschalen. VDI Verlag, Fortschritt-Berichte VDI, Reihe 1.

Zimmermann, R. and Rolfes, R. (2006). POSICOSS — Improved postbuckling simulation for design of fibre composite stiffened fuselage structures. *Composite Structures*, 73, 171–176.

Author Index

Subject Index

Printed in the United States
By Bookmasters